collisions through thickness t

$\pi b^2 . N . t$ where $N =$ number particles per unit volume

of collision for a big deflection

$$= \frac{a^2}{b^2}$$

no of big deflection in distance t

$$= \pi a^2 . N . t$$

for chance even for a distance a such that $\frac{N_0 e}{a}$ is force²

$$\frac{N_0 e}{a} = \frac{4\pi}{3}\rho t . \tfrac{1}{2}mv^2 . K \quad \text{when } K \text{ is } < 1$$

$$\frac{N_0 e}{a} = \frac{4 \times 10^4}{3} . K$$

$$a = \frac{3 N_0 e}{4 \times 10^4 . K} = \frac{3 \times 200 \times 4.65}{4 \times 10^4 . \times 10^{10} . K}$$

$$= \frac{7}{10^{12} K}$$

no of big deflection $= \pi \times \frac{49}{10^{24} . K} . N . t$

3.1
155

$$= \frac{1.5}{10^{22} . K} N . t$$

$$= 6.2 \times 10^{22} = \frac{9.4}{10^{22} . K} . 10^{22} . t$$

$$= \frac{9.4 t}{K} \quad \therefore \text{chance in level field}$$

$$\text{unity for even pair} = \frac{9.4 . a}{10^5} = \frac{1.9}{10^4} = \frac{2}{10000}$$

$$\text{for air} = .00006 \text{ atm}$$

SCIENCE

IN THE

MAKING

Volume 3: 1900–1950

SCIENCE

IN THE

MAKING

**SCIENTIFIC DEVELOPMENT
AS CHRONICLED BY
HISTORIC PAPERS IN THE
PHILOSOPHICAL MAGAZINE —
WITH COMMENTARIES
AND ILLUSTRATIONS**

Volume 3: 1900–1950

EDITED BY
E. A. DAVIS

Taylor & Francis
ALERE FLAMMAM
· Founded 1798 ·

Dedicated to
my Mother and Father

UK Taylor & Francis Ltd., 1 Gunpowder Square, London, EC4A
3DE

USA Taylor & Francis Inc., 1900 Frost Road, Suite 101,
Bristol, PA 19007

British Library Cataloguing in Publication Data

A catalogue record for this book is available from the British
Library.

ISBN 0 7484 07669

**Library of Congress Cataloguing in Publication Data are
available**

Typeset by Keyword Typesetting Services Ltd

Printed in Great Britain by
Biddles Ltd., Guildford and King's Lynn

Contents

Introduction and Acknowledgements

Chronologically significant years, such as the end or the middle of a century, offer historians of science the opportunity to reflect on major advances that have taken place during preceding periods of time and to identify changes of emphasis or direction in various lines of investigation. As far as physics is concerned, 1900 is a particularly favourite year, mainly because of the division it provided between the classical conquests of the nineteenth century — in mechanics, thermodynamics and electromagnetic theory and the major accomplishments of atomic, nuclear and quantum theories during the present century.

In 1900 physics was heralded as the queen of sciences, not only because it seemed to have provided the answers to most of the fundamental and basic questions concerning the natural world, but also because of its undoubted successes in bringing electrical technology, combustion engines, telegraphic communications and other recognizable benefits to mankind. The first indications that physics was about to enter a new phase could be discerned in the closing years of the last century. From 1895, a number of spectacular discoveries were made in rapid succession, each of which suggested that all was not right with classical treatments and that radical new insights were going to be required to understand them. The most significant of these discoveries were: X-rays by Röntgen in 1895; the spontaneous disintegration of uranium (radioactivity) by Becquerel in 1896; the magnetic splitting of spectral lines by Zeeman in 1896; the identification of cathode rays with negatively charged corpuscles (electrons) by J. J. Thomson in 1897; and the isolation, by painstaking labour, of two new radioactive elements, radium and polonium, by the Curies in 1898. These extraordinary new discoveries were to provide challenges as great as anything that had hitherto been met in the history of scientific enquiry.

It could not have been foreseen that the problems faced by physicists were not to be overcome by the rejection of all that had been achieved in the previous centuries, from the time of Newton onwards. Two revolutionary concepts — quantum mechanics and relativity — theories introduced and developed in the first decades of the present century, were not only able to throw light on the puzzling new phenomena but more importantly to provide a framework in which they could be understood without abandoning the traditional doctrines. Classical physics was found to be the limiting case of the new theories — a methodology still applicable in the familiar world in which objects move with speeds considerably less than that at which light propagates and with sizes that are large compared to those of atoms.

The architects of the new theories, notably Planck and Einstein, were adamant that the principle of conservation of energy be retained, but the doctrine of the conservation of mass was destined to fall, as was that of the indivisibility of the atom. Einstein's celebrated equation $E = mc^2$ demonstrated the equivalence of mass and energy, so energy conservation took on a more general meaning. J. J. Thomson's electron was soon accepted as a basic constituent of all matter, but it was Rutherford who revealed the internal structure of the atom with its central positive core carrying most of the mass. It was also Rutherford who first realized that radioactivity involved disintegration of the atomic nucleus and subsequently showed how this could be achieved by artificial means.

Quantum physics had a very uneasy start. Revolutionary concepts always take time to be assimilated and Planck's hypothesis in 1900 that the vibrations of the 'resonators' responsible for the emission of electromagnetic radiation from a heated body were quantized, lay dormant for 5 years — in spite of its success in explaining the energy distribution of the radiation. The most influential advocate of Planck's ideas was Einstein, who introduced the concept of light quanta and, in so doing, appeared to challenge the conclusion from a century of work that electromagnetic radiation was wavelike in nature. Although other pioneers pursued the concepts of quantum theory, its principles were not generally accepted until sometime after Bohr applied them with great success to the Rutherford model of the atom in 1913, and 10 years later Compton used them to explain the change of wavelength of X-rays during their scattering from solids. In 1924 de Broglie made his startling prediction that wave-particle duality need not be restricted to light but could be applied equally well to matter itself. It was left to Schrödinger to give meaning and mathematical articulation to de Broglie waves, and others — notably Heisenberg, Pauli, Born and Dirac — developed more fully the many theoretical and philosophical ideas associated with the new quantum mechanics.

The present volume begins, however, with some celebrated papers relating to the electron. The fundamental electronic charge was recognized even before the discovery of the electron as an important universal constant. Efforts to determine its magnitude were pursued with vigour, culminating in the oil drop experiments of Millikan who in 1917 obtained a value that was within half a percent of today's accepted value. Long before the significant part played by electrons as constituents of matter was appreciated, their role as independent particles, when produced by thermionic emission, was being utilized in vacuum tubes such as the Fleming diode and later the triode — devices which when incorporated into electrical circuits considerably advanced the technology necessary for the transmission and reception of radio waves. Solid-state electronics was to make its impact after the discovery of the transistor in 1947, but cathode ray tubes and electron microscopes, in which electron beams are steered and focussed by magnetic and electric fields, came first and, although now highly developed, operate in ways fundamentally the same as those of the early instruments. As in the last century, fundamental science spawned useful practical applications.

The following books have proved useful during the writing of the introductory sections to each Part and the commentaries on the papers selected for this volume.

R. Spangenburg and D. K. Moser, *The History of Science from 1895 to 1945* (Facts on File 1994)

E. N. da C. Andrade, *Rutherford and the Nature of the Atom* (Peter Smith, Gloucester, Mass. 1978)

A. S. Eve, *Rutherford* (Cambridge University Press 1939)

Per. F. Dahl, *Flash of the Cathode Rays* (Institute of Physics 1997)

T. J. Trenn, *The Self-Splitting Atom* (Taylor & Francis 1977)

M. R. Wehr, J. A. Richards and T. W. Adair, *Physics of the Atom* (Addison-Wesley 1978)

L. M. Brown, A. Pais and A. B. Pippard (eds), *Twentieth Century Physics Volumes I–III* (Institute of Physics and American Institute of Physics 1995)

R. A. Millikan, *The Electron* (University of Chicago Press 1924)

S. F. Mason, *A History of the Sciences* (Macmillan 1962)

A. Keller, *The Infancy of Atomic Physics* (Clarendon Press, Oxford 1983)

M. Bunge and W. R. Shea (eds) *Rutherford and Physics at the Turn of the Century* (Wm. Dawson and Sons, 1979)

Classical Scientific Papers: Physics (Mills & Boon 1964)

G. P. Thomson, *J. J. Thomson and the Cavendish Laboratory* (Nelson 1964)

S. Rozental (ed.) *Niels Bohr* (North-Holland 1968)

B. Schonland, *The Atomists (1805–1933)* (Clarendon Press, Oxford 1968)

W. H. Cropper, *The Quantum Physicists* (Oxford University Press 1970)

E. Whittaker, *A History of the Theories of Aether & Electricity* (Dover 1989)

G. C. Gillespie (ed.) *Dictionary of Scientific Biography* (Charles Scribner's Sons 1976)

Acknowledgements

It is a pleasure to record my thanks to Dr E. G. Thomas of the Department of Physics and Astronomy, University of Leicester for reading and commenting on several parts of this volume, to Mrs Stephanie Brooks for typing the manuscript and to staff at Taylor & Francis for assisting with production.

The endpapers of this volume are facsimiles of pages from one of Rutherford's notebooks and are reproduced by permission of the Syndics of Cambridge University Library.

EDWARD A. DAVIS
Department of Physics and Astronomy
University of Leicester

Note to the Reader

Throughout this volume facsimile pages have been reproduced in their entirety. Where only part of the reproduced page is relevant, a 'pointing hand' in the margin indicates the start of the chosen extract. In these cases, disregard the text above the pointer.

1. Expansion apparatus of the type used by H. A. Wilson in the determination of the electronic charge. Courtesy of the Cavendish Laboratory, University of Cambridge.

2. Marie and Pièrre Curie in their laboratory.

3. Ernest Rutherford at McGill University 1905.

4. Frederick Soddy.

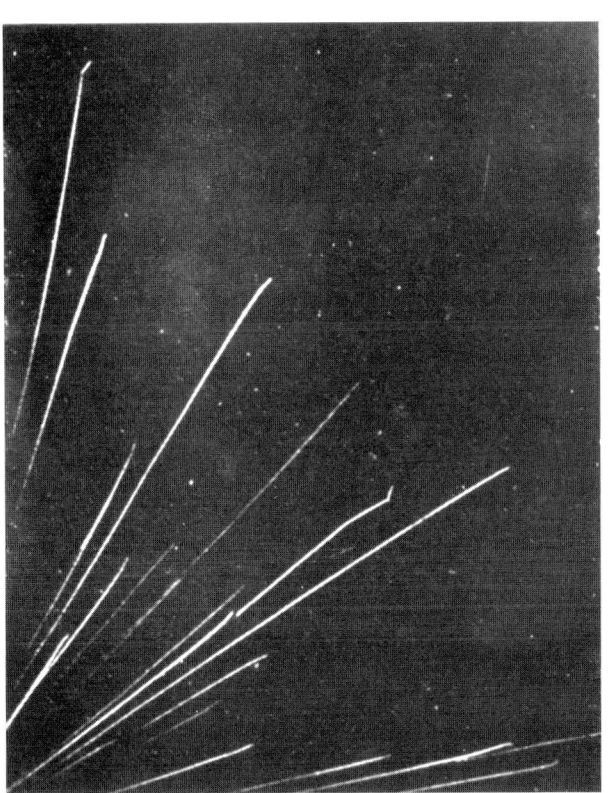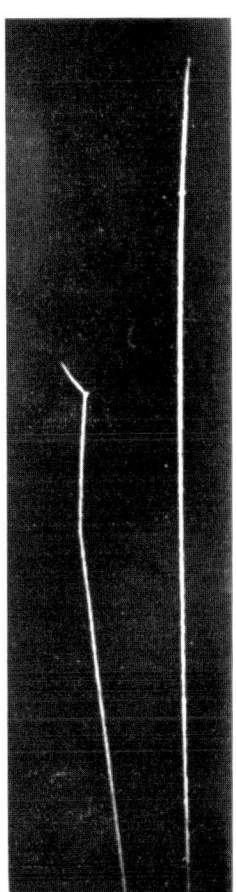

5. Cloud chamber photographs of α-particle tracks.

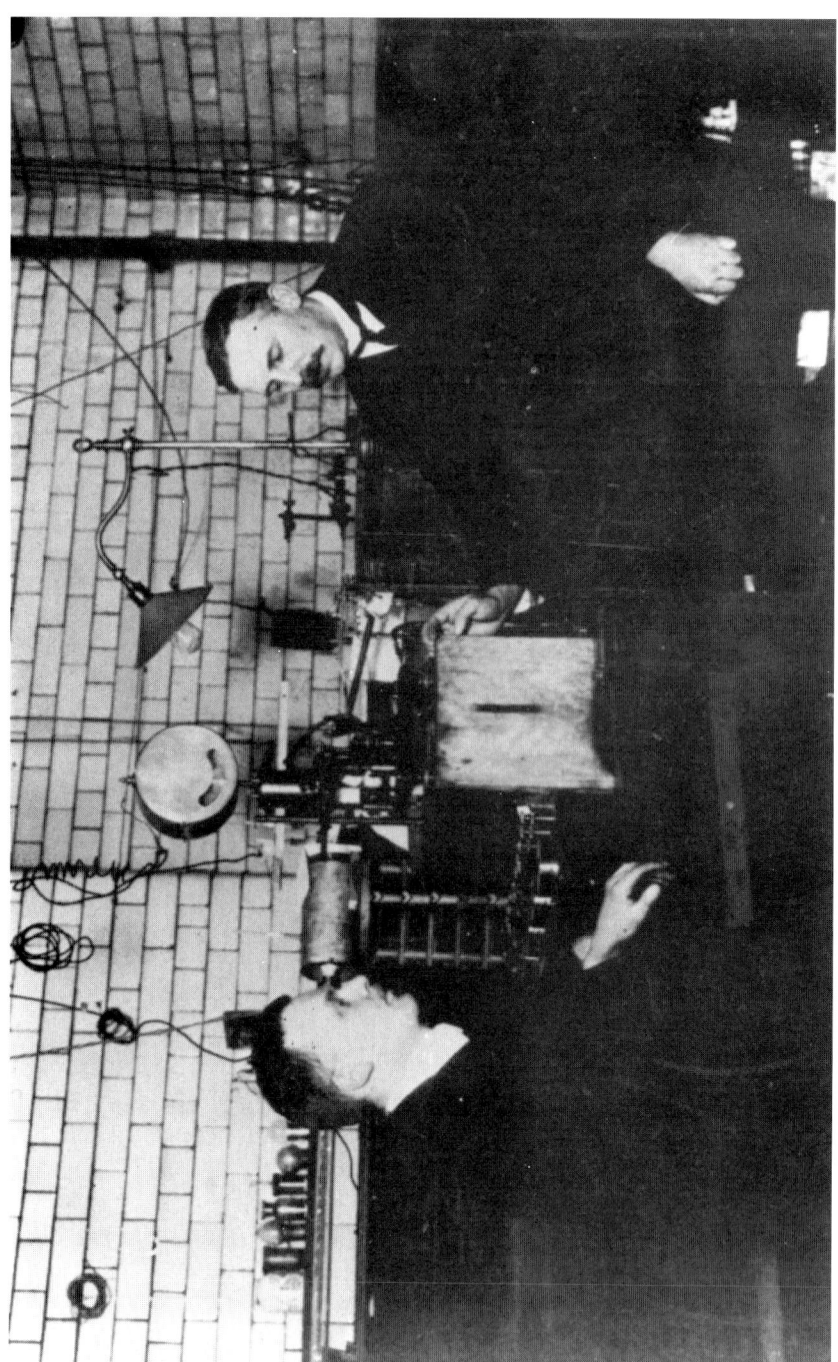

6. Geiger and Rutherford at Manchester University.

7. J. J. Thomson with his positive ray apparatus. Courtesy of the Cavendish Laboratory, University of Cambridge.

8. Positive ray tube circa 1910. Courtesy of the Cavendish Laboratory, University of Cambridge.

9. Ernest Rutherford (1871-1937). Photograph taken in 1920. Courtesy of Getty Images.

10. Niels Bohr (1855-1962).

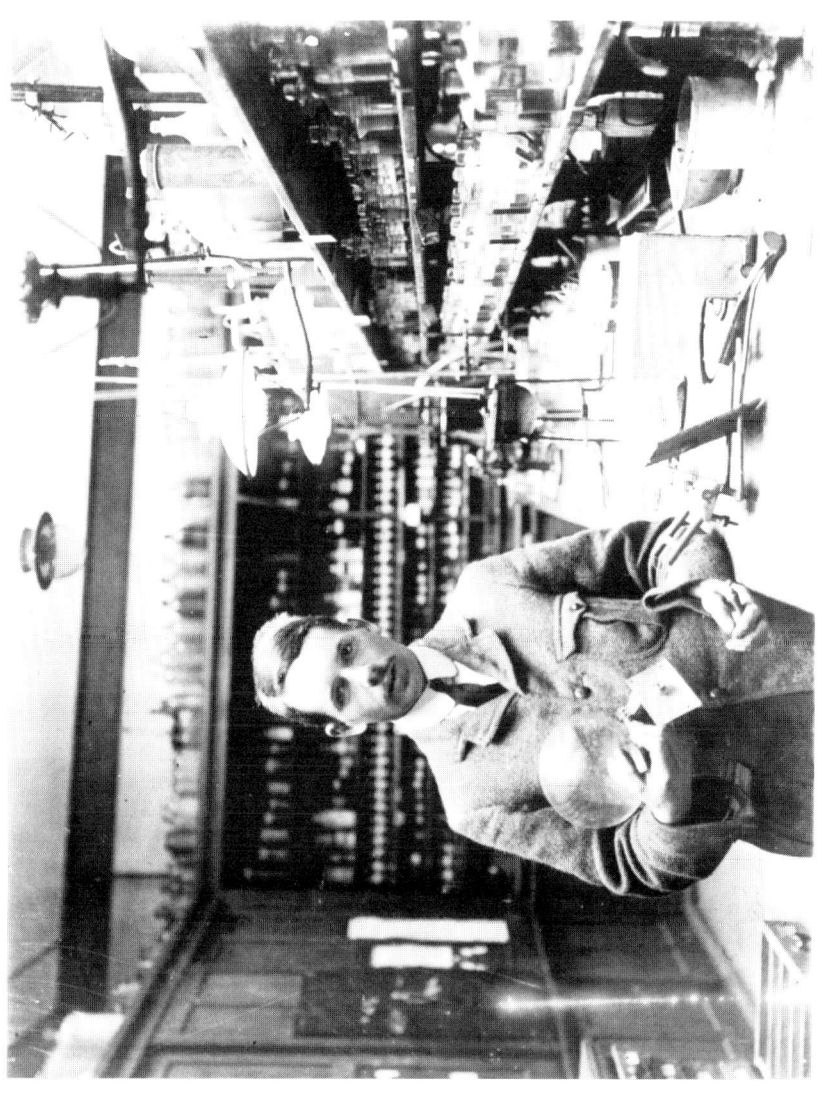

11. Harry Moseley in his Oxford Laboratory. © Courtesy of the Museum of the History of Science, Oxford. Inventory no. 18,874.

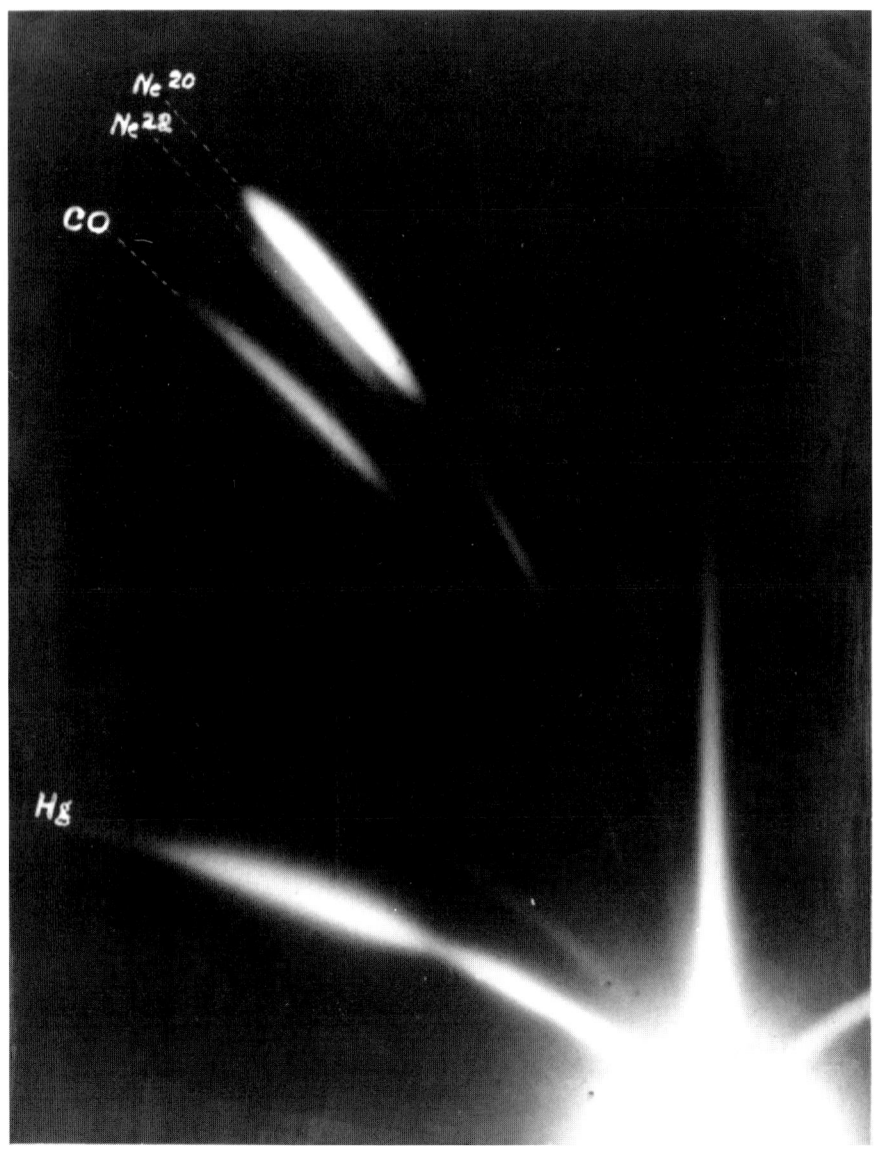

12. Positive ray parabolae. Courtesy of the Cavendish Laboratory, University of Cambridge.

13. Aston's mass spectrograph. Courtesy of the Cavendish Laboratory, University of Cambridge.

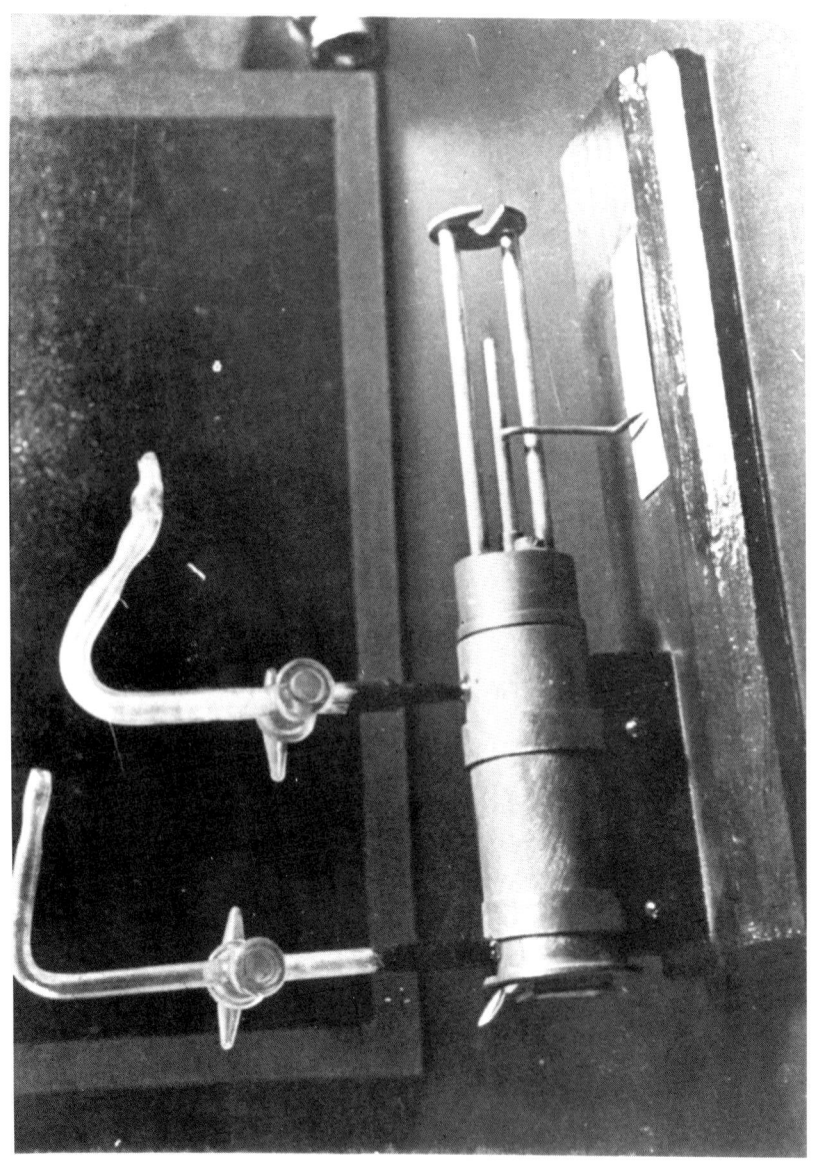

14. Rutherford's apparatus used for the artificial disintegration of nitrogen nuclei. Courtesy of the Cavendish Laboratory, University of Cambridge.

15. Delegates at the 1927 Solvay Congress in Brussels. From left to right: front row: I. Langmuir, M. Planck, Madame Curie, H. A. Lorentz, A. Einstein, P. Langevin, Ch E. Guye, C. T. R. Wilson, O. W. Richardson: second row: P. Debye, Mr Knudsen, W. L. Bragg, H. A. Kramers, P. A. M. Dirac, A. H. Compton, L. V. de Broglie, M. Born, N. Bohr: back row: A. Piccard, E. Henriot, P. Ehrenfest, E. D. Herzen, T. H. de Donder, E. Schrödinger, E. Verschaffelt, W. Pauli, W. Heisenberg, R. H. Fowler, L. Brillouin. Institut International de Physique Solvay, courtesy AIP Emilio Segrè Visual Archives.

16. Max Planck (1858-1947).

Part One

The Electron and Thermionics

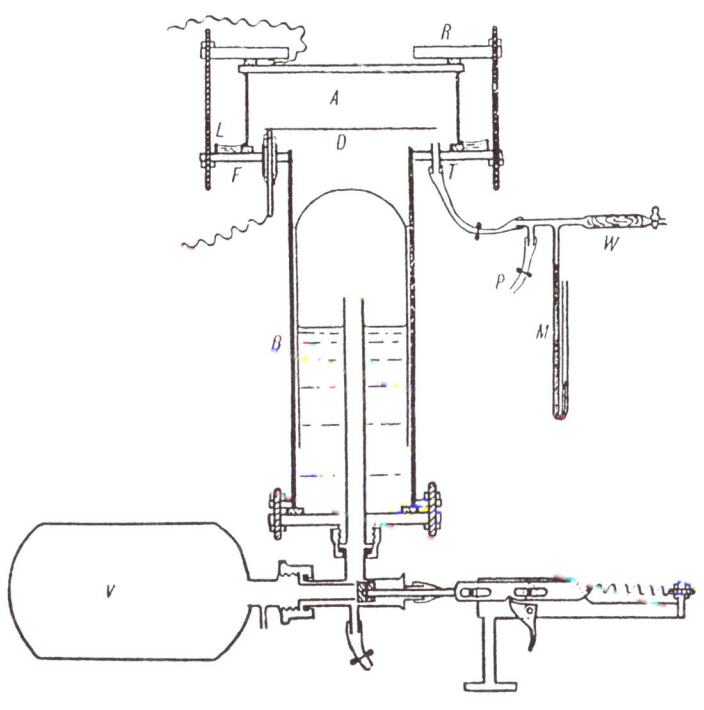

Part One
The Electron and
Thermionics

The discovery of the electron — the first sub-atomic particle to be isolated and identified — was an extraordinary event in science, not only for the incentive it provided for further explorations of the nature and structure of atoms, but also for the stimulus it gave to the development of thermionic valves and other electronic devices. The name 'electron' was in fact coined by Johnstone Stoney six years before the discovery of the particle now known by that name. Stoney proposed the term to describe the 'natural unit of electricity' carried by ions in the electrolysis of liquids, i.e. the amount of charge which must pass through a solution in order to liberate one atom of hydrogen or other univalent substance. Neither Stoney nor his contemporaries believed that the charge was capable of independent existence. When J. J. Thomson demonstrated experimentally in 1897 that the cathode rays in a gaseous discharge were negatively charged particles with a ratio of charge to mass (e/m) about one thousand times larger (or, for the same charge, a mass one thousand times smaller) than the smallest known electrolytic ion, namely hydrogen, he christened the particles 'corpuscles'. Only much later was Stoney's term, the electron, adopted for the particle itself.

The first non-electrolytic determination of the magnitude of the electronic charge was made in 1897 by Townsend using a method which involved the condensation of water droplets around gaseous ions to form a charged cloud. The total charge of the cloud was measured with a quadrant electrometer and the average charge carried by each drop deduced from a separate determination of the number of drops. Townsend calculated this number by dividing the total mass of the cloud (determined by passing it through drying tubes and measuring the increase in their weight) by the average mass of each drop, this being computed from their density and mean radius. The latter he obtained by observing the rate of fall of the water droplets and application of Stoke's law. In spite of the many different steps and assumptions involved, Townsend's determination of e by this method was surprisingly good, being about 3×10^{-10} esu in comparison with today's value of 4.8×10^{-10} esu.

In 1898, J. J. Thomson utilized a similar method to determine the charge on gaseous ions produced by X-rays and one year later the charge on electrons generated by the photoelectric effect. A new feature of these experiments, however, was the use of an expansion device to produce cooling of air supersaturated with water vapour to facilitate the condensation of droplets on the charged particles. This approach was also used in 1903 by H. A. Wilson, whose paper on the determination of e is reproduced in this Part, along with a paper by C. T. R. Wilson, who pioneered the expansion technique and later developed it into the Wilson cloud chamber for the detection of tracks produced by charged particles of many kinds.

The paper by J. A. Fleming, reproduced next, illustrates the extent to which devices based on the flow of electrons between electrodes in vacuum tubes were being developed in response to the need for reliable and sensitive detectors of radio waves. The first transmission and reception of high-frequency electromagnetic waves through space was accomplished by Hertz in 1886, an achievement which confirmed Maxwell's theoretical predictions of 1861 (see *Science in the Making*, Volume 2). Signalling with these wire-less waves was suggested by Crookes in 1892 and a year later Lodge used a 'coherer'[1], one of the most important of the early detectors, as a receiver.

Public interest was not aroused, however, until Marconi filed a patent in 1896 for circuits and devices for 'the effective telegraphic transmission and intelligible reception of signals produced by artificially formed Hertz oscillators' and achieved signalling over many miles. After experimenting with coherers of various geometries, Marconi used a magnetic detector, an early form of which had been invented by Rutherford in 1895 (see p 85). The first vacuum tube rectifier for the reception of signals was made by Fleming in 1904. Its development, together with the invention of the triode by Lee de Forest in 1906 — a valve which could amplify electrical signals — made radio telephony practical on a large scale.

The emission of electrons from heated metals was a phenomenon of considerable importance in vacuum tubes, but a real understanding of its nature and what controlled its characteristics occupied physicists for several decades. Pioneering work, noteworthy for its thoroughness, was carried out by Owen Richardson, first at the Cavendish Laboratory and then at Princeton University where he became a professor in 1906. An equation describing the variation of emission current with voltage and temperature, was named after him. One of Richardson's early papers, co-authored with Brown, is reproduced on p. 42

The final paper in this part is by Millikan, whose experimental determination of the fundamental electronic charge by the balanced oil drop method is known to every student of physics, many of whom have repeated Millikan's experiment as an undergraduate laboratory exercise. Robert Millikan (1868–1953) graduated from Oberlin College, Michigan, in 1891 and two years later entered Columbia University as its sole graduate student in physics. After spending a year in Europe, where he undertook research with Nernst at Göttingen, he joined the faculty at the University of Chicago in 1896. His promotion in 1907 to Associate Professor followed research on the photoelectric effect, work to which he was to return in 1912. By 1915 Millikan had verified in every detail Einstein's celebrated equation for the photoelectric effect and obtained the best value of Planck's constant then available. However, it is his successive determinations of the fundamental charge, each one more accurate than the last, for which Millikan will always be remembered. Before his success in making measurements on singly charged drops, there was always the possibility that the electronic charge was a statistical average. Millikan convincingly dispelled this notion. His determinations of e, described in the paper on p 66, were made with a precision of 0.1% and his final value served the scientific world for a generation. Although it was about 0.5% lower than today's value, most of the discrepancy can be attributed to his use of a value for the viscosity of air that was later shown to be inaccurate.

1903 **5**   *A Determination of the Charge on the Ions produced in Air by Röntgen Rays. By Harold A. Wilson, Fellow of Trinity College, Cambridge*

The fact that the electronic charge was a fundamental constant, comparable in importance to the velocity of light and Avogadro's number, had been recognized during the nineteenth century, even before the discovery of the electron in 1897. Stoney had drawn attention to the significance of the 'elementary unit of charge' carried by the hydrogen ion in electrolysis in 1891. In 1898 J. J. Thomson measured the charge on ions created in a gas by X-rays and obtained the value 6.5×10^{-10} esu. One year later he determined the charge on 'corpuscles' (electrons) ejected from a metal by the photoelectric effect and found this to have a similar value, namely 6.8×10^{-10} esu, albeit with an uncertainty of about $\pm$ 15%. (See *Science in the Making*, Volume 2, for facsimiles and commentaries on these earlier papers.)

In 1903 Thomson repeated his gaseous ionization experiments using radioactive radium instead of X-rays as the ionizing agent (*Philosophical Magazine* 1903 **5** 354) and obtained a value for the fundamental charge carried by ions of 3.4×10^{-10} esu, which he gave reasons for preferring over his former value.

The present paper by H. A. Wilson describes an important modification to the method used by Thomson for the determination of the
charge on a single ion, although the apparatus he used was similar.
Designed by C. T. R. Wilson (who was later to develop it into the
'Wilson cloud chamber'), it consisted of a vacuum-driven expansion
device (see the figure on page 431 of the paper and Plate 1 of this
volume) which produced supersaturation conditions in water vapour.
Ions created in the gap between two parallel electrodes by X-rays from
a Röntgen tube then acted as nuclei for the condensation of water
droplets. Whereas Thomson had used the electrodes merely to determine the total charge carried by the droplets as they fell under gravity,
H. A. Wilson applied electric fields so as to increase or retard the rate of
fall of the charged drops. Indeed some drops which had acquired a
high charge could be made to travel upwards.

By way of explanation, the second equation in the paper relating the
mass of a droplet to its velocity, in the absence of an electric field,
involves application of Stoke's law which gives the terminal velocity of
a spherical particle falling in a medium of known viscosity. The constant 3.1×10^{-9} thereby involved values of the density of the drops and
the viscosity of air, which were in fact not accurately known for saturated air at the temperature existing immediately after the expansion.

The major advance in the determination of e by the method
described in the paper was that the measurements were made on
ions that were *singly* charged. Ions carrying multiple charges were
driven down more rapidly by the field and could therefore be distinguished. Wilson timed the rate of fall of the top of the cloud of droplets
which formed a fairly well defined plane.

The average of eleven different determinations of the charge yielded
3.1×10^{-10} esu which was close to Thomson's later value referred to
above. Although more confidence could be given to Wilson's value, the
accuracy obtained (around $\pm$ 20%) was in fact no better than that
obtained by Thomson and it was to be left to Millikan to improve on
this with further refinements of the same general method.

Today's accepted value for the electronic charge e is 4.8×10^{-10} esu.

1904   **7**   *The Condensation Method of Demonstrating the*
         Ionisation of Air under Normal Conditions. By C. T. R.
         Wilson, M.A., F.R.S., Fellow of Sidney Sussex College,
         Cambridge

The Wilson cloud chamber, which was eventually developed into a
powerful research tool for the visualization and photographing of ionization tracks of fundamental particles, had its origins in the type of

apparatus illustrated and described in this paper. By the time the account was published, early forms of the apparatus had been used by J. J. Thomson and H. A. Wilson (see preceding commentary) in their determinations of the fundamental charge carried by ions.

Here C. T. R. Wilson reports on a later version of the apparatus built to study the effects of an electric field on drop formation produced by expansion of supersaturated air. Its construction and mode of operation are described in detail in the paper. The most significant changes from earlier models were the replacement of the outer glass cylinder of the expansion vessel by one made of brass and a larger cloud chamber, the latter being introduced to reduce the effects of loss of ions by diffusion to the walls.

Without any source of ionization, a critical volume expansion rate of 1.25 was found to be required to produce rain-like condensation. Beyond a second limit of 1.38, a shower of fine mist appeared. Between these two values the effect of ionizing rays increased the number of drops enormously. The reason for this (not given in the paper) is that very small drops, which would normally evaporate, grow in size when they are charged, because of the electrostatic forces opposing the effect of surface tension. Once such drops grow beyond a critical size, they become an efficient nucleus for further condensation.

Even without a source of ionization such as X-rays, and after repeated expansions to remove dust particles which act as nuclei for condensation, Wilson observed that a number of drops still formed but that an applied electric field was effective in removing these from the cloud chamber. Clearly therefore they carried a charge. It was some years later that cosmic rays were recognized as the principal cause of this residual ionization.

1906 **11** *The Construction and Use of Oscillation Valves for Rectifying High-Frequency Electric Currents. By J. A. Fleming, M.A., D.Sc., F.R.S., Professor of Electrical Engineering in University College, London*

John Ambrose Fleming was born in 1849 and led an active scientific life until his death aged 95. He started his research career under James Clerk Maxwell at the Cavendish Laboratory, Cambridge but remained there for only four years before his appointment in 1881 as Professor of Physics and Mathematics at University College, Nottingham — a post he resigned after one year in order to become a consultant to the Edison Electric Light Company in London. In 1885 Fleming accepted a Chair of Electrical Technology at University College, London where he remained for 41 years. He cooperated with Marconi and helped to design the transmitter used to span the Atlantic by radiotelegraphy in 1901.

Fleming is best known for his invention of the first thermionic valve — a diode based on the Edison effect, the unilateral flow of current between two electrodes in a vacuum tube. His objective was to produce a rectifier to replace the rather fickle 'coherer' and carborundum point-contact detectors then used in radiotelegraphy. British and United States patents for the vacuum diode (or glow-lamp detector as it was then called) were issued to Fleming in 1904 and 1905 respectively.

The present paper followed a much longer one by Fleming published 10 years earlier (*Philosophical Magazine* 1896 **42** 52–102) in which he described investigations of various geometrical arrangements for the electrodes and reported on their rectifying characteristics. Here he outlines the design and use of devices for rectifying high-frequency oscillatory signals such 'as are employed in the receiving circuits of wireless telegraph apparatus'. Figures 1 and 2 illustrate the valve and the manner in which it is used for this purpose. Oscillations in the primary circuit, produced by an induction coil and two Leyden jars (capacitors), generate Hertzian waves which are detected some distance away by the secondary circuit. The valve rectifies this signal, as confirmed by the steady deflection of a mirror galvanometer.

Fleming used this arrangement to investigate the effects of changing various features, for example the separation and the material of the spark-balls in the primary circuit. He also experimented with different filaments in the valve and found certain advantages of carbon over platinum coated with calcium or barium oxide, although the latter type produced larger emission currents.

The development of the vacuum valve, particularly by Lee de Forest in 1906 — who inserted a third electrode (a grid) and demonstrated the ability of this *triode* to amplify electrical signals — was of immense importance to the growth of wireless communication and electronic instrumentation during the first decades of this century.

1908   **16**   *The Kinetic Energy of the Negative Electrons emitted by Hot Bodies. By O. W. Richardson, Professor of Physics, and F. C. Brown, Experimental Science Fellow, Princeton University*

The copious emission of electrons from heated metal filaments is now such a commonplace phenomenon (used, for example, as the source of electron beams in television tubes, computer monitors, electron microscopes, etc.) that it is difficult to appreciate the painstaking work that went into establishing its characteristics and physical origin.

The discharge of electricity from hot metals had been observed in the eighteenth century but more than a hundred years passed before any systematic studies of the effect were undertaken. In 1880 the German physicists, Elster and Geitel, showed that both positive and negative emissions could occur and that the pressure and nature of the surrounding gas were important parameters. J. J. Thomson, in a classic paper (*Philosophical Magazine* 1899 **48** 547) — the main purpose of which was to determine the charge of the 'corpuscles' (later called electrons) which he had identified with cathode rays two years earlier — demonstrated that the negative emission from a hot carbon filament into a vacuum was associated with the release of particles. Furthermore the path traced out by these particles could be deflected in a magnetic field in exactly the same way as the electrons emitted by the photoelectric effect from a zinc plate. Using a combination of electric and magnetic fields, J J determined the charge to mass ratio (e/m) of the particles to be the same in the two cases and furthermore to have the same value as the cathode ray particles. These experiments clearly demonstrated that the negative emission from heated metals was the release of electrons.

From 1900 detailed investigations of the thermal emission of electrons from metals were carried out at the Cavendish Laboratory, Cambridge, by Owen Richardson, an experimentalist of first rank, who was later appointed to Chairs of Physics at Princeton (1906) and at King's College, London (1914). Richardson received the Nobel Prize for Physics in 1928 for his work on 'thermionics' — a name he coined in 1909 (*Philosophical Magazine* 1909 **17** 813).

The principal theoretical expression derived by Richardson and Brown in this paper is equation (10) which expresses the emission current i as a function of the voltage V acquired by a nearby plate and the filament temperature θ: $i = i_0 \exp(-Ve/R\theta)$ where R is the gas constant and e the electronic charge. This relation was obtained by considering the component of the electron's kinetic energy normal to the emitting surface, under the assumption that the distribution of their velocities is the same as that of molecules in a gas at the same temperature.

The emission of electrons from a heated filament, even in a good vacuum, depends critically on the condition of the surface of the emitter; prolonged heating is required to drive off occluded gases and impurities in order to obtain reproducible results. Taking this precaution, and more besides, Richardson and Brown obtained experimental verification of the above theoretical formula — at least in terms of the voltage dependence. The predicted variation with temperature could not be verified in these experiments.

Later work by Richardson in 1913 using ductile tungsten as a filament enabled him to use larger currents and to establish further details of his thermionic theory. He laid to rest a rival view that the emission was associated with chemical reactions between the hot filament and residual gases. By 1916 he had discovered that when a positive voltage was *applied* to the nearby plate the current *increased* with voltage, but that above a certain voltage the current remained vitrually constant. Under the latter conditions (corresponding to the removal of the negative space charge which builds up around the filament), the current was then found to obey the relation: $i = AT^2 \exp(-E/kT)$ where A is a constant, T the absolute temperature, and E the height of the barrier at the filament surface which the electrons must surmount in order to escape.

The development of quantum mechanics in the 1920s led to a more detailed understanding of the 'electron gas' in metals and amongst those who re-examined Richardson's findings in the light of the new concepts, Dushman, a Russian–American electronics expert, made the most significant contributions. The emission law is now commonly known as the Richardson–Dushman equation. It was a triumph of quantum mechanics that the universal constant of proportionality A was expressed in terms of fundamental quantities, namely the mass and charge of the electron, and the Boltzmann and Planck constants. Richardson himself was quick to recognize the importance of de Broglie matter waves and Schrödinger's wave mechanics, and applied them to a study of lines of the hydrogen spectrum in the 1930s.

1917 **34** *A new Determination of e, N and Related Constants. By R. A. Millikan*

Robert Millikan recognized in 1907 that a likely source of error in previous determinations of the electronic charge by the method of monitoring the fall of charged water drops could be attributed to evaporation loss during the time of observation. While undertaking experiments similar to those reported by H. A. Wilson (see p 12), he noticed, by chance, that in a high electric field a few individual drops remained stationary, the gravitational and electrical forces being exactly balanced. By keeping such drops in the field of view of the observing telescope, Millikan was able to select those drops that were not evaporating to a noticeable extent and to make measurements of their motion with and without an applied field in the way that Wilson had done earlier for the top of the cloud. He made the important discovery (*Philosophical Magazine* 1910 **19** 209) that individual drops always carried an exact multiple of the smallest charge measured and the latter he calculated to have a weighted mean of 4.65×10^{-10} esu — a significant increase over previously determined values of e, but in excellent agreement with half the value Rutherford and Geiger had obtained for the charge on α particles in 1908 (see p 87).

The crucial idea of using oil drops instead of water occurred to Millikan in 1909. In his 1950 autobiography he recounts of himself 'What a fool I have been to try in this crude way to eliminate the evaporation of water droplets when mankind has spent the last three hundred years in improving clock oils for the very purpose of obtaining a lubricant that will scarcely evaporate at all.' In 1910 he reported the first results using charged oil drops and by 1913, after further refining his techniques and methods of analysis, he gave a new value for e of 4.774×10^{-10} esu with an experimental error of about 1 part in 500.

The present paper has been selected because it is his last publication on this topic and includes a detailed description of the apparatus and the precautions he took in determining a value of e with a precision of about 0.1%. (The Appendix of the paper has been cut to include only raw data from one drop; twenty-two drops were in fact measured.) Millikan's final quoted value for e relied, as he recognized, on measurements by others of the viscosity of air. A new determination of this in the 1920s showed that the value Millikan had calculated for e was too low by about one-half of a percent. Re-evaluation of his result yields $e = 4.799 \times 10^{-10}$ esu, agreeing to within 0.2% of today's accepted value.

Other features of interest in the paper are, firstly, the footnote on page 2, which recalls the fact that J. J. Thomson, the 'discoverer' of the electron in 1897, was still referring 20 years later to the particles as 'corpuscles' — the name he had originally assigned to them; and, secondly, the evaluation at the end of the paper of other fundamental constants using the value of e.

Notes

1   Coherers were devices consisting of a quantity of metal filings contained between two metal plugs in a small evacuated glass tube. When a radio-frequency signal from an aerial was applied to the plugs, the device changed from a high to a low resistance state, a condition which remained until the tube was mechanically shaken. A precise understanding of the physics involved in their operation has never been obtained.

[429]

XLII. *A Determination of the Charge on the Ions produced in Air by Röntgen Rays.* By HAROLD A. WILSON, *Fellow of Trinity College, Cambridge* *.

THE experiments described in this paper were undertaken with the object of making a fresh determination of the charge on one ion. This charge will throughout this paper be denoted by e.

Prof. Townsend (Phil. Mag. Feb. 1898), in a paper on the "Electrical Properties of Newly Prepared Gases," has described a determination of the average charge on the droplets composing the cloud formed when newly prepared oxygen is bubbled through water. This charge was found to be about 3×10^{-10} electrostatic units of electricity. There are some reasons for supposing that each droplet contains one ion, and consequently Townsend's result may be regarded as a determination of the charge on one ion. The result which I have obtained is in very good agreement with his.

Prof. J. J. Thomson (Phil. Mag. Dec. 1898 and 1899) has given two estimates of e, the first depending on a determination of the average charge on the droplets of a cloud formed by condensation of water-vapour on the ions produced in air by Röntgen rays, and the second on a similar determination for the ions given off by a zinc plate under the action of ultra-violet light. The mean result of the first research was $e = 6 \cdot 5 \times 10^{-10}$ and of the second $e = 6 \cdot 8 \times 10^{-10}$ †.

Since from the value of e the number of molecules in a cubic centimetre of a gas can be immediately deduced, and also since the absolute value of e is of considerable interest in itself, a fresh determination by a different method appeared to be worth making.

The method I have used depends, like Prof. Thomson's, on the fact discovered by C. T. R. Wilson ‡, that the ions produced in air by Röntgen rays act as nuclei for the cloudy condensation of water-vapour when supersaturation exceeding a definite amount is produced by a sudden expansion.

* Communicated by Prof. J. J. Thomson.

† Since this paper was written Prof. Thomson has informed me that he has lately made a fresh determination of e by his original method, but with an improved apparatus, and he has very kindly consented to my mentioning the result he has obtained, here. It is $e = 3 \cdot 8 \times 10^{-10}$, and so agrees very well with the mean result of my experiments, viz. $3 \cdot 1 \times 10^{-10}$. It appears that in his earlier experiments the cloud was formed mainly on the negative ions and not on both positive and negative ions as was supposed at the time, consequently the result obtained was nearly twice too big.

‡ Phil. Trans. A. 1897, p. 265, and A. 1899, p. 403.

430   Mr. H. A. Wilson: *Determination of the Charge*

The droplets of the cloud produced presumably each contain one or more ions. Let a droplet containing one ion, and consequently having a charge e, have a mass m which can be determined by observing its rate of fall (v_1 say) in air. If now a vertical electrostatic field of strength X is applied to this droplet, there will be a vertical force on the droplet equal to Xe due to the field, so that the total force on the droplet will be $Xe + mg$, where g is the acceleration due to gravity, and reckoning Xe positive when it is in the same direction as the weight mg. Now the rate of steady motion of a sphere in a viscous fluid is proportional to the force acting on it, so that the rate of fall of the droplet will be altered by the electric field. Let it be now v_2. Then we have

$$\frac{mg}{mg + Xe} = \frac{v_1}{v_2}.$$

The relation between m and v_1 is given by the equation

$$m = 3 \cdot 1 \times 10^{-9} \times v_1^{\frac{3}{2}} \text{ *}$$

so that

$$e = 3 \cdot 1 \times 10^{-9} \frac{g}{X} (v_2 - v_1) v_1^{\frac{1}{2}}.$$

Thus if X is known measurements of v_1 and v_2 are sufficient to determine e absolutely. This is the method which I have employed.

It was found that, using strong Röntgen rays, some of the droplets in the cloud had bigger charges than others. In fact there sometimes appeared to be several sets of droplets having charges nearly in the ratios $1 : 2 : 3$. It appeared, therefore, that some of the droplets contained one ion, some two ions, and so on. This agrees with Prof. Thomson's observation that when the strength of the Röntgen rays was increased beyond a certain amount, the number of droplets in his clouds did not increase proportionally to the number of ions present at the moment of expansion. Prof. Thomson therefore used weak rays so that in his experiments each droplet probably only contained one ion, which is a necessary condition for the success of the method he employed.

The principal advantages of my method are that it is not necessary to estimate either the number of drops in the cloud, or the number of ions present at the moment of its formation, or to make the assumption that each droplet contains only one ion. Both these estimations involve assumptions which in practice can only be approximately true, and there is

---

* J. J. Thomson, Phil. Mag. Dec. 1899, p. 561.

always a danger that some of the drops in the cloud contain more than one ion.

The apparatus used is shown in the accompanying diagram.

It consisted of a glass tube AB about 4 cms. in diameter and 10 cms. long. Its lower end was closed by an india-rubber stopper and its upper end joined on to a short length of narrow tubing. Two circular brass disks, C and D, each 3·5 cms. in diameter, were supported one above the other in

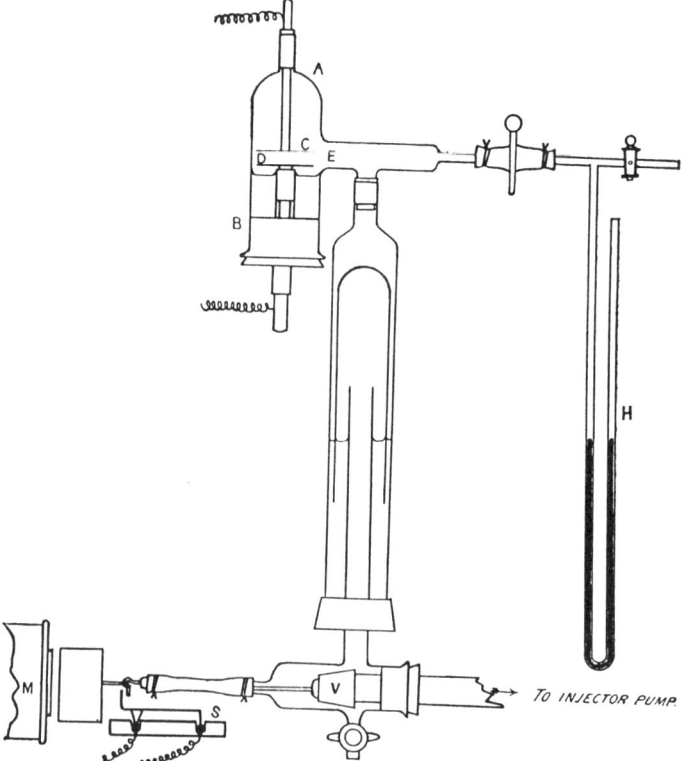

this tube as shown; the cloud on which the observations were made was formed between them, and they could be maintained at any required difference of potential up to 2000 volts by means of a battery of small secondary cells. A glass tube E was sealed on to the side of AB and served to connect the space between the disks with an apparatus for producing a sudden expansion of any desired amount. A small mercury manometer (H) was used to measure the expansion.

432   Mr. H. A. Wilson : *Determination of the Charge*

The expansion apparatus used was kindly lent to me by Mr. C. T. R. Wilson, and it was similar to those he has described in the papers referred to above. The apparatus was arranged so that the valve V, the opening of which produces the sudden expansion, could be pulled back suddenly by means of an electromagnet (M). This enabled the valve to be pulled away every time in exactly the same way.

The space in the tube AB below the disks was filled with water so that the air between the disks was thoroughly saturated with moisture. This air was rendered " dust free " in the usual way, by repeated expansions with intervals in between to allow the clouds formed to settle. The apparatus was then ready for a measurement of e.

A Röntgen-ray bulb was worked near AB, so that the rays passed between the disks. Then the battery circuit through the magnet was closed and a sudden expansion so produced. A cloud was thus formed between the disks, and the time which its upper surface took in falling from the upper disk to the lower disk was measured. This gave v_1 the rate of fall without an electric field. The experiment was then repeated, but immediately after the expansion the disks were connected to the battery, and so v_2, the rate of fall in an electric field, was obtained.

It was found that if the rays were kept on all the time during an experiment, then very large values for the charge on each droplet were obtained. A field of a few hundred volts per centimetre was then sufficient to cause many of the droplets to rise instead of falling. It soon became clear that the fresh ions formed after the expansion attached themselves to the droplets, so that the longer the rays were kept on after the expansion the bigger the charge on the droplets became. A switch S was therefore put in the primary circuit of the induction-coil used to excite the Röntgen bulb and arranged so that the armature of the magnet turned the switch, broke the circuit, and so stopped the rays a small fraction of a second before the expansion was produced.

The disks C and D were also connected to a commutator which first connected them together, and then on being turned connected them to the large battery used to charge them up.

A narrow beam of light was passed between the disks C and D to illuminate the cloud and enable its upper surface to be observed. The falling of the cloud was watched through a small hole on a level with the disks, and about twenty centimetres away from them in a direction nearly perpendicular to the beam of light. A second screen was put up close to the apparatus having a vertical slit in it through which only the central portion of the illuminated part of the cloud could be seen. This slit and the beam of light were each about half a centimetre wide, so that the portion of the cloud which was observed was that occupying a vertical prism half a centimetre square at the axis of the tube AB between the disks.

The disks were never more than one centimetre apart, and consequently very little circulation of the air could take place between them. When a cloud is formed by expansion in a large vessel, the walls of the vessel heat up the air near them which produces a circulation of the air upwards near the walls and downwards in the middle. If this sort of thing happened in these experiments, v_1 and v_2 would both be obtained too high ; but it was found that when the disks were not more than a centimetre apart the circulation which occurred near the glass walls of the tube did not extend to the centre, and the surface of the cloud between the disks remained plane as the cloud fell.

The disks were always connected together until the expansion had taken place, when, if it was desired to determine v_2, they were immediately connected to the battery by turning the commutator lever. If they were connected to the battery before the expansion took place no cloud was obtained because the field removed the ions as fast as they were formed.

In making a measurement of e the time of fall from the upper disk to the lower one was measured with a stop-watch, alternately with and without the electric field. v_1 and v_2 were then calculated from the mean results for the times of fall.

C. T. R. Wilson (Phil. Trans. A, 1899, p. 440) found that with an expansion of nearly 15 cms. of mercury only the negative ions acted as nuclei, but with greater expansions condensation occurred on both positive and negative ions. These results were easily verified with my apparatus. With an expansion of 15 cms. charging the upper disk negatively caused the whole cloud to fall more quickly than it fell when the disks were uncharged, while charging the upper disk positively reduced the rate of fall of the cloud. It was clear, therefore, that the droplets were negatively charged.

Also with considerably larger expansions than 15 cms. some of the droplets fell more slowly and some more quickly when the disks were charged than when they were not charged, showing that both positively and negatively charged droplets were present. However, there always seemed to be more negatively charged droplets present than positively

434　　Mr. H. A. Wilson : *Determination of the Charge*

charged ones, and unless the expansion used was nearly enough to produce a fog even in the absence of any ions, the positively charged droplets were not very easy to observe. There seemed, in fact, to be a large excess of negative ions present between the disks and not equal numbers of positive and negative ions. The explanation of this is, no doubt, to be found in the secondary radiation emitted by brass under the action of Röntgen rays. This secondary radiation has been proved to consist of negatively charged corpuscles, so that an excess of negative ions in the air near the disks might have been expected. The excess would, however, not have been expected to be as great as appeared to be the case. It is perhaps possible that when both positive and negative ions are present together, condensation takes place mainly on the negative ions, although when either kind are present alone, there is very little difference between the supersaturations required to produce condensation on the positive or negative ions.

An expansion of from 16 to 17 cms. of mercury was always used in the experiments described below, and all the results are for the charge on the negative ions.

All the droplets did not always fall at the same rate when the electric field was applied. This appeared to be nearly always the case, but was especially so when strong rays were used. There appeared to be several sets of droplets, each set falling all at the same rate. The rate of fall of the most numerous set indicated that the droplets in it had the smallest charges. The observations given below refer to this set only, the other sets will be considered later.

Since the cloud begins to evaporate soon after it is formed, it is very important to get the measurement of its rate of fall over as quickly as possible. I therefore generally only allowed it to fall about half a centimetre, and applied the electric field in the direction which increased the rate of fall. Another reason why a very small distance between the disks had to be used, was that the available P.D. was only 2000 volts, so that unless the disks were near together, the electric field between them was not strong enough to appreciably alter their rate of fall. For these various reasons nearly all the observations were made with the disks as near together as possible, because it was clear that reliable results could not otherwise be obtained. For the same reason the maximum P.D. available was used in nearly every case. It would of course have been more satisfactory if observations could have been made with a greater variety of distances between the *on the Ions produced in Air by Röntgen Rays.*   435

plates, and through a larger range of P.D., but to accomplish this with the battery available was not possible.

The following table contains the results of a set of observations: —

Distance between the disks $d = 0.45$ cm.
Potential-difference used 1800 volts.

Time of Fall without P.D. $X = 0$.	Ditto with P.D. $X = 13.3$ E.S. units.
secs.	secs.
(1) 23·6	(2) 17·8
(3) 23·3	(4) 16·9
(5) 23·9	(6) 17·0
(7) 23·8	(8) 17·2
Mean…23·65	Mean…17·22

The numbers in brackets refer to the order in which the observations were made. The above results give

$$v_1 = \frac{0.45}{23.65} = 0.0190 \ \frac{\text{cm.}}{\text{sec.}}$$

$$v_2 = \frac{0.45}{17.22} = 0.0262 \ \frac{\text{cm.}}{\text{sec.}}$$

Consequently, since

$$e = 3.1 \times 10^{-9} \frac{g}{X} (v_2 - v_1) v_1^{\frac{1}{2}},$$

we have

$$e = \frac{3.1 \times 10^{-9} \times 981 \times 0.0072 \times (0.019)^{\frac{1}{2}}}{13.3}$$

$$= 2.3 \times 10^{-10} \text{ E.S. units.}$$

Also

$$m = 3.1 \times 10^{-9} \times v_1^{\frac{3}{2}} = 8.1 \times 10^{-12} \text{ gram.}$$

The method of experimenting can be varied by measuring the velocity of fall first with the field in one direction and then with the field in the opposite direction. If v_2 and v_3 are the velocities, then

$$\frac{mg + Xe}{mg - Xe} = \frac{v_2}{v_3},$$

taking v_2 to be the velocity when Xe acts in the downward direction. The mean of v_2 and v_3 gives the velocity when $X = 0$.

436   Mr. H. A. Wilson : *Determination of the Charge*

In an experiment made in this way the following numbers were obtained :—

Distance between the disks 1·0 cm.
P.D. = 2000 volts.

Time of Fall.

X = +6·7.	X = −6·7.
secs.	secs.
(1) 18·4	(2) 21·6
(3) 25·7	(4) 27·1
(5) 19·4	(6) 22·4
(7) 24·2	(8) 27·8
(9) 27·0	(10) 32·6
Mean...22·94	Mean...26·3

These numbers give

$$v_2 = 0.0436 \frac{\text{cm.}}{\text{sec.}}$$

$$v_3 = 0.0380 \frac{\text{cm.}}{\text{sec.}}$$

Also

$$e = \frac{3.1 \times 10^{-9}}{2^{\frac{3}{2}}} \frac{g}{X} (v_2 - v_3)(v_2 + v_3)^{\frac{1}{2}},$$

so that

$$e = 2.6 \times 10^{-10} \text{ E.S. units.}$$

and

$$m = 3.1 \times 10^{-9} \times \left(\frac{v_2 + v_3}{2}\right)^{\frac{3}{2}} = 2.5 \times 10^{-11} \text{ gram.}$$

As already mentioned the cloud soon begins to evaporate after it is formed, so it is important to get the measurement of its rate of fall over as quickly as possible. It was, therefore, found most satisfactory to use the rates of fall with X = 0 and with X positive, making the rate of fall greater than when X = 0.

In making a series of measurements an observation with X positive was always made as quickly as possible after one with X = 0, in order that the strength of the rays and other conditions should be as nearly as possible the same in both cases. Although the individual observations in a series, say with X = 0, often vary a good deal, yet there is usually a corresponding variation in the observations with X positive, so that the value of *e* obtained from the mean results for the

on the Ions produced in Air by Röntgen-Rays. 437

series is not necessarily affected by any error due to these variations.

The following tables contain all the other results obtained except a few done at an early stage, before the apparatus had been got to work satisfactorily, none of which are included.

$d = 0·50$ cm.
P.D. $= 2000$ volts.

t_1. (X=0.) secs.	t_2. (X=+13·3.) secs.
12·2	9·6
11·1	9·3
11·4	9·3
12·0	9·2
10·6	9·6

Mean 11·4   Mean 9·4

$v_1 = 0·0439$ $v_2 = 0·0530$
$m = 2·86 \times 10^{-11}$
$e = 4·37 \times 10^{-10}$.

$d = 0·50$ cm.
P.D. $= 2000$ volts.

t_1. (X=0.) secs.	t_2. (X=+13·3.) secs.
18·3	10·8
20·3	15·6
18·2	17·6
18·0	13·8
18·4	15·4

Mean 18·64   Mean 14·64

$v_1 = 0·0268$ $v_2 = 0·0341$
$m = 1·36 \times 10^{-11}$
$e = 2·73 \times 10^{-10}$

$d = 0·5$ cm.
P.D. $= 2000$ volts.

t_1. (X=0.) secs.	t_2. (X=+13·3.) secs.
14·9	13·2
15·0	11·4
14·9	12·2
14·0	10·7

Mean 14·7   Mean 11·87

$v_1 = 0·034$ $v_2 = 0·042$
$m = 1·95 \times 10^{-11}$
$e = 3·4 \times 10^{-10}$.

$d = 0·55$ cm.
P.D. $= 2000$ volts.

t_1. (X=0.) secs.	t_2. (X=+12·1.) secs.
15·6	13·0
17·2	12·8
16·0	12·4
17·0	13·8
17·4	14·1
18·4	15·0
17·2	14·1
16·0	12·8
16·9	13·1
16·7	12·6

Mean 16·84   Mean 13·37

$v_1 = 0·0327$ $v_2 = 0·0411$
$m = 1·83 \times 10^{-11}$
$e = 3·81 \times 10^{-10}$.

438   Mr. H. A. Wilson: *Determination of the Charge*

<table>
<tr><td colspan="2">$d = 0\cdot4$ cm.
P.D. $= 2000$ volts.</td><td colspan="2">$d = 0\cdot40$ cm.
P.D. $= 2000$ volts.</td></tr>
<tr><td>t_1.
(X=0.)
secs.</td><td>t_2.
(X=+16·7.)
secs.</td><td>t_1.
(X=0.)
secs.</td><td>t_2.
(X=+16·7.)
secs.</td></tr>
<tr><td>21·5</td><td>10·1</td><td>21·0</td><td>12·4</td></tr>
<tr><td>21·9</td><td>13·0</td><td>20·4</td><td>13·2</td></tr>
<tr><td>20·9</td><td>12·0</td><td>20·6</td><td>13·0</td></tr>
<tr><td>21·0</td><td>12·0</td><td></td><td></td></tr>
<tr><td>19·4</td><td>12·0</td><td>Mean 20·7</td><td>Mean 12·9</td></tr>
<tr><td>21·6</td><td>11·6</td><td></td><td></td></tr>
</table>

Mean 21·05   Mean 11·80
$v_1 = 0\cdot0190$ $v_2 = 0\cdot0340$
$m = 8\cdot1 \times 10^{-12}$
$e = 3\cdot8 \times 10^{-10}$.

$v_1 = 0\cdot0193$ $v_2 = 0\cdot0310$
$m = 8\cdot3 \times 10^{-12}$
$e = 3\cdot0 \times 10^{-10}$.

$d = 0\cdot4$ cm.
P.D. $= 2000$ volts.

t_1. (X=0.) secs.	t_2. (X=+16·7.) secs.
20·0	12·0
20·4	12·0

Mean 20·2   Mean 12·0
$v_1 = 0\cdot0198$ $v_2 = 0\cdot0334$
$m = 8\cdot64 \times 10^{-12}$
$e = 3\cdot5 \times 10^{-10}$.

$d = +0\cdot40$ cm.
P.D. $= +1500$ volts.

t. (X=0.) secs.	t_2. (X=+12·5.) secs.
33·6	20·0
33·0	20·0
30·5	20·0
31·6	20·4
29·4	19·6

Mean 31·6   Mean 20·0
$v_1 = 0\cdot0126$ $v_2 = 0\cdot0200$
$m = 4\cdot4 \times 10^{-12}$
$e = 2\cdot04 \times 10^{-10}$.

$d = 0\cdot44$ cm.   P.D. $= 2000$ volts.

t_1. (X=0.) secs.	t_2. (X=+15·2.) secs.
21·8	15·4
22·6	18·2
23·4	16·7
23·6	17·2
	17·6
	16 4
	14·6
	14·4
	14·9

Mean...22·85   Mean...16·10
$v_1 = 0\cdot0193$ $v_2 = 0\cdot0272$ $m = 8\cdot3 \times 10^{-12}$ $e = 2\cdot3 \times 10^{-10}$.

The following table contains a summary of the above results :—

d.	X.	$v_1.$	$v_2.$	*m.*	*e.*
0·45	13·3	$1·9 \times 10^{-2}$	$2·62 \times 10^{-2}$	$8·1 \times 10^{-12}$	$2·3 \times 10^{-10}$
1·00	±6·7	......	4·36 ,,	25 ,,	2·6 ,,
0·50	13·3	4·39 ,,	5·3 ,,	28·6 ,,	4·4 ,,
0·50	13·3	2·68 ,,	3·41 ,,	13·6 ,,	2·7 ,,
0·50	13·3	3·4 ,,	4·2 ,,	19·5 ,,	3·4 ,,
0·55	12·1	3·27 ,,	4·11 ,,	18·3 ,,	3·8 ,,
0·40	16·7	1·9 ,,	3·4 ,,	8·1 ,,	3·8 ,,
0·40	16·7	1·93 ,,	3·1 ,,	8·3 ,,	3·0 ,,
0·40	16·7	1·98 ,,	3·34 ,,	8·6 ,,	3·5 ,,
0·40	12·5	1·26 ,,	2·00 ,,	4·4 ,,	2·0 ,,
0·44	15·2	1·93 ,,	2·72 ,,	8·3 ,,	2·3 ,,
				Mean	......$3·1 + 10^{-10}$

It will now be convenient to consider the less numerous sets of droplets which fell quicker than the principal set on which the above observations were made.

When no field was applied the whole cloud fell at the same rate and its upper surface was sharp. No sign of any separation into sets could be detected. When the field was applied the cloud fell quicker than before, but otherwise its appearance was at first the same. After a few seconds, however, the surface of the cloud began to separate into two ; apparently some of the cloud falling quicker than the rest. The line of separation between the two sets was fairly sharp. Sometimes three such sets were observed.

The following numbers were obtained in one series of experiments with the disks 0·4 cm. apart :—

X=0.	X = +12·5. Principal set.	X = +12·5. Second set.	X = +12·5. First set.
secs.	secs.	secs.	secs.
33·6	20·0	15·4	11·0
33·0	20·0	15·0	10·6
30·5	20·0	14·0	10·0
31·6	20·4	...	10·8
29·4	...	...	10·4
Mean 31·62	Mean 20·1	Mean 14·8	Mean 10·56
$v_1 = 0·0126$	$v_2 = 0·020$	$v_3 = 0·027$	$v_4 = 0·038$

We might suppose that the subsidiary sets are produced by two droplets coalescing under the influence of the field, but it is easy to show by calculation, that a droplet with twice the mass and twice the charge of the others ought to have

2 G 2

440 *Charge on the Ions produced in Air by Röntgen Rays.*

fallen in about 6 secs. in the above experiment. If we suppose that two droplets, one with a positive charge and the other with a negative charge, coalesced, which of course is a probable thing to happen, the resulting droplet with twice the mass and no charge ought to have fallen in 11·2 secs., which is very nearly the mean time (10·6 secs.) taken by the quickest set to fall. However, it is not easy to see how droplets coalescing could produce a set of drops having a sharp upper limit, for we should expect coalescence to occur from time to time during the existence of the cloud. The existence of a sharp upper surface to the set seems to show that all the droplets forming it were formed at the moment of the expansion.

Another possible explanation of these sets seems to be that when the cloud is formed some of the droplets contain more than one ion. If two ions were very near together during the expansion they might easily give rise to only one droplet. An objection to this view is that such a droplet ought to be larger than one containing a single ion. This objection, however, falls to the ground when the magnitude of the effect of the charge on the equilibrium size of the droplets is remembered, for it is known to be very small.

If we suppose that the droplets in the three sets are all of the same size, but have different charges, then it is easy to calculate these charges. The results of this calculation for the observations given above are

Principal set, charge per droplet	2·04 × 10⁻¹⁰	
Second set,	,,	3·94 ,,
First set,	,,	6·94 ,,

If, then, the principal set has one ion per droplet, the second has two, and the first about three.

It has been shown by Townsend (Phil. Trans. A. 1899, p. 129) that the charge on an ion produced in air by Röntgen rays or by other forms of radiation is equal to the charge on the hydrogen ion or atom in solutions. According to the result of the present experiments it consequently follows that the charge on one hydrogen atom is $3\cdot1 \times 10^{-10}$ E.S. unit or 10^{-20} of an electromagnetic unit. One E.M. unit of electricity deposits from a solution 0·01118 gram of silver in electrolysis, and consequently $\dfrac{0\cdot01118}{107\cdot11} = 1\cdot043 \times 10^{-4}$ gram of hydrogen. It follows that the mass of an atom of hydrogen is approximately $10^{-4} \times 10^{-20} = 10^{-24}$ of a gram. The mass of a molecule of hydrogen is therefore 2×10^{-24} of a gram, so that since the mass of one cubic centimetre of hydrogen at 0° C. and 760 mms. of mercury pressure is 9×10^{-5} gram, the number of molecules (N) in one cubic centimetre of hydrogen is $\dfrac{9 \times 10^{-5}}{2 \times 10^{-24}}$ or approximately $N = 4 \times 10^{19}$.

The mean result of the present experiments, viz. $e = 3\cdot1 \times 10^{-10}$ of an electrostatic unit, cannot be very far from the truth. I think that it may be considered established by these experiments that e lies between 2×10^{-10} and 4×10^{-10} E.S. unit.

The values of N which have been obtained from the kinetic theory of gases vary between rather wide limits. The value obtained depending usually on the radius assigned to a molecule of the gas under consideration. O. E. Meyer ('Kinetic Theory of Gases,' p. 333) gives the value $N = 6\cdot1 \times 10^{19}$, which is based on the assumption that the average radius of a molecule of air is 10^{-8} cm. If $N = 4 \times 10^{19}$ then the average radius of a molecule of air must be $1\cdot2 \times 10^{-8}$ cm.

A great many different lines of argument (see Meyer's 'Kinetic Theory of Gases') lead to values for the radius of a molecule or sphere of molecular action near to 10^{-8} cm., but the magnitude of this quantity certainly cannot be considered to be established except within limits not very near together. The agreement between the value of N obtained from the present experiments, and the values deduced from the kinetic theory of gases, may consequently be considered as good as could have been expected.

In conclusion I wish to say that my best thanks are due to Prof. J. J. Thomson for much valuable advice during the carrying out of these experiments in the Cavendish Laboratory.

Ionisation of Air under Normal Conditions.   681

pressures and corresponding dimensions of vessels for different gases $\left(\text{according to } h=1; \quad n=\alpha; \quad b=\dfrac{\beta}{\sqrt{\alpha}}\right).$

Also the thermic variability of conductivity—not yet known with desirable precision—may be investigated in an analogous manner, by application of similar motions. If we make use, for the higher temperatures, of vessels with dimensions increased in proportion of the first, and of pressure increased in proportion of the $(\epsilon-\tfrac{1}{2})$th power of temperature, the quantity of heat transferred must be proportional to θ^{ϵ}, whence ϵ may be determined. The method of heating wires by electric currents may be easily adapted to this way of experimenting.

We confine ourselves to these few examples on this sort of similarity, since its range of applications is less extensive and since there is little experimental work hitherto done which could serve as a basis for further speculations.

LXXV. *The Condensation Method of Demonstrating the Ioni-sation of Air under Normal Conditions.* By C. T. R. WILSON, *M.A., F.R.S., Fellow of Sidney Sussex College, Cambridge**.

SOME years ago I described experiments † which proved that when air saturated with water-vapour has been freed from dust particles, it will still give condensation in the form of drops on sudden expansion provided the expansion exceeds a definite limit. If $v_1\,v_2$ be the volume of the air before and after the sudden expansion, then if v_2/v_1 be less than 1·25 no drops are produced on expansion, but if this critical expansion be exceeded a rainlike condensation results. The drops remain comparatively few if v_2/v_1 does not exceed a second limit about 1·38. It was found that exposure of the air to Röntgen or other ionising rays increased enormously the number of drops produced by expansions between these limits, the least expansion required to cause the formation of drops remaining, however, the same. It was concluded that the nuclei giving the clouds in air exposed to Röntgen rays are to be identified with the ions to which its conducting power under the action of the rays is attributed, and that the few drops always produced with expansions exceeding the critical value are due to ions of the same nature continually being produced even in the absence of the rays.

* Communicated by the Author.
† Phil. Trans. vol. clxxxix. p. 265 (1897).

682   Mr. C. T. R. Wilson *on the Condensation Method*

Further experiments showed * that the number of drops produced by expansions between the above-mentioned limits in air exposed to Röntgen rays, is reduced in a very striking manner when a sufficiently strong electric field is maintained across the air before expansion, thus proving that the nuclei move in an electric field and are therefore electrically charged, and presumably identical with the ions to which the conducting power is due. On the other hand, similar experiments made in the absence of ionising agents failed to show any diminution of the number of drops by the action of even very strong fields. The absolute identity of the degree of supersaturation required to cause condensation upon ions and upon the nuclei to which the rainlike condensation is due, made it difficult to believe that the latter are not ions also, and to explain their non-removal by an electric field it was suggested that they might be ions produced in some way as a result of the expansion. When, however, subsequent experiments † on the leakage of electricity from conductors suspended within closed vessels, showed that a continual slight ionisation of the air is always going on in such vessels, it appeared more likely that the rainlike condensation really is due to this ionisation, and that the failure to detect any diminution in the number of drops under the action of an electric field is due to some defect in the conditions of the experiments. In the experiments thus far made the vessels used had been small, and to permit of a strong electric field being applied the air was enclosed between conducting surfaces generally only a centimetre or less apart ; in many cases one of the conductors was a layer of water at the bottom of the vessel, the other being a horizontal metal plate coated with wet filter-paper. The drops were under these conditions very few whether an electric field was applied or not ; it was thought that if a much larger volume of air were used there would be more chance of detecting the diminution in number when an electric field was applied. This expectation has been realized. With the large apparatus described below the effect of an electric field in removing the nuclei which gave rise to the rainlike condensation is very striking.

The construction of the apparatus ‡ (shown in the figure) is the same in principle as in the experiments on condensation nuclei which I have described in previous papers. On

---

* Phil. Trans. vol. cxcii. p. 403 (1899).

† Geitel, *Physikalische Zeitschrift,* vol. ii. p. 116 ; C. T. R. Wilson, Roy. Soc. Proc. vol. lxviii. p. 151.

‡ The apparatus was made by Messrs. W. G. Pye & Co., Cambridge. To Mr. Pye I am indebted for many suggestions as to the mechanical details.

of Demonstrating the Ionisation of Air.   683

account of the much larger size of the new apparatus, the mechanism by which the sudden expansion is produced was

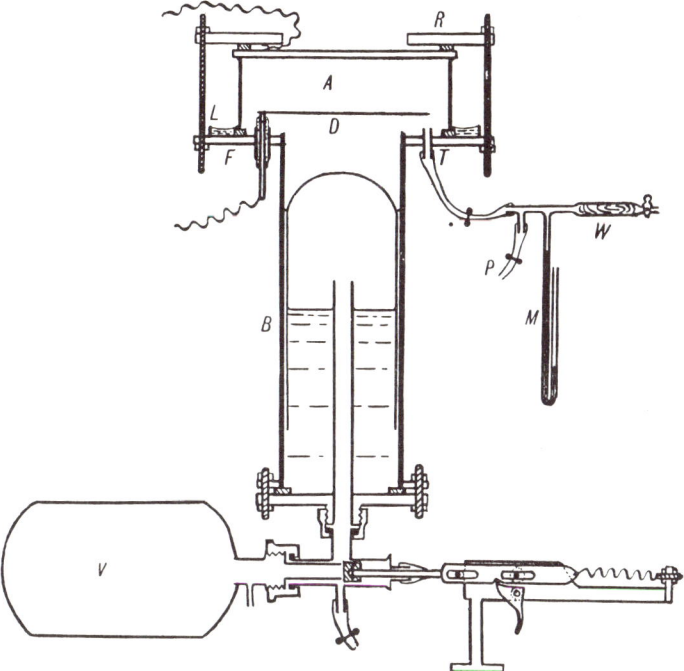

constructed, however, of brass, not, as in the older experiments, of glass. The cloud chamber A, in which the drops formed by expansion are viewed, is a glass cylinder 18·5 cms. in internal diameter and 5·9 cms. high. Its roof consists of a thick brass disk cemented to it by means of sealing-wax. The cylinder rests on an indiarubber ring lying on an annular brass plate F, which forms a flange at the top of the expansion cylinder B. The glass cylinder is squeezed down on the indiarubber by means of an upper annular brass plate R resting on the roof of the cloud chamber, from which it is separated by a second indiarubber ring ; the upper and lower annular plates are connected by six bolts, by means of which the necessary pressure can be applied. The external diameter of the annular plates is 26 cms. ; about one cm. from the edge of the lower one on its upper surface a thin ring L of brass 1·2 cm. high is soldered. This serves to contain water, all risk of air leaking in below the edge of the glass cylinder being thus removed. Through three symmetrically placed tubes penetrating the lower plate of the cloud chamber are

684　　Mr. C. T. R. Wilson *on the Condensation Method*

sealed three insulated brass rods supporting a horizontal brass disk D, 15·3 cms. in diameter. Between this disk and the roof of the cloud chamber, 4·7 cms. above it, any desired difference of potential could be maintained by means of a battery of storage-cells. Both this disk and that forming the roof were covered on the surfaces facing one another with wet filter-paper. In addition to the three tubes through which pass the supports of the brass disk, the floor of the vessel is pierced by a fourth smaller tube T, by means of which air can be removed from or admitted into the apparatus.

Below the cloud chamber and supporting it is a vertical brass expansion cylinder B, 10 cms. in internal diameter and 30 cms. long. Sliding freely in this and serving as a piston is a thin-walled brass cylinder open below and with a hemispherical top, the length of the cylindrical part being 18·75 cms., the thickness of the walls being less than one millimetre. The expansion cylinder is bolted by means of a flange at its lower end against a thick brass disk, an india-rubber ring, of which the internal diameter is considerably less and the external diameter greater than that of the cylinder, being inserted between them. Rising up from the centre of the disk is a brass tube 18 cms. long and 1·3 cms. in internal diameter. The cylinder is filled with water to within a few cms. from the top of this tube. By means of the mechanism to be presently described, the central tube can be put into sudden communication with a vacuum chamber V, thus causing the piston to fly sharply down against the indiarubber at the bottom of the cylinder, and to remain pressed tightly against this so that no air or water can escape. It is in this way that the sudden expansion is produced : on putting the central tube in communication with the atmosphere instead of the vacuum chamber, the piston rises to its original position.

The thick brass disk to which the expansion cylinder is attached rests upon an iron tripod (not shown in the figure) to the top of which it is firmly fixed by three screws. The feet of the tripod are screwed down to a board. The tubes for making connexion with the vacuum chamber are shown below the expansion cylinder. For convenience the connexions are made with screw-joints, indiarubber washers being inserted to prevent leakage. The vacuum chamber was a brass cylinder 22 cms. long and 14 cms. in diameter, with rounded ends ; it was maintained at low pressure by a water-jet pump. A gauge was connected to avoid the risk of making an expansion while the vacuum was not sufficiently good. The construction of the mechanism for making sudden

of Demonstrating the Ionisation of Air.   685

communication with the vacuum chamber is the same as in
the smaller apparatus described in previous papers, but of
brass instead of glass ; its mode of working will be understood from the figure.   An indiarubber stopper held tightly
by the pressure of the atmosphere against the end of the
tube leading to the vacuum-chamber V can be suddenly
pulled away by a spring released by the trigger arrangement
shown.   In this way the sudden motion of the piston, and
the consequent expansion of the air in the cloud chamber,
are effected.

The final volume of the air after expansion is always
the same, for the piston is then held against the indiarubber
at the bottom, owing to the low pressure below.   To vary the
expansion the initial volume has to be varied.   The air before
expansion is always at very nearly atmospheric pressure—
really at a pressure less than that of the atmosphere by the
pressure required to balance the weight of the partially immersed piston, the air below the piston being always before
expansion at atmospheric pressure.   The amount of any
expansion is determined by adjusting the pressure after opening communication with the gauge M, before allowing the
piston to rise after the previous expansion, a sufficiently long
interval being allowed to elapse for the temperature to return
to that of the surroundings ; the constancy of the pressure
serves as a test of this condition being fulfilled.   The pressure
thus determined will be the same as the pressure after the
next expansion has taken place and the temperature has
again become steady.   The ratio of the air-pressure before
expansion (*i. e.*, of the whole pressure in the cloud chamber less
the saturation pressure of the aqueous vapour) to that after
expansion will then be equal to v_2/v_1, the ratio of the final to
the initial volume.   To adjust the final pressure to any desired
value, air can be admitted from the atmosphere through the
cotton-wool filter W, or removed by opening communication
with the water-pump through the tube P.

The drops resulting from expansion are illuminated by a
narrow beam of light converging to a focus at the centre of
the cloud chamber ; the source was in most cases an arc or
lime-light ; the effects were, however, quite easily observed
with the light from an ordinary luminous flame.   It was
found convenient to coat the outside of the glass with black
enamel over half the circumference, leaving, however, a
vertical strip about one cm. wide in the middle for the light
to enter ; the glass immediately opposite this slit was also
blackened over a width of a few cms.   Any drops produced
were then well seen on looking towards the centre of

686   Mr. C. T. R. Wilson *on the Condensation Method*

the cylinder through the unblackened portions of the glass*.

The apparatus gave a value for the least expansion required in order that rainlike condensation might result, which agreed well with that obtained in previous experiments with much smaller apparatus.

The results of one series of measurements, those of May 15th, 1903, are given below.

p =gauge-reading when piston is at bottom.

w =pressure required to support weight of piston.

B=barometer-reading.

π =maximum vapour-pressure of water at temperature of experiment.

Then
$$v_2/v_1 = \frac{B-w-\pi}{B-p-\pi}.$$

In the experiment temperature $=15°$ C. Barometer$=766$ mm. $\pi=13$ mm. $w=5$ mm.

In the Table which follows the result of the expansion corresponding to various values of the gauge-reading is given (*a*) when the upper and lower plates were metallically connected; (*b*) when a difference of potential of 160 volts was maintained between them.

Gauge-reading in millimetres.	P.D.=0.	P.D.=160 volts.
159	Shower.	
153	Very few drops.	
151	No drops.	
152	No drops.	
153	Very few drops.	No drops.
155	Shower.	No drops.
160·5	Shower.	No drops.
171	Shower.	Very few drops.
177	Dense shower.	Very few drops.

Value of v_2/v_1 when rainlike condensation begins (in the absence of an electrical field)

$$v_2/v_1 = \frac{766-5-13}{766-153-13} = 1·247.$$

In the early experiments† the value found for the least expansion required for rainlike condensation was $v_2/v_1 = 1·252$.

* The apparatus was exhibited at the British Association meeting at Southport last September, and the removal of the ions by an electric field demonstrated.

† Phil. Trans. vol. clxxxix. p. 265.

of Demonstrating the Ionisation of Air.   687

The above observations show also the very marked effect of an electric field. It will be observed that when v_2/v_1 only slightly exceeds the critical value, a difference of potential of 160 volts entirely prevents the formation of drops. When the expansion is made larger a few drops are seen even in presence of the field ; the drops were found to increase in number as the expansion was increased. That some drops should be formed when the expansion considerably exceeds the critical value, even in presence of a strong field, is not surprising ; for the supersaturation when nuclei are few or absent will exceed the critical value for a finite time, the longer the greater the expansion. Any ions, set free while the supersaturation exceeds the critical value, will come into action as condensation nuclei and give rise to drops before the electric field has had time to remove them.

A difference of potential of 40 volts between the plates was found to reduce the number of drops formed on expansion to sensibly the same degree as 1000 volts. Even a potential-difference of two volts produced a noticeable diminution. In the older small-scale experiments, in which the distance between the plates was often considerably less than one cm., the strength of the field for a given difference of potential would be about five times as great as in the present experiments, and the maximum distance the ions had to travel only one-fifth as great ; the effect of a given difference of potential in removing ions would then be much greater. It is possible that accidental differences of potential may in some cases have already largely reduced the number of ions present, so that the additional reduction following the application of much stronger fields was not noticeable. In the small apparatus also the number of ions present per c.c. in the absence of an electric field would be less on account of the much greater rate of loss of ions by diffusion to the walls of the vessel. The number of drops being small a larger share of water would fall to each, and they would fall too rapidly for variations in their number to be readily detected.

The total number of ions present when a steady state is reached, in the absence of an electric field, is such that the number of ions removed per second by recombination and by diffusions to the sides is equal to the number produced per second. On account of the small rate of production of ions, diffusion rather than recombination is in these experiments the more important factor in limiting the number of ions even with the large apparatus ; with the small apparatus the loss of ions by recombination is negligible in comparison with that due to diffusion to the walls. Let us assume, for example,

688   Mr. C. T. R. Wilson *on the Condensation Method*

that 40 ions of either sign are produced per second in each c.c. This is the rate of ionization deduced from leakage experiments in a small vessel of silvered glass*. (The number 20, given in the original paper, was deduced from the experiments by using J. J. Thomson's first value for the ionic charge, since shown by him and by H. A. Wilson to be about twice too large †.)   If q be the rate of production of ions per c.c. per second, then the number of ions of either sign in each c.c., if loss by diffusion be ignored, will be given by

$$N = \sqrt{q/\alpha},$$

where α is the coefficient of recombination.   Putting $q=40$, $\alpha=3\cdot3 \times 10^3 e$ (where e is the ionic charge) and $e=3 \times 10^{-10}$ we have $N=6 \times 10^3$.   If, on the other hand, we neglect the loss by recombination and consider only the loss by diffusion, then in a column of one sq. cm. in cross section extending from plate to plate and perpendicular to the plates, the number of ions when a steady state is reached is equal to $\dfrac{2}{3}\dfrac{q}{D}l^3$,

where $2l$ is the distance between the plates and D is the coefficient of diffusion of the ions through the gas (J. J. Thomson, 'Electrical Properties of Gases,' p. 21).   The average number for every c.c. of air between the plates will be

$$N = \frac{1}{3}\frac{q}{D}l^2.$$

For negative ions in moist air D is equal to 0·035 ‡.   Thus when q is 40, $N=4 \times 10^2 l^2$.   In an apparatus with plates less than a centimetre apart, as in many of the older experiments, so that l is less than $\frac{1}{2}$, N is less than 100 per c.c. instead of the 6×10^3 obtained when only loss by recombination is considered : thus in this case the final number of ions is determined by diffusion, the loss by recombination being negligible.   With the larger apparatus of the present paper $2l=5$cms., $N=4 \times 10^2 \times 2\cdot5^2=2\cdot5 \times 10^3$ when we consider only the loss of ions by diffusion.   The loss by recombination is now no longer negligible, but the loss by diffusion is the more important factor ; the total number of ions when a steady condition is reached will be less than the value obtained when either recombination or diffusion is ignored, it must be somewhat less than $2\cdot5 \times 10^3$ per c.c.

* C. T. R. Wilson, Roy. Soc. Proc. vol. lxviii. p. 151.
† J. J. Thomson, Phil. Mag. vol. v. p. 346 ; H. A. Wilson, Phil. Mag. vol. v. p. 429 (1903).
‡ Townsend, Phil. Trans. A. cxcv. p. 259 (1900).

of Demonstrating the Ionisation of Air.   689

The value of q, the rate of production of ions, used above was deduced from measurements of the leakage of electricity through air contained in a small vessel of silvered glass. As a part of the effect is almost certainly due to a somewhat easily absorbable radiation from the walls, a smaller value of q is to be expected in a larger vessel such as that used in the experiment now described ; q has, moreover, been shown by several observers to depend on the material of which the walls are composed. The experiments of H. L. Cooke*, who used a brass vessel of 1100 c.c. capacity, are more nearly comparable with those of the present investigation. The value of q found by him (without special shielding of the apparatus) was about one-third of the value used in the above calculations. If we use this value the effect of recombination becomes still less important in comparison with that of diffusion ; and if we ignore recombination, the maximum number of negative ions in the absence of an electrical field, being proportional to q, is reduced to less than 900 per c.c.

I have not yet succeeded in making any direct determination of the number of drops actually produced on expansion. A superior limit to the number was, however, obtained indirectly by observing the rate of fall of the drops—the method adopted by J. J. Thomson in his determination of the charge carried by an ion. In this method of finding the number of the drops, the total quantity of water which separates out from each c.c. as a result of a given adiabatic expansion is calculated and assumed to be equally distributed among the drops ; while the radius of the drops is obtained from the rate of fall by the use of Stokes' formula. When the drops are so few and the fall so rapid as in the present experiments, one cannot assume that they attain their maximum size; in other words, that sensibly all the available water is condensed upon the drops. The value found for the number of drops by dividing the total available water by the volume of each drop (as obtained from the rate of fall) will therefore be too high, but may be considered as a superior limit below which the actual number of drops really lies.

A series of observations, in which comparatively rough measurements of the time taken by the drops to fall were attempted, gave the following results. Expansions capable of catching all the negative ions produced drops all of which had fallen to the lower plate in less than three seconds ; the temperature being 14° C. The distance between the

---

* H. L. Cooke, Phil. Mag. vol. vi. p. 403 (1903).

Phil. Mag. S. 6. Vol. 7. No. 42. *June* 1904.   3 A

690

plates, *i. e.* the maximum distance fallen, was 4·7 cms.; the rate of fall when negative ions alone came into action thus did not exceed 1·6 cm. per second. Treating these data in the manner described by Professor Thomson*, we find that the number of negative ions present, in the absence of an electric field, is less than 1000 per c.c. : this is in agreement with the value calculated above from the data afforded by leakage experiments.

[659]

LIX. *The Construction and Use of Oscillation Valves for Rectifying High-Frequency Electric Currents.* By J. A. FLEMING, *M.A., D.Sc., F.R.S., Professor of Electrical Engineering in University College, London*.*

ATTENTION was directed by the author in 1890 to the fact that if two carbon filaments are sealed into a single vacuous glass bulb so as to make an incandescent lamp with two separate carbon loops, the resistance between these filaments, though infinite when the carbon is cold, becomes quite small as soon as the loops are made incandescent †. Moreover, if a metal plate is sealed into an incandescent lamp it was shown that the space between the metal plate and the incandescent carbon filament possesses a unilateral conductivity, negative electricity being able to pass freely from the hot carbon to the plate, but not in the opposite direction ‡. More recently the author discovered that such an arrangement may be used as a valve to permit the passage of one constituent current only of a high-frequency current or to rectify an electric oscillation §. The reason for this action is now recognized to be the copious emission of negative ions or electrons from the incandescent carbon. This operation has been studied quantitatively by the present writer and many other observers.

For the purpose of rectifying electrical oscillations and thus be able to detect them by an ordinary galvanometer, these *oscillation valves* are now made as follows :—A carbon-filament glow-lamp is constructed, the carbon loop of which is upheld in the centre of a highly exhausted glass bulb (see fig. 1). Around the loop is fixed a small cylinder of nickel, C, which is connected to a platinum wire sealed through the side of the bulb. The valve is used as follows:— The carbon loop is made incandescent by a suitable battery of secondary cells, a sliding rheostat being added to adjust the voltage on the terminals of the lamp. The circuit in which oscillations are to be detected is joined in series with a dead-beat mirror-galvanometer, and the valve connected with the circuit by wires joined respectively to the terminal of the nickel cylinder and the negative terminal of the carbon

* Communicated by the Physical Society : read March 23, 1906.
† See J. A. Fleming, "On Electric Discharge between Electrodes at Different Temperatures in Air and High Vacua," Proc. Roy. Soc. Lond. vol. xlvii. p. 122 (1890); also " Problems on the Physics of an Electric Lamp," Proc. Royal Institution, vol. xiii. part 34, p. 45 (1890).
‡ See J. A. Fleming, "On a Further Examination of the Edison Effect in Glow-Lamps," Phil. Mag. July 1896.
§ See also Proc. Roy. Soc. Lond. vol. lxxiv. p. 476, 1905, "On the Conversion of Electrical Oscillations into Continuous Currents by means of a Vacuum Valve."

660   Prof. Fleming *on Oscillation Valves*

loop. The oscillation valve is most conveniently mounted for this purpose on a special form of stand (see fig. 1). In using the valve the carbon filament must be brought to bright

Fig. 1.

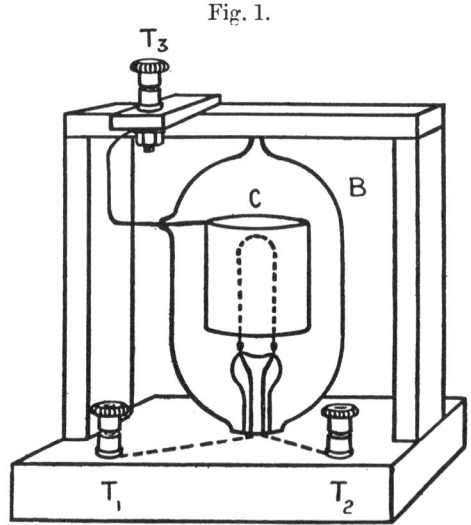

B, Exhausted glass bulb.   C, Nickel cylinder. T_1 T_2, Carbon filament terminals. T_3, Insulated cylinder terminal.

incandescence, about equal to that which in a carbon glow-lamp would correspond to a so-called "efficiency" of 3 watts per candle. So used, the valve enables us to employ a sensitive

Fig. 2.

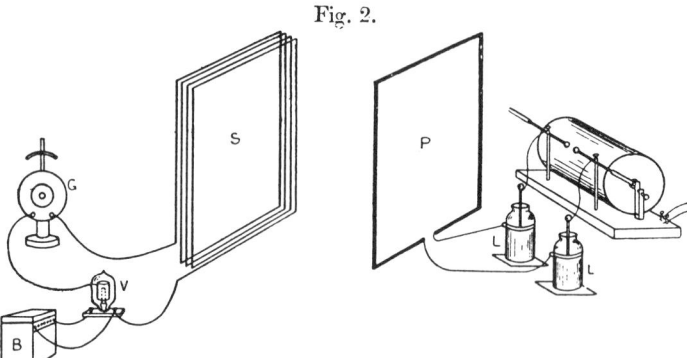

P, Primary oscillation circuit.   S, Secondary oscillation circuit. G, Galvanometer.   V, Valve.   B, 12-volt battery for incandescing filament of valve.

mirror-galvanometer of the ordinary type to detect the presence of electric oscillations in a circuit and to institute comparative measurements.

for Rectifying High-Frequency Electric Currents.   661

Thus, for instance, we form an oscillatory circuit (see fig. 2) by connecting a Leyden jar in series with a square coil of wire of a few turns P, and join the condenser and inductance across a spark-ball discharger connected to the secondary terminals of an induction-coil. At a certain distance we place another square coil of wire S in series with a galvanometer G and oscillation valve V. We then find that when oscillations are set up in the primary circuit, we obtain a steady deflexion of the galvanometer indicating that its coils are being traversed by a series of discharges in the same direction, all those in the opposite direction being practically stopped.

The author has already described the methods by which the amount of rectification produced by the valve can be ascertained (see Proc. Roy. Soc. vol. lxxiv. p. 484, 1905). Perfect rectification does not exist, but, as shown, the number expressing the percentage which the actual unilateral electric flow is of that which would flow if the unilateral conductivity were perfect, can be ascertained by sending the current which passes through the vacuous space of the valve through a calibrated galvanometer and electrodynamometer placed in series with each other. In valves as described this rectification may amount to 90 per cent.

We may use the above arrangements to investigate the effect of different kinds of discharge-balls and different lengths of spark. If we employ a fairly long spark in the primary condenser-circuit we may find that we obtain a very small effect on the galvanometer in the secondary circuit, but if we shorten the spark-gap until the spark at the balls is hardly visible, the galvanometer deflexion is generally increased. The reason for this is partly because the oscillations are damped out much more by the long spark than by the short one, and partly because with a short spark the condenser discharges occur more frequently. Hence, although in the latter case the condenser is charged to a less voltage owing to the lower discharge potential, the decreased damping and greater charge frequency causes the galvanometer to be traversed by a larger quantity of electricity per second, and therefore to give a greater deflexion.

In the same manner, we can exhibit the difference in the damping due to variations in the material of the spark-balls. Thus, using iron, brass, and zinc spark-balls of the same diameter and a spark-length of 0·1 mm., the galvanometer deflexions in one case were respectively 40, 57, and 70 scale-divisions, thus showing the smaller damping of zinc spark-balls. The writer has found by this means that carbon in the state used for arc-lamp carbons presents many advantages

662   ꞏ Prof. Fleming *on Oscillation Valves*

as a discharge surface. All who have experimented much
with Hertzian oscillators know how the state of the polish of
the surface of the metal balls (generally brass) affects the
electric wave-producing power. It can be shown by the use
of an oscillation valve that for quantitative work a discharger
made of carbon rods, as follows, presents many advantages:—
A row of arc-lamp carbons C, C, C (see fig. 3) are fixed like
posts in a piece of ebonite and another row of slightly conical
carbon rods B, B are inserted transversely between them, the
distances between the rods being fixed so that very small

Fig. 3.

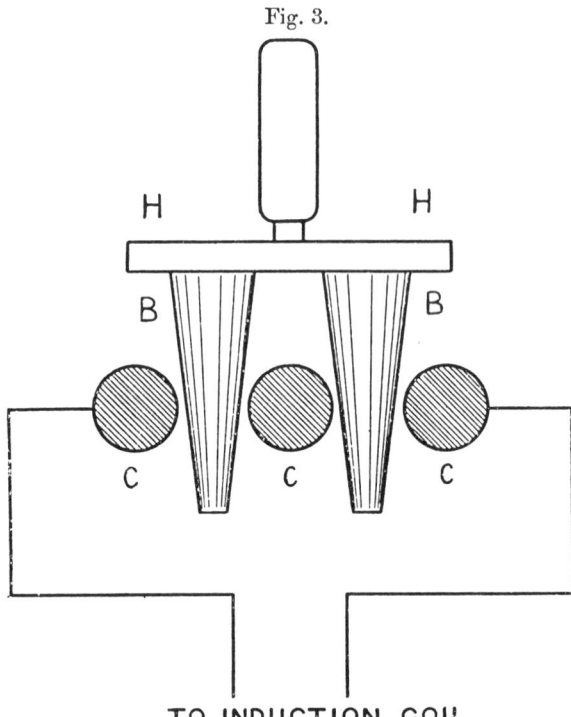

TO INDUCTION COIL

air-gaps are left between carbon and carbon. We thus construct a multiple spark-gap of carbon surfaces which has small
damping and great constancy. By enclosing the rods in a
non-oxidizing atmosphere we can prevent the rods burning
away. Another advantage of the arrangement is the ease
with which new surfaces can be brought into use.

We can also investigate by the same means whether the
use of spark-balls immersed in oil presents any advantages.
Also the same arrangements may be used to exhibit the
screening action of conductors for high frequency magnetic

fields. For if we interpose between the primary and secondary circuits a sheet of tinfoil or zinc, we see a notable decrease in the galvanometer deflexion, thus making the screening action of the metal very evident. Employed in this manner, it enables us to show strikingly the rapid rate at which the field due to a current in a square or circular circuit decreases with distance from the circuit, and therefore to illustrate one of the disadvantages under which wireless telegraphy by electromagnetic induction labours when compared with space telegraphy by electric waves. When using the valve to detect the oscillations in an antenna produced by the impact of Hertzian waves, an oscillation transformer is inserted in the circuit of the receiving antenna, and its secondary circuit is connected through a valve with a dead-beat mirror-galvanometer. We are thus able to receive signals over short distances by the direct effect of the rectified oscillations themselves on the galvanometer.

The action of other substances besides incandescent carbon as a cathode in a vacuum-valve has also been studied. It has been found by G. Owen[*] and by A. Wehnelt[†] that glowing metallic oxides, including the rare oxides employed in the manufacture of the Nernst lamp-glowers, copiously emit negative ions when incandescent both at atmospheric and at reduced pressures. Wehnelt found that the incandescent oxides of calcium, barium, and strontium produce an abnormally powerful electronic discharge, and, following the recommendations of the author, he has proposed to employ vacuum-tubes with one electrode covered with such oxides and heated, as rectifying valves for alternating currents.

As far back as 1890 the writer showed in a lecture experiment at the Royal Institution that the so-called Edison effect, that is the passage of negative electricity across space from an incandescent carbon filament to a metallic plate near it, could take place at atmospheric pressure if the plate was very near the filament. It is easy to show a similar experiment with a Nernst electric glow-lamp. If a Nernst lamp is supported with the bare glower horizontal and placed within a few millimetres of a vertical insulated metal tube kept cold by being filled with water, it is found that negative electricity will pass freely across the glower to the cold metal, but not in the opposite direction. Hence if the glower and metal tube are inserted as a gap in an electric circuit containing a sensitive galvanometer, and if secondary

[*] See G. Owen, Phil. Mag. vol. viii. p. 230 (1904). "On the Discharge of Electricity from a Nernst Filament."
[†] See A. Wehnelt, Phil. Mag. vol. x. p. 80 (1905). "On the Discharge of Negative Ions by Glowing Metallic Oxides and Allied Phenomena."

oscillations are created by induction in this circuit, we find that the galvanometer gives a steady deflexion showing the passage of a continuous current through it, and therefore of the unilateral conductivity of the space between the glower and the metal tube.

The distance over which this transference of electricity can take place depends very much upon the temperature of the glower, and the amount of rectification of the alternating current obtained upon success in keeping down the temperature of the metallic electrode. This is best achieved by circulating water through it.

It follows as a consequence from the above facts, that there is a considerable emission of negative ions or electrons from the incandescent portion of the lime cylinder used with the oxy-coal gas-burner to produce the lime-light, and that the space near the incandescent portion of the lime cylinder as well as the space near the Nernst lamp-glower is highly conductive by reason of the presence there of negative ions emitted from the oxide surface.

In the practical construction of oscillation valves, the advantage of placing the heated and non-heated electrodes in a vacuum is that the plate which acts as an anode can be placed at a greater distance from the incandescent surface and thereby kept cool, since the electrons ejected from the heated surface are projected to a much greater distance when the atmospheric pressure is reduced. Although platinum coated with calcium or barium oxides undoubtedly emits a much larger electronic current per square millimetre than carbon at the same temperature and under the same surrounding conditions as to gas pressure, I find that for rectifying electric oscillations the carbon-filament oscillation-valve as I have designed it, affords more conveniently all that is required. There are some advantages in employing a thick carbon filament and constructing it to be worked at about 12 volts and take a fairly large current of 2 or 3 amperes. For one thing, the filament is much less likely to be destroyed by overheating in working, and hence the valve lasts longer. In some cases an advantage may ensue from working valves in parallel, that is joining up a number of such carbon-filament valves with their carbon filaments in parallel on the same heating battery, connecting together the insulated metal cylinders contained in each bulb together, and then using the multiple arrangement as if it were a single valve.

When used, however, to rectify such oscillations as are employed in the receiving circuits of wireless telegraph apparatus, a single valve will do all that is required because

665

the quantity of electricity which has to be carried is small and the electronic emission from even a 4-volt 1-ampere carbon filament is amply sufficient to carry the negative component of the feeble oscillations used across the vacuous space.

It should be noted that such oscillation-valves as are here described have quite a different range of use from other rectifying arrangements such as the Cooper-Hewitt mercury-vapour tube, and the electrolytic aluminium-carbon valve of Nodon and others.

The electrolytic valves produce no rectifying effects with high-frequency alternating currents, because the time element enters into the formation of the aluminic hydroxide film on which their action depends. On the other hand, the mercury-vapour tubes which have been proposed for use with high-tension alternating currents will not operate below a certain minimum potential-difference between the electrodes. The vacuum-valve as here described, however, will pass current unilaterally with a fraction of a volt difference of potential between the incandescent and the cold electrode, and there is no minimum potential difference below which they will not act ; hence their use is conditioned solely by the sensitiveness of the galvanometer employed with them.

By its simplicity and ease of use the carbon-filament vacuum-valve recommends itself as a useful addition to our resources for experimental work in connexion with electric oscillations and electric-wave telegraphy.

THE

LONDON, EDINBURGH, and DUBLIN

PHILOSOPHICAL MAGAZINE

AND

JOURNAL OF SCIENCE.

[SIXTH SERIES.]

SEPTEMBER 1908.

XXX. *The Kinetic Energy of the Negative Electrons emitted by Hot Bodies.* By O. W. RICHARDSON, *Professor of Physics, and* F. C. BROWN, *Experimental Science Fellow, Princeton University* *.

THAT the carriers of negative electricity emitted by hot metals are electrons was first proved by the experiments of J. J. Thomson† on their deflexion in a magnetic field. This result alone did not compel any definite view of the origin of this ionization. Somewhat later one of the authors‡ showed that the phenomena then known were such as would arise if the electrons originate in the metals, from which they were able to escape when their velocity normal to the surface exceeded a certain value. This method of looking at the question was found to give a particularly satisfactory account of the thermal relations of the phenomena which were then accurately investigated for the first time. In Richardson's method of developing the subject (*loc. cit.*) the assumption is justified by theoretical considerations, that the translational kinetic energy of the electrons inside the metal has the same value as that of the molecules of a gas at the same temperature as that of the metal. From the principles there laid down it also follows that the translational kinetic energy of the electrons outside the metal possesses the same value. This equality applies not only to the average value,

* Communicated by the Authors.
† Phil. Mag. [5] vol. xlviii. p. 547 (1899).
‡ O. W. Richardson, Camb. Phil. Proc. vol. xi. p. 286 (1901); Phil. Trans. A, vol. cci. p. 497 (1903).

354   Prof. Richardson *and* Mr. Brown *on the Kinetic*

but to the way in which the energy is distributed among the different electrons as well. Although a knowledge of the kinetic energy of the emitted electrons is of obvious importance, no attempt appears to have been made to determine it. The present paper embodies the results of an investigation of the portion of the kinetic energy which depends upon the component of the velocity normal to the emitting surface. What is determined is the value of $\frac{1}{2}mu^2$, where m is the mass of an electron and u is its component of velocity perpendicular to the surface from which it is emitted. Both the average value of this portion of the energy and the law according to which it is distributed among the different particles given off are examined. The method employed gives no information about the part of the energy which depends on the component of velocity parallel to the surface of emission: the sideways energy will form the subject of another communication by one of the authors.

Theory of the Method.

Stated briefly the method used consisted in measuring the rate at which an insulated plate A charged up when a portion of another plate B parallel to A consisted of metal sufficiently hot to emit ions. It is clear that as the metal in B emits negative ions the plate A will become negatively charged, so that a difference of potential will be established tending to stop the flow of electricity. From the way in which the current between the two plates varies with the time or with the difference of potential between them the desired information can be obtained.

Consider two parallel planes of indefinite extent perpendicular to the axis of x. The lower plane, determined by $x=0$, has a small portion of its central region heated so that it is emitting ions. The potential of this plane is maintained at zero. The potential of the upper plane, determined by $x=x_0$, has the value V at the instant considered. Consider an ion whose charge is e and mass m situated at a point between the planes whose coordinates are x, y, z. Its equation of motion will be

$$m\frac{d^2x}{dt^2} = -e\frac{dV}{dx}, \quad m\frac{d^2y}{dt^2} = m\frac{d^2z}{dt^2} = 0. \quad \cdot \quad \cdot \quad (1)$$

From these equations it follows (see Phil. Trans. A, vol. cci. p. 499) that the electron will only arrive at the upper plate provided

$$\tfrac{1}{2}mu_0^2 \geq Ve, \quad \cdot \quad \cdot \quad \cdot \quad \cdot \quad \cdot \quad (2)$$

where u_0 is the initial velocity perpendicular to the plate,

Energy of Negative Electrons emitted by Hot Bodies.   355

and V is the difference of potential between the two plates. There is no reason why all the ions should be emitted from the hot metal with the same velocity. If we consider a sufficiently great number the velocity or energy will be distributed among them according to some regular law. Out of any large number of ions emitted by the plate let us denote the fraction having velocity components between u_0 and $u_0 + du_0$ by $F(u_0)du_0$, those between v_0 and $v_0 + dv_0$, and w_0 and $w_0 + dw_0$ by $f(v_0)dv_0$ and $f(w_0)dw_0$ respectively. Here v_0 and w_0 are the initial components of velocity parallel to y and z respectively; if the planes are of sufficient extent the condition that the ions should reach the upper plane will be independent of these components of velocity. If n is the total number of ions emitted by the lower plane per second the current i to the upper plane is clearly

$$i = c\frac{dV}{dt} = ne \int_{\sqrt{2\frac{e}{m}V}}^{\infty} F(u_0)du_0 \int_{-\infty}^{\infty} f(v_0)dv_0 \int_{-\infty}^{\infty} f(w_0)dw_0, \quad (3)$$

where c is the capacity of the system.

Circular Plate.

In practice it is impossible to use planes of indefinite extent, so that it is important to determine the effect of the finite size of the two plates. In the experiments the plates were circular, the lower being somewhat larger than the upper one. The upper plate was surrounded by a guard-ring. The hot metal could be treated with sufficient approximation as a point at the centre of the lower plate. We shall now calculate the number of ions which reach the upper plate, everything being supposed to be symmetrical about the axis of the two circles. Consider a charged particle whose distance from this axis is ρ, and whose distance from the lower plate is x. Its equations of motion are

$$m\frac{d^2x}{dt^2} = -e\frac{\partial V}{\partial x} \quad \text{and} \quad m\frac{d^2\rho}{dt^2} = 0. \quad . \quad . \quad . \quad (4)$$

Integrating these subject to the conditions that when

$$t = 0, \quad x = 0, \quad \frac{dx}{dt} = u_0, \quad \rho = 0,$$

and

$$\frac{d\rho}{dt} = W,$$

2 B 2

356   Prof. Richardson *and* Mr. Brown *on the Kinetic*

we get

$$mx = +\tfrac{1}{2}Xet^2 + mu_0 t \Big\} ,\qquad \cdots \quad (5)$$
$$\rho = Wt$$

where $X = -\dfrac{\partial V}{\partial x}$ is supposed to be constant.

The equation to the path is thus

$$\tfrac{1}{2}Xe\frac{\rho^2}{W^2} + m\frac{u_0}{W}\rho = mx. \quad \cdots \quad (6)$$

If x_0 is the distance between the two plates and ρ_0 the radius of the upper one the condition that the particle should just reach the boundary of the upper plate is

$$mx_0 W^2 - mu_0\rho_0 W = \tfrac{1}{2}Xe\rho_0^2. \quad \cdots \quad (7)$$

If the value of W considered as a function of u_0, x_0, and ρ_0 is $\gtreqless$ that given by equation (7) the particle will reach the upper plate, provided u_0 also satisfies the inequality (2), otherwise it will either be returned to the lower plate or it will reach the guard-ring. Solving the quadratic equation we see that the particle will just reach the edge of the plate if

$$W = \tfrac{1}{2}\frac{\rho_0}{x_0}\left(u_0 \pm \sqrt{u_0^2 - 2\frac{e}{m}V}\right),$$

when

$$V = 0,\quad \frac{W}{\rho_0} = \frac{u_0}{x_0},$$

so that the positive sign is the one to take. Hence for the particle to reach the upper plate W must lie between the limits

0 and $\tfrac{1}{2}\dfrac{\rho_0}{x_0}\left(u_0 + \sqrt{u_0^2 - 2\dfrac{e}{m}V}\right)$, whilst u_0 must lie between

∞ and $\sqrt{2\dfrac{e}{m}V}$. If the fraction of the ions emitted whose velocity parallel to the plates lies between W and $W + dW$ is $F(W)dW$, the current received by the upper plate will be

$$i = ne\int_{\sqrt{2\frac{e}{m}V}}^{\infty} F(u_0)du_0 \int_0^{\frac{1}{2}\frac{\rho_0}{x_0}\left(u_0 + \sqrt{u_0^2 - 2\frac{e}{m}V}\right)} F'(W)dW. \quad \cdots \quad (8)$$

Particular Case of Maxwell's Law of Distribution.

If we assume tentatively that the law of distribution of velocity among the emitted ions is the same as that among

the molecules of a gas which start from any surface bounding
it or within it, then the above functions may be calculated
by the methods of the kinetic theory of gases.   They are

$$F(u_0)du_0 = 2km\,u_0\,\epsilon^{-kmu_0{}^2}du_0,\; f(v_0)\,dv_0 = \sqrt{\frac{km}{\pi}}\,\epsilon^{-kmv_0{}^2}dv_0$$

$$f(w_0)dw_0 = \sqrt{\frac{km}{\pi}}\,\epsilon^{-kmw_0{}^2}dw_0 \text{ and } F'(W)dW = 2km\,W\epsilon^{-kmW^2}dW \Bigg\}, (9)$$

where $\frac{3}{4}k$ is the average energy of translation of a molecule
at the temperature of the hot body.   It is to be borne in
mind that the above functions are expressed as fractions of
the total number of ions leaving an element of area perpendicular to the axis x in unit time, and not in terms of the
number in unit volume as is usually done.

If we substitute these values of the functions in the preceding formulæ and carry out the integrations, we shall
obtain the current to the upper plate as a function of the
potential-difference, provided the law of distribution of velocity
among the emitted electrons is Maxwell's law.   Under these
circumstances, in the case where the planes are of indefinite
extent, the current to the upper plate becomes

$$i = n\,e\,\frac{2(km)^2}{\pi}\int_{\sqrt{2\frac{e}{m}V}}^{\infty} du\,u\,\epsilon^{-kmu^2}\int_{-\infty}^{\infty} dv\,\epsilon^{-kmv^2}\int_{-\infty}^{\infty} dw\,\epsilon^{-kmw^2} \Bigg\} (10)$$

$$= ne\epsilon^{-2kVe} = i_0\,\epsilon^{-2kVe}$$

if i_0 is the value of the current at the initial instant when
$V = 0$.   Since $k = \dfrac{1}{2R_1\theta}$ where R_1 is the constant in the gas
equation $pv = R_1\theta$, calculated for a single molecule, and θ is
the absolute temperature, we have, taking logarithms

$$\log_\epsilon \frac{i}{i_0} = -\frac{Ve}{R_1\theta} = -\frac{ve}{\nu R_1\theta}V = -\frac{ve}{R\theta}V,\; \cdot \;\; \cdot \;\; (11)$$

where ν is the number of molecules in 1 c.c. of gas at $0°$ C.
and 760 mms. pressure, and R is the constant in the equation
$pv = R\theta$ taken to refer to the quantity of gas occupying unit
volume under these standard conditions.   Assuming what is
now fairly well established, that the charge e on an electron
is equal to that carried by a monovalent ion during electrolysis, νe is equal to the quantity of electricity required to
liberate half a cubic centimetre of hydrogen in a water
voltameter under standard conditions of temperature and

358   Prof. Richardson *and* Mr. Brown *on the Kinetic*

pressure, since hydrogen is a monovalent element having a diatomic molecule in the gaseous state. Thus·on this view both *ve* and R are well known physical constants.

The preceding relations may be used to determine the way in which the potential of the upper plate varies with the time *t*, during which the current from the hot body has been flowing. If C is the capacity of the upper plate and its connexions we have

$$C\frac{dV}{dt} = i = i_0 \epsilon^{-\frac{ve}{R\theta}V},$$

so that

$$\epsilon^{\frac{ve}{R\theta}V}\, dV = \frac{i_0}{C}\, dt;$$

integrating this, subject to the condition that V=0 when *t*=0, we have

$$e^{\frac{ve}{R\theta}V} = 1 + \frac{ve}{R\theta}\frac{i_0}{C}t,$$

or

$$V = \frac{R\theta}{ve}\log_\epsilon\left(1 + \frac{ve}{R\theta}\cdot\frac{i_0}{C}t\right), \quad \cdot \quad \cdot \quad \cdot \quad \cdot \quad (12)$$

and

$$i = C\frac{dV}{dt} = i_0\Big/\left(1 + \frac{ve}{R\theta}\cdot\frac{i_0}{C}t\right). \quad - \quad \cdot \quad \cdot \quad (13)$$

The current, therefore, is always finite and vanishes when *t*=∞. Nevertheless the potential is infinite when *t* is infinite. This approach to an infinite value of the potential is not observed in practice. This is probably due to the fact that the current falls off with the time so rapidly that it soon becomes comparable with the small leaks inherent in the apparatus, and with the discharging current carried by ions of the opposite sign. For these reasons a limit is soon found to the potential to which the upper plate can be charged in this way.

The two formulæ (11) and (12), which give the current as a function of the potential-difference and the potential-difference as a function of the time respectively, are not independent of one another, since the former can be obtained from the latter by differentiation with respect to the time. To test the theory, therefore, it is only necessary to examine the truth of one of the two formulæ. This has been done for the formula $\log_\epsilon\frac{i}{i_0} = -\frac{ve}{R\theta}V$ in a manner which will now be described.

Energy of Negative Electrons emitted by Hot Bodies.   359

Experimental Arrangements.

The general arrangements of the apparatus used to investigate the kinetic energy of the negative ions from hot platinum is shown diagrammatically in fig. 1. The central

Fig. 1.

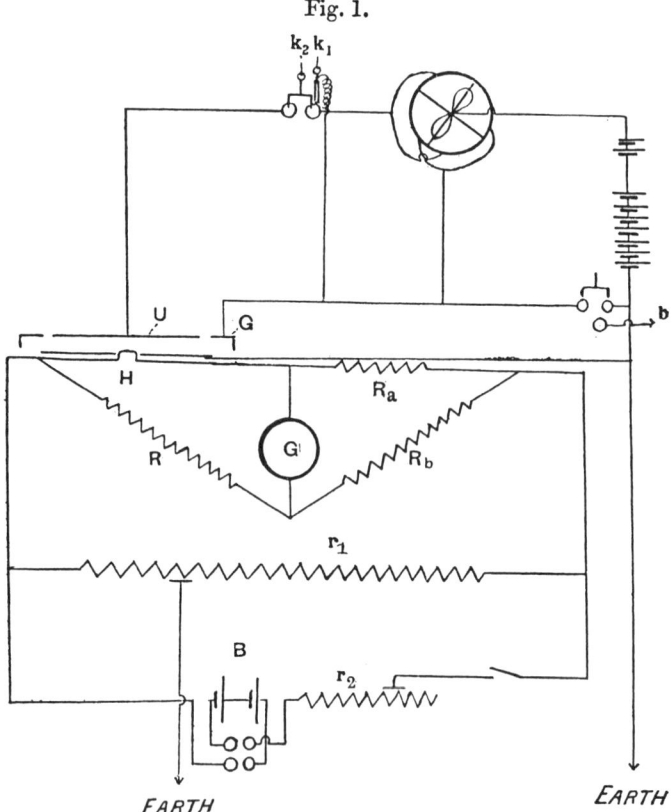

portion of a narrow platinum strip H was bent upwards as shown through a square hole at the centre of the lower of two circular plates, so as to be flush with the upper surface of the latter. The upper plate U was somewhat smaller than the lower, and was surrounded by a guard-ring G. Both plates were perpendicular to the common axis passing through their centres. The upper plate was connected to one pair of the quadrants of a Dolazalek electrometer by means of the key k_1k_2. The plug k_2 was connected with the guard-ring and with the second pair of quadrants. By means of the key b it could be connected either with the earth or any desired potential. In this way the rate of charging of the upper plate could be determined both with the two plates initially at the same potential and with any desired difference of potential between them. The charging quadrants could also be connected to a subdivided standard condenser not shown in the diagram.

The platinum strip was heated by a current furnished by the battery B and regulated by two sliding rheostats in parallel (shown as one in the diagram) at r_2. One of these had a much higher resistance than the other, and served as a fine adjustment. The temperature was determined from the resistance, which was measured by the Wheatstone's bridge arrangement, of which H, R_a, R_b, and R formed the four arms. This is the method previously used by Richardson. The essential conditions that R and R_b should both be large compared with R_a and H, and that R_a should carry the current without heating, were satisfied.

It is important that the middle point of the exposed portion of the hot strip should be at zero potential. If a fine wire was welded to the middle portion of H and soldered to the lower plate it was found that this gave rise to local variations in the heating, and also that any slight displacement of the strip during the experiments put it out of adjustment. These difficulties were overcome by shunting the whole Wheatstone's bridge circuit with a high resistance r_1, any point of which could be connected to earth. By trial a point in r_1 was found so that when it was connected to earth the initial rate of leak to the upper plate was unchanged on reversing the main heating current. The condition for this is evidently that the centre of the hot strip should be at zero potential. By simply reversing the main current from time to time this adjustment could be tested and a readjustment made, if it were required during the course of the experiments. The lower plate was permanently connected to earth.

A section through the platinum strip, showing the plates and arrangement of apparatus in their immediate neighbourhood, is shown in fig. 2. The detailed construction of the plates will be described later. The lower plate consisted of two sections screwed together. The platinum strip was held between them and insulated from them by strips of mica. The thickness of the platinum strip was ·0018 cm. Its other dimensions were ·2 × ·5 cm. Its ends were welded to heavier platinum leads which dipped into glass mercury-cups sealed into the heavy brass base-plate B. These served to introduce the heating current. The resistance of the portion

Energy of Negative Electrons emitted by Hot Bodies.   361

of the platinum strip which emitted the ions was between
·04 and ·006 ohm.   That of the leads about ·1 ohm.   The
heating current was usually in the neighbourhood of 2
amperes.   The lower plate L was supported by a brass ring
soldered to the brass plate B.   The upper plate and guard-
ring were rigidly connected together and insulated from each
other by ebonite pieces.   These are not shown in fig. 2.

Fig. 2.

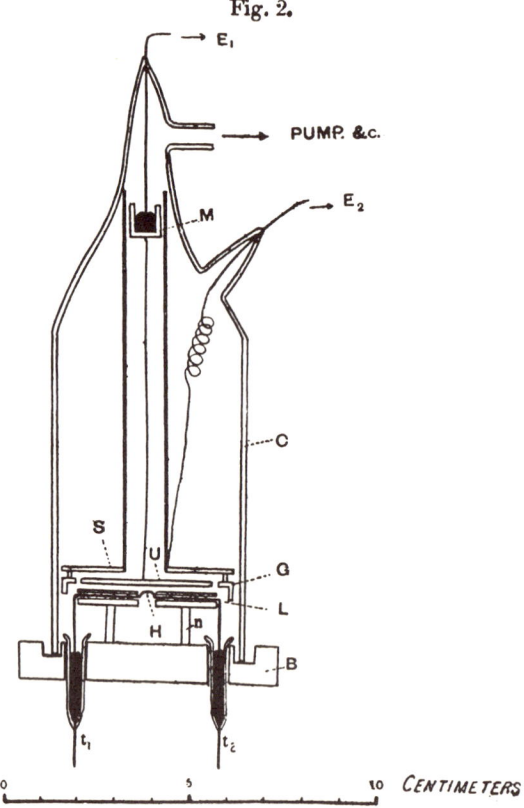

Additional ebonite supports, also not shown, enabled it to
rest on the lower plate, and also kept it insulated from the
latter.   These supports also ensured that the plates were
always at a constant distance apart and parallel to each other.
Connexion with the electrometer was made by means of a
stout wire soldered to the top of the upper plate, which
supported a mercury-cup M insulated from the shield S by
ebonite.   The wire connecting it to the electrometer system
dipped into this mercury-cup.

362   Prof. Richardson *and* Mr. Brown *on the Kinetic*

The whole of this portion of the apparatus was enclosed by a glass tube C, so that any desired degree of evacuation could be obtained.   The bottom of this tube was ground flat and rested in a groove in the base-plate, to which it was joined with sealing-wax done over with soft wax.   All sealing-wax joints were made air-tight in this way.   The platinum wires E_1 and E_2 were of course fused into the glass.   An outlet tube connected with the mercury-pump, McLeod gauge, drying-apparatus, and tap for letting in gases. The drying agent used was phosphorus pentoxide redistilled in oxygen.   In order to make certain that effects were not being caused by charges accumulating on the glass the upper plate and the wire leading from it were shielded by the flanged brass tube S.   This was connected to the guard-ring and also to earth by means of the extensible wire E_2.   It will be seen that the system was very thoroughly protected from any effects which might arise from charging up of insulation.

The particular design of apparatus was adopted so that it might very readily be taken apart and a new platinum strip substituted whenever that became desirable.   It was admirable from that point of view.

The detailed construction of the two plates is shown in fig. 3.   The three ebonite pieces which served to insulate the

Fig. 3.

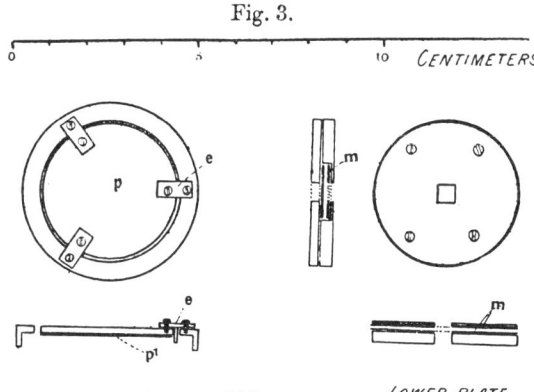

UPPER PLATE AND GUARD RING   LOWER PLATE

upper plate from the guard-ring and the lower plate, and also to support both on the lower plate, are denoted by *e*. The plan of the lower plate shows the square hole where the hot strip came up flush with the surface.   The two sections show the way in which the platinum strip was supported. The mica insulating strips are denoted by *m*.   A sheet of thin platinum foil was soldered to the lower side of the upper plate so as to avoid any effects which might conceivably be due to anything of the nature of contact electromotive force.

The diameter of the upper plate used in the experiments was 3·6 cms. and the distance between the two plates was two millimetres. In discussing the results the formulæ for infinite planes will be used. Strictly speaking we ought to apply formula (8), substituting the values of $F(u_0)$ and $F'(W)$ in (9). When this is done an integral is obtained which cannot be evaluated in finite terms. It is, however, easy to show from expression (8) that the fractional error introduced by neglecting the finiteness of the radius ρ_0 of the upper plate will be comparable with a quantity lying

between $\epsilon^{-k\frac{\rho_0^2 Ve}{2x_0^2}}$ and $\frac{x_0^2}{x_0^2+\rho_0^2}\epsilon^{-2kVe\frac{\rho_0^2}{x_0^2}}$, depending on the

value of V. A preliminary calculation showed that with the dimensions of the apparatus chosen this error would always be smaller than the expected error of observation. This conclusion was subsequently confirmed by experiments in which the current to the upper plate, to the guard-ring, and to the two together were measured and compared. Under all conditions the current to the guard-ring was small compared with that to the plate, the ratio of the two being smaller than the probable observational error. Since the guard-ring was constituted so as to intercept all the ions from the lower plate which were not received by the upper plate, it is clear that practically all the ions from the metal strip which were not returned to the lower plate by the field reached the upper plate.

The fiducial points used in calibrating the temperature-resistance curve were the temperature of the room, the melting-point of potassium sulphate (1066° C.) and the melting-point of platinum (1820° C.) The temperature varied perceptibly along the strip owing chiefly to the conduction of heat from the ends. In standardizing it, therefore, the potassium sulphate was placed on the hottest portion, as this, owing to the tremendous rapidity with which the emission of ions increases with the temperature, would be the region which gave rise to the greater proportion of the total. In determining the resistance at the melting-point of platinum, that at which the strip melted was observed. This would obviously give the temperature of the hottest portion. When the three temperatures were plotted against resistance they were found to lie as nearly as possible on a straight line.

364   Prof. Richardson *and* Mr. Brown *on the Kinetic*

This is not in agreement with the known variation of the resistance of platinum with temperature. Presumably the discrepancy arises from a change in the relative distribution of the heat along the strip as the temperature changes. The experimental temperatures were obtained from the measured resistances by reference to a chart constructed in the above way. We believe that this method of getting the temperature is trustworthy though not very accurate.

The electrometer and everything connected with it were insulated on blocks of paraffin or ebonite, so that the rate of leak to or from the upper plate could be measured when it was charged to any desired potential. Potentials as high as 400 volts were used in some of the experiments.

Results of the Experiments.

In the earlier experiments trouble was experienced on account of the occurrence of positive ionization as well as negative. With a new wire, as is well known, the positive ionization is large compared with the negative, but decays away with time, so that after long continued heating the positive ionization becomes small compared with the negative. The latter was found to remain practically constant under comparable conditions throughout the experiments. In order to get rid of the positive ionization it was found necessary to heat the platinum strip from ten to thirty hours before taking readings. In these experiments it was never found possible to get rid of the positive ionization completely ; a considerable leak was always obtained if the upper plate was charged to a high negative potential. We think that most of this positive ionization was caused by gas evolved from the apparatus when the metal strip was heated. Fortunately the positive leak was small when the negative potential applied to the upper plate was small. In the experiments it was found that the negative ionization emitted by the strip was never able to charge the upper plate to a potential greater than one volt. If, therefore, the positive leak when the upper plate was charged to a potential of -2 volts was inappreciable compared with the negative leaks against the potential which were measured, it was taken that the positive leak had been sufficiently got rid of. Great care was taken to ensure that this condition was always fulfilled in practice.

One of the authors (Mr. F. C. Brown) observed that the positive leak could be got rid of more quickly by charging the upper plate to a high negative potential, for example 200 volts. This procedure was, however, found to entirely alter the nature of the subsequent discharge of the negative

Energy of Negative Electrons emitted by Hot Bodies. 365

ionization, the emitted ions being capable of going against a much higher potential than formerly. A detailed investigation of this interesting effect has been carried out by Mr, Brown, and will shortly be published, so that it is not necessary to say much about it here. It is evident that after such treatment the hot metal is in a peculiar state. In order to ensure that the metal was in a normal state we were careful to always keep the potential of the upper plate in the neighbourhood of zero so as to make sure that the hot metal strip was never placed in a strong electrostatic field. Incidentally it may be mentioned that it was found that the abnormal state induced in this way could be got rid of by heating the strip for a short time to a very high temperature. This property was made use of in some of the experiments which follow.

The pressure of the gas (air) in the different experiments varied considerably from ·001 to as much as ·06 mm. So far as the authors were able to judge, the actual value of the pressure had no effect on the phenomena investigated, provided it was as low as the above limits indicate.

The chief experimental problem in hand was the determination of the current to the upper plate, when this was allowed to charge up, as a function of the potential-difference between the two plates. This relation when obtained enables the applicability of the formulæ (10) and (11) to be tested. The platinum strip and the lower plate were always maintained at zero potential. The potential of the upper plate at any instant was determined by the reading of the electrometer to which it was connected. The electrometer was arranged to give 115 divisions deflexion for a volt, as this degree of sensitiveness was found to be most convenient for these experiments. The deflexions could be estimated to one-tenth of a division. In measuring the currents a suitable capacity was connected to the quadrants of the electrometer. The readings of the latter, which was nearly dead beat, were recorded at definite intervals of time, and the current was obtained from the formula $i = c \dfrac{\Delta V}{\Delta t}$, where c is the capacity of the electrometer and its connexions, including the condenser, Δt is the interval of time between two readings, and ΔV is the corresponding increment of potential. Strictly, of course, this formula is only true for infinitesimal intervals, but by inserting a sufficiently large capacity in the system it was found in practice that the rate of increase of potential with the intervals used did not diminish very much during any one interval. The error thus introduced was also averaged

366   Prof. Richardson *and* Mr. Brown *on the Kinetic*

out to some extent by taking the current thus obtained to correspond to the potential at the middle of the interval. In some of the experiments a different method of reducing the observations was made use of. The observations of potential and time were used to plot a curve connecting these two variables. The values of $\dfrac{dV}{dt}$ were then obtained from this curve by geometrical differentiation. This method was not found to give results which were either more consistent or more accurate than those given by the other, so that as it was much more laborious it was discarded.

The maximum potential of the upper plate against which a measurable current would go was always about ·6 of a volt. With potentials of this magnitude the current was small, and the rate of change of voltage was only measurable when either a very small capacity or no capacity at all was added to that of the electrometer. Generally speaking it was found that the insertion of two capacities which changed the total capacity in a ratio of about 50 to 1 enabled the currents to be measured conveniently throughout the whole range. Two series of readings were usually taken with the heating current in opposite directions. By taking the mean of two deflexions corresponding to any given potential any error arising from the central point of the hot strip not being connected to earth was eliminated.

As the potential-difference between the two plates was always small, the fall of potential along the hot strip itself is of considerable importance. It was not possible to determine what this amounted to, but it was estimated to be between ·08 and ·012 volt, and was probably nearer the lower than the higher of these two limits. Assuming that the middle of the strip was at zero potential the greatest potential at any point of the hot metal would lie between the limits $\pm\,·04$ volt and $\pm\,·006$ volt. It is difficult to be quite certain how the results would be affected by the existence of this external field along the strip.

A large number of series of observations were taken in the manner indicated with slightly varied conditions. The numbers recorded in one of them are shown in the accompanying table. It will be noticed that with the same mean potential-difference (·11 volt) between the plates the currents measured were independent of the capacity used. The actual determinations were $21·5 \times 10^{-12}$ ampere with ·001 microfarad and 22×10^{-12} ampere with ·1 microfarad. The temperature in this experiment was $1283°$ C.

Energy of Negative Electrons emitted by Hot Bodies.   367

Negative Ionization.

Capacity, microfarads.	Pressure, mm.	Time.	Electrometer Readings, mm.			Increment of potential, scale divisions.	Interval of time, seconds.	Current, amperes ×10².	Mean potential, volts.
			Heating current						
			direct	reversed	mean				
0·001	0·015	10 sec.	26	24	25	25	10	21·5	0·11
		20 sec.	34	32	33	8	10	6·9	·25
		30 sec.	40	38	39	6	10	5·1	·31
		1 min.	51	50	50·5	11·5	30	3·0	·39
		1·5 min.	58	56	57	6·5	30	1·9	·47
		2 min.	62·5	61	61·7	4·7	30	1·36	·53
		2·5 min.	66·0	64·5	65·2	3·5	30	1·00	·55
		3 min.	69	67·5	68·2	3·0	30	·84	·58
		4 min.	72·8	...	...	3·8	60	·54	·62
0·10	0·015	30 sec.	1·8	1·2	1·5	1·5	30	42·8	·014
		2 min.	4·0	5·2	4·6	3·1	90	33·0	·035
		3 min.	6 3	7·8	7·05	2·4	60	35·0	·055
		1 min.	2·6	2·6	2·6	2·6	60	37·3	·018
		5 min.	9·9	11·9	10·9	8·3	240	27·0	·08
		7 min.	13·0	15·0	14·0	3·1	120	22·0	·11
		10 min.	17·1	19·7	18·4	4·4	180	18·0	·14
		13 min.	20·2	...	...	3·1	180	14·0	·16
		17 min.	24·9	...	...	3·8	240	13·5	·19
		20 min.	26·3	...	...	2·3	180	10·7	·22

Resistance 7570, temperature 1556 absolute scale.   Gas constant 4·1×10¹.

A series of observations similar to those in the table have been plotted in fig. 4, so as to exhibit the potential to which

Fig. 4.

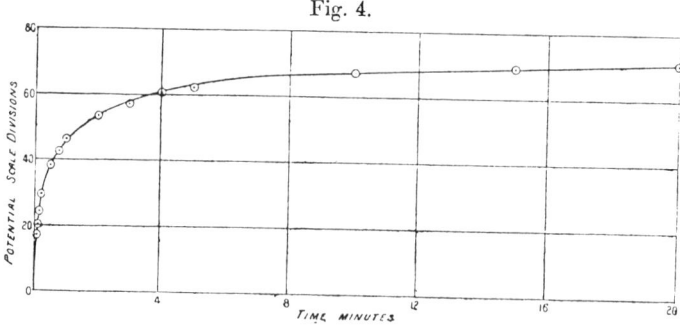

the upper plate charges as a function of the time. ꣸ The general form of this curve is in agreement with the equation (12) which was deduced theoretically.

Fig. 5 exhibits the current to the upper plate as a function of the potential-difference tending to stop it.   The points

368   Prof. Richardson *and* Mr. Brown *on the Kinetic*

with circles round them denote the values of the current
recorded in the preceding table at the potentials given by
the abscissæ. These points are seen to be distributed fairly
evenly about the smooth curve shown. In order to test the
theoretical formulæ (10) and (11) a series of points on the

Fig. 5.

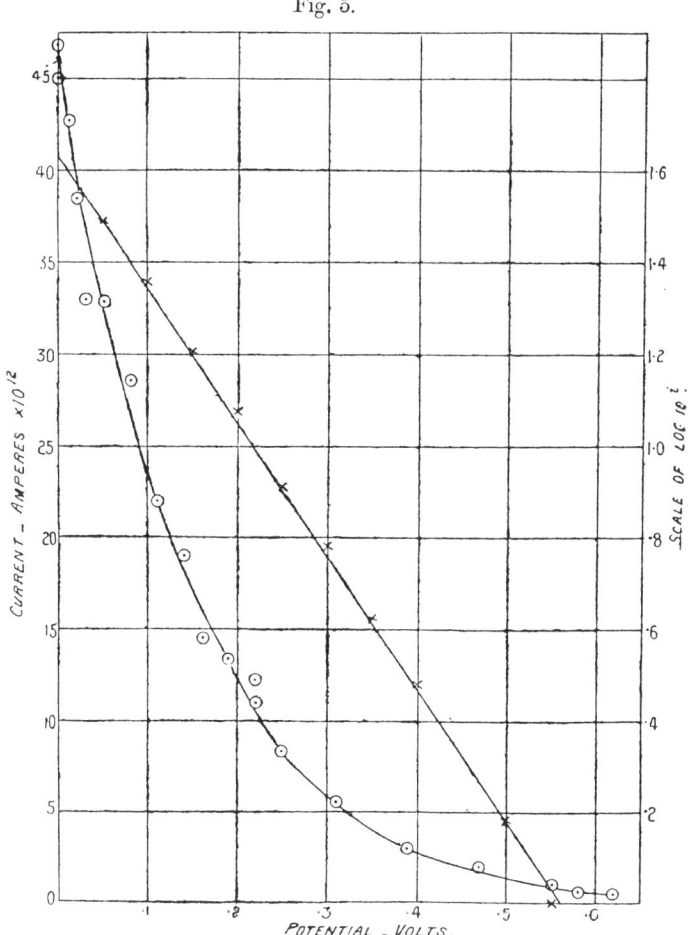

smooth curve were taken, and the logarithms of the corresponding values of the current were obtained. These are
plotted against the potential-difference in the figure. The
scale of the logarithmic curve is shown on the right-hand
side of the diagram.

Energy of Negative Electrons emitted by Hot Bodies.   369

If we use logarithms to the base 10 formula (11) becomes

$$\log i_0 - \log_{10} i = \cdot 432 \frac{ve}{R\theta} V.$$

As i_0, v, e, θ, and R are constant in any one experiment, the curve obtained by plotting $\log_{10} i$ against V should be a straight line. The accuracy with which the linear relation was fulfilled is shown by the diagram. From the slope of this line the value of the coefficient $-\frac{ve}{R\theta}$ could be deduced. Substituting the known value of ve, the quantity of electricity required to liberate half a cubic centimetre of hydrogen at 0° C. and 760 mms. by electrolysis, and the value of θ, the absolute temperature, found experimentally, a value of R the gas constant could be determined by these experiments and compared with the well-known value of this constant. This particular experiment gave for R the value $4\cdot1 \times 10^3$ c.g.s. units compared with the standard value $3\cdot7 \times 10^5$. The agreement is fairly good when all the difficulties of the investigation are taken into account.

A large number of experiments were made with platinum under varied conditions, and this linear relation was always verified provided the general method of treatment of the platinum which has been described was adhered to. In most of the experiments the surrounding gas was air at a low pressure; but the effect of replacing the air by hydrogen at a similar pressure was examined. In another set of experiments the platinum was covered with calcium oxide by heating a drop of calcium nitrate solution placed on it. In both these cases it was noticed that for some time after the change in the conditions had been made the above law connecting the current with the potential-difference was obeyed just as with platinum alone surrounded by air. It was also noticed that on first heating the strip after letting in hydrogen or after coating it with lime, the value of the current at any given temperature was practically unchanged from that which had obtained for pure platinum at the same temperature in air. The subsequent increase in the emission of ions, whether caused by hydrogen or by lime, appeared to take some time to establish itself, and as soon as a marked increase in the initial absolute value of the current occurred, it was found that the law was no longer obeyed. Thus the character of the phenomena appears to change when the metal is heated for a long time either in a hydrogen vacuum or when coated with lime. The results which we obtained after heating for some time in hydrogen were too irregular to draw very definite

370   Prof. Richardson *and* Mr. Brown *on the Kinetic*

conclusions from ; but in the case of the lime-covered strip more consistency was observed. The main feature of the change consisted in a reduction of the curvature of the curve connecting the current and the potential. This reduction was greater the higher the temperature of the lime-coated strip, and also therefore the greater the absolute value of the current. At higher temperatures the current appeared within the limits of experimental error to be a linear function of the potential instead of an exponential function. In one experiment the current was $3\cdot4 \times 10^{-8}$ amp. for $V=0$ and diminished as a linear function of the potential-difference to the value zero for $V=1\cdot22$ volts. Usually the current reached zero for a somewhat smaller voltage than this. The general effect of both hydrogen and lime on the hot platinum appears to be to change the law of distribution of energy among the emitted electrons entirely, and also to change the average value of the kinetic energy to some extent. In every case the average value of the kinetic energy was found to be greater than what it would have been for pure platinum at the same temperature in air. This change is, comparatively speaking, not very great. The greatest increase recorded in our experiments amounted to about twice the value for platinum alone.

It has been pointed out that the normal behaviour of the strip was also deranged if the hot metal was subjected to the action of a strong electric field at any time. In this case both the law of distribution of energy and the average value of the latter appear to be entirely changed. This peculiar state of the metal can be got rid of very rapidly by heating it for a short time to a high temperature. After this treatment the current against the potential again obeys the formula

$$\log_\epsilon i_0 - \log_\epsilon i = \frac{\nu e}{R\theta} V$$ with the same value of the coefficient $\frac{\nu e}{R}$ as in the normal case.

The value of R calculated from eight different series of experiments which have been selected as illustrating all the different conditions under which the linear relation between $\log i$ and V was found to be satisfied are given in the accompanying table. It will be observed that other conditions besides the previous treatment of the hot metal were varied during the experiments. The pressure of the gas varied from ·006 to ·06 mm., the temperature from 1473 to 1840 ablute, and the maximum current from 3×10^{-12} to $4\cdot7 \times 10^{-12}$ ampere. The last series in the table was obtained after the hot metal had been put into the peculiar state already described by charging the upper plate with a high negative

potential, and this peculiar state had afterwards been destroyed
by strongly heating the strip. The last observation but one
refers to similar conditions, except that the peculiar state
was induced by charging the upper plate negatively. It is
possible that some change took place in the strip which made
the recorded temperatures too high in these two experiments.

The values of R calculated from the different experiments
are shown in the last column of the table. These range from
$2·9 \times 10^3$ to $4·2 \times 10^3$ with a mean value of $3·5 \times 10^3$. The
disagreement of these numbers among each other is probably
greater than could arise from errors in the measurements of
any of the physical quantities involved, such as the temperature for example. When we consider the number of things
which appear to affect this phenomenon in a way which is
not yet understood, the agreement is probably as satisfactory
as could be expected. The agreement of these numbers with
the theoretical value $R = 3·7 \times 10^3$ is very striking, and seems
scarcely likely to be a chance coincidence.

Date.	Treatment of Platinum before Observations.	Pressure, mm.	Absolute Tempe- rature.	Maximum Current, amperes.	R.
Jan. 29.	16 hours' heating	·015	1556	$4·7 \times 10^{-11}$	$4·1 \times 10^3$
Jan. 31.	....	{ ·008 ·009	} 1473	$1·2 \times 10^{-11}$	$4·2 \times 10^3$
b. 5.	{ Just after lime was placed on { platinum	{ ·006 { ·06	} 1503	3×10^{-11}	$3·5 \times 10^3$
Feb. 10.	{ Just after hydrogen was let { into vacuum	} ·04	1553	4×10^{-11}	$3·6 \times 10^3$
Mar. 8.	About 35 hours' heating	·015	1660	$1·4 \times 10^{-11}$	$2·9 \times 10^3$
Mar. 9.	About 30 hours' heating	·01	1560	3×10^{-12}	$3·1 \times 10^3$
Mar. 13.	{ Highly charged with negative { electricity and strongly { heated subsequently	} ·02	1840	4×10^{-11}	$3·2 \times 10^3$
Mar. 14.	{ Highly charged with positive { electricity and strongly { heated subsequently	}	1813	1×10^{-11}	$3·4 \times 10^3$

In order to test the theory further the obvious thing to
try was whether the coefficient $\dfrac{ve}{R\theta}$ deduced from the log i, V
diagram really was universally proportional to the absolute
temperature. A glance at the preceding table will show
that in the case of platinum this is a difficult if not impossible task. The disagreement between the different determinations of R shows a greater ratio of variation than the
fractional change in the absolute temperature over the whole
range of temperature during which the effect could conveniently be measured. For this reason it seemed likely that

372   Prof. Richardson *and* Mr. Brown *on the Kinetic*

it would prove impossible to disentangle any change in the coefficient due to θ being changed from changes due to unknown causes. We have therefore not attempted to test this part of the theory by experiments on platinum itself.

The negative ionization from the liquid alloy of sodium and potassium suggested itself as a more likely way of attacking this question. With this substance the current can conveniently be measured at temperatures as low as 500° absolute; so that if the theory were obeyed by this substance the value of $\frac{ve}{R\theta}$ ought to be about three times as great as in the experiments on platinum. In making experiments with this substance we have had considerable experimental difficulties to contend with, and so far have only been able to obtain results of a qualitative character. So far as our experiments go, they indicate that the nature of the negative ionization from the substance is not at all what was expected. Instead of the current falling off more rapidly with the potential than with hot platinum, it falls off less rapidly, indicating that the emitted electrons have a much greater quantity of kinetic energy than those emitted from platinum at a much higher temperature. The experiments that we have been able to make with the alloy so far are not sufficiently accurate to decide whether the formula $\log i - \log i_0 = kV$ is obeyed with a different value of k or not. We hope to be able to resume this part of the investigation in the autumn.

Discussion of Results.

In presenting the results of this investigation the method has been adopted of making certain hypotheses as to the distribution of energy among the emitted electrons. From these hypotheses formulæ have been deduced which were then compared with the experimental observations. This method is justifiable on the ground that these theoretical considerations first suggested the investigation, and also because in the case of platinum the phenomena are in accordance with the theory. We have seen, however, that a number of cases have arisen where the law of distribution of energy among the particles does not coincide with Maxwell's law, so that to analyse these cases it is important to have a more general method of deducing the mode of distribution of the energy than that which has been made use of. This may always be done in the following way:—First of all construct the curve giving the current-densities as ordinates in terms of the opposing potentials as abscissæ. Our interpretation of this curve is that the current C_V corresponding to any potential V is equal to e the charge on an ion multiplied by the number of ions shot off in unit time for which $\frac{1}{2}mu^2$ is greater than eV. Calling this number $N_{(V)}$ we have then $C_V = eN_{(V)}$. But if the number of ions emitted per second for which the normal component of the energy lies between eV and $e(V+dV)$, (*i. e.* between $\frac{1}{2}mu^2$ and $\frac{1}{2}mu^2 + mu\,du$) is denoted by $eN'_{(eV)}dV$ we shall have

$$N_{(V)} = e\int_{eV}^{\infty} N'_{(eV)}dV.$$

So that

$$N'_{(eV)} = \frac{d}{d(eV)}\,N(V) = -\frac{1}{e^2}\cdot\frac{dC_V}{dV}.$$

To obtain the number which have velocity components perpendicular to the emitting surface lying between u and $u+du$ we have simply to replace eV by its kinetic equivalent $\frac{1}{2}mu^2$. We thus get

$$mu N'(\tfrac{1}{2}mu^2)du = -\frac{1}{e}\frac{dC_V}{dV}dV.$$

The number of particles whose normal velocity or energy lies between given limits can thus always be calculated from the tangent to the CV curve.

If we apply this method to the experimental numbers obtained for platinum in what we have called the normal condition, the function giving the number having energy between assigned limits is that required by Maxwell's law. This is sufficiently obvious, since otherwise the equations obtained previously would not have been satisfied. In the case where the electrons were emitted from platinum covered with lime the CV curve lost its exponential character and became a straight line at high temperatures. In this case $\frac{dC_V}{dV}$ is constant, so that the number of particles whose energy lies between x and $x+dx$ is proportional to dx and independent of x, or, in other words, the number of particles having an amount of $\frac{1}{2}mu^2$ lying within a given range is independent of the amount itself. This is only true within certain limits; in the case referred to the number of particles which had an amount of energy greater than that which corresponds to 1·22 volts was too small to be detected.

The measurements that we have made in the cases in which the distribution of energy is abnormal are too meagre

374   Prof. Richardson *and* Mr. Brown *on the Kinetic*

as yet to enable much to be said positively as to the causes which make the distribution of the energy differ from the Maxwell type.   There are, however, certain obvious causes which might change the distribution of energy.   If a double layer formed outside the metal, and the direction of the electric force in the double layer was such that it tended to drive the ions away from the metal, the distribution of energy would be altered.   Admitting that the free electrons inside the metal have the distribution of velocity given by Maxwell's law, those which escaped into the double layer would also have this distribution provided they were enabled to get out by virtue of their kinetic energy overcoming the surface forces.   All the ions which reached the double layer would escape into the region outside, but the value of $\frac{1}{2}mu^2$ for each of them would be increased by the work done in passing through the layer.   Thus, in this simple case, the distribution of energy among the emitted particles would be that given by Maxwell's law + a constant.   If the electrons escaping were deflected by collisions with atoms inside the double layer itself this simple law of distribution would be altered and would become very complex, but in any case we should expect that any variation from the normal would result in an increase in the average energy of the particles. This explanation is supported by the facts, so far as this particular conclusion is concerned ; for in all the cases of deviation from the type which we have examined the mean kinetic energy appeared to be greater than that required by Maxwell's law.

The most reasonable way of interpreting the results which have been obtained so far appears to us to be to suppose that generally speaking the distribution of velocity among the free electrons inside the metal is that given by Maxwell's law for a molecule of gas at the same temperature as that of the metal.   That when the electrons which escape simply have to do a certain amount of work against surface forces this law also holds for the distribution of energy among the electrons which have escaped.   It seems probable, however, that there are a large number of cases of the escape of electrons from hot metals when the mechanism is not so simple as this.   There may for instance be a double layer like the one already alluded to ; there is some evidence that the very large change produced by absorbed hydrogen on the leak from hot platinum may be due to an effect of this kind.   It is possible also that the expulsion of these electrons in some cases is due to a more indirect process.   It might for

Energy of Negative Electrons emitted by Hot Bodies.   375

instance be an effect of the radiant energy of the metal analogous to the photoelectric effect. In this case the kinetic energy would probably be much greater than the thermal value. We hope that further research will throw light on this point.

It seems to us an important point to have established that in one case at any rate, that of platinum heated in air at low pressure, the distribution of the square of the velocity component normal to the surface among the electrons emitted is identical with Maxwell's law of distribution of the same quantity for a gas at the temperature of the metal. It has been suggested to us that experiments of this kind do not necessarily enable us to deduce the law of distribution of velocity among the emitted particles, for the reason that formulæ similar to those on which our conclusion is based might be deduced by a purely hydrodynamical kind of treatment assuming that the particles exerted a pressure which was related to the temperature according to the law $pv = \mathrm{R}\theta$. Since this kind of treatment supposes the matter concerned to behave as though it were continuous it would follow that the experiments would not warrant any conclusion as to the distribution of velocity among the particles. It appears to us, however, that this is an unfair view of the question to take. It is now well established that the electric currents under investigation are carried by charged particles whose charge and mass are known. Admitting for the moment our interpretation of the experiments, it follows that at 1650° absolute the mean value of u, the component of the velocity, perpendicular to the plate, of the ions emitted is about $1\cdot5 \times 10^7$ cm. per sec. The distance between the plates being 2 millims., the average time occupied by the ions in crossing under zero field would be $1\cdot3 \times 10^{-8}$ sec. The maximum current in any of the experiments was $4\cdot7 \times 10^{-11}$ ampere, which corresponds to an emission of $3\cdot6 \times 10^8$ ions per sec. The number of ions present at any instant between the two plates would therefore be comparable with 5. The average distance between them would be so great that their mutual forces would be entirely negligible. On these grounds it appears to us that the only reasonable view to take is that the current is carried by discrete charged particles whose motion after they have left the plate is determined solely by the magnitude of the electric field and their initial velocity. Unless we are prepared to deny the atomic theory of electricity there appears to be no escape from the conclusion that the distribution of velocity among

376 *Kinetic Energy of Negative Electrons emitted by Hot Bodies.*

the emitted particles is that which has been deduced from these experiments.

This method does not enable us to determine by experiment the distribution of velocity among the electrons in a closed space including a piece of hot metal when the final state of statistical equilibrium has been reached. All that we are able to do is to examine the distribution of velocity among the particles emitted from the hot metal at any instant, and to show that in the case of platinum at least the results are consistent with what would be required if in the state of statistical equilibrium the distribution of velocity among the electrons outside the metal were determined by Maxwell's law. This leads to a strong presumption that the distribution of velocity among the external electrons in the steady state would be given by Maxwell's law, with the mean translational kinetic energy identical with that of the molecules of a gas at the temperature of the metal. This involves the further conclusion that the distribution of velocity among the free electrons inside the metal is also determined by Maxwell's law. For if the free electrons inside the metal are free in the sense of the kinetic theory of gases, the only difference between those inside and those outside the metal will be due to the difference of their potential energy. There is a well-known theorem in the kinetic theory of gases which proves that when two regions of the same gas at the same temperature are compared, the regions being such that the mean potential energy of the molecules is different in the two regions, the mean translational kinetic energy is the same in both, and is distributed according to the same law. The only effect of the difference of potential energy is to make the concentration of the molecules different in the two regions. Applying this theorem to the case of the electrons inside and outside a piece of hot metal, it follows that the mean translational energy and the way in which it is distributed among the electrons will be the same both inside and outside the metal. The conclusion that the average translational kinetic energy and the law of distribution of velocity of the electrons inside a metal are identical with those among the molecules of a gas at the same temperature as that of the metal is of great importance in the electron theory of metallic conduction and thermal radiation.

Princeton, N. J., June 5, 1908.

THE

LONDON, EDINBURGH, AND DUBLIN

PHILOSOPHICAL MAGAZINE

AND

JOURNAL OF SCIENCE.

[SIXTH SERIES.]

JULY 1917.

I. *A new Determination of* e, N, *and Related Constants.* *By* R. A. Millikan *.

1. *Introductory.*

THE only preceding determination of *e* and N for which a high degree of precision has been claimed was completed in 1912 and published in full in 1913 †. This determination yielded, for the electronic charge and the number of molecules in a gram-molecule, the following values :—

$$e = 4.774 \times 10^{-10} + .0095 \text{ electrostatic unit,}$$
$$N = 6.062 \times 10^{23} + .012.$$

Although these values, as well as the method by which they were obtained, received quite general recognition, it soon became imperative to reopen the problem and to attempt to make a new and, if possible, more convincing determination.

For, first, Professor Ehrenhaft and his pupils began publishing in 1914 ‡ a series of results which, though obtained by a modification of my method, were wholly irreconcilable with the results which I had found. I wished, therefore, to see whether I could check their conclusions and find conditions under which my method failed.

Secondly, there developed a tendency, especially among

* Communicated by the Author.
† Physical Review, ii. pp. 109–143 (1913).
‡ *Ann. der Physik,* xliv. p. 657 (1914), xlvi. p. 261 (1915) ; also *Phys. Zeit.* xvi. p. 10 (1915).

Phil. Mag. S. 6. Vol. 34. No. 199. *July* 1917.   B

2   Prof. R. A. Millikan *on a new*

British physicists, to regard the value of *e* given above as somewhat too high, a value being commonly adopted which was about 2 per cent. lower. As this was much greater than the necessary error in the method, I was anxious to see by entirely new work whether a numerical error could have crept into the former determination.

Finally, the constant *e* has recently taken on added importance, since not only does it now carry with it, as formerly, the knowledge of the most important molecular, atomic, and radioactive magnitudes (such as the exact number of molecules in a given weight of any substance, the absolute weight of any atom or molecule, etc.), but all of the most significant of the radiation constants as well (such as Planck's *h*, the Stefan-Boltzmann constant σ, the Wien constant C_2, all the X-ray constants, *i. e.*, the wave-lengths of characteristic X-rays) have recently been found to depend for their most reliable evaluation * upon the value of *e*. Further, if electricity exists in nature only in exact multiples of *e*, then *e* is in a more complete sense than any other physical quantity a natural unit, having none of the arbitrariness about it which inheres in so-called absolute units like the centimetre, the gram, and the second. In a word, *e* is increasingly coming to be regarded both as the most fundamental of physical or chemical constants, and also as the one of most supreme importance for the solution of the practical numerical problems of modern physics. It seemed worth while, therefore, to drive the present method for its evaluation—a method which is certainly exceedingly exact if its validity is granted—to the utmost limits of its possible precision.

Accordingly, early in 1914, the work herewith reported was begun.

2. *The Method.*

For the sake of completeness, it may be stated again that the method consists in capturing electrons † on an oil-drop

---

* Physical Review, Millikan, vii. pp. 353-388 (1916), and Webster, vii. p. 607 (1916).

† The word electron is used with the meaning originally given to it by Dr. G. Johnstone Stoney, viz., "the natural unit of electricity." This use has been consistently followed by the most authoritative writers, like Sir J. J. Thomson, Sir E. Rutherford, O. W. Richardson, N. Campbell, etc., all of whom speak in recent books or articles of positive as well as negative electrons, though the mass associated with the former is never less than that of the hydrogen atom. When an electron is found associated with a mass but $\frac{1}{1845}$ of that of the hydrogen atom it may be called "a free negative electron," or following Sir J. J. Thomson "a corpuscle."

situated between the plates of an air condenser between which a constant electrical field may be thrown on parallel to gravity. If v_1 is the constant speed of descent under gravity, and v_2 the speed of ascent under the influence of the constant field F, then it is found by experiment that when the charge on the drop is changed through the capture of electrons, or their loss through the direct incidence of X-rays or ultra-violet light, the series of speeds imparted to it by the field—namely, the series of values of $v_1 + v_2$ constitutes an exact arithmetical progression the greatest common divisor $(v_1 - v_2)_0$ of which is the value of the electron measured in terms of a velocity. This is reduced to apparent electrical units by means of the equation derived from Stokes's law,

$$e_1 = \tfrac{4}{3}\pi \left(\frac{9\eta}{2}\right)^{\frac{3}{2}} \left(\frac{1}{g(\sigma-\rho)}\right)^{\frac{1}{2}} \frac{(v_1+v_2)_0 v_1^{\frac{1}{2}}}{F}, \quad \cdot \quad \cdot \quad (1)$$

in which η is the coefficient of viscosity of air, σ the density of the particle, and ρ that of the air. The radius, a, of the drop is then found to a sufficiently close approximation by inserting an approximate value of e in the equation

$$a = \sqrt[3]{\frac{3Fe}{4\pi g(\sigma-\rho)} \frac{v_1}{(v_1+v_2)_0}}, \quad \cdot \quad \cdot \quad \cdot \quad (2)$$

this equation being derived from

$$\frac{v_1}{v_2} = \frac{mg}{Fe-mg} \quad \text{or} \quad e = \frac{mg}{Fv_1}(v_1+v_2)_0$$

$$\text{and} \quad m = \tfrac{4}{3}\pi a^3(\sigma-\rho).$$

Such observations are made on a considerable number of drops at various pressures or on the same drop at different pressures, and, for the sake of obtaining a linear relation, the values of $e_1^{\frac{2}{3}}$ are obtained and then plotted against the corresponding values of $\frac{1}{pa}$. This procedure amounts to adding a first-order correction term to Stokes's law, and writing it in the form

$$v_1 = \frac{2}{9}\frac{ga^2}{\eta}(\sigma-\rho)\left(1+\frac{b}{pa}\right). \quad \cdot \quad \cdot \quad \cdot \quad (3)$$

The relation between e_1 and e then takes the form

$$e^{\frac{2}{3}}\left(1+\frac{b}{pa}\right) = e_1^{\frac{2}{3}}. \quad \cdot \quad \cdot \quad \cdot \quad \cdot \quad (4)$$

4   Prof. R. A. Millikan *on a new*

From this equation * it is seen at once that the intercept of the $e_1^{\frac{2}{3}}\dfrac{1}{pa}$ line on the $e_1^{\frac{2}{3}}$ axis is the value of $e^{\frac{2}{3}}$, and that the slope of this line divided by the above-mentioned intercept is the constant b, the significance of which has been pointed out before † and will be more fully discussed in a following paper.

3. *The Apparatus.*

The apparatus is new throughout, and every constant involved in this method of determing e has been re-evaluated with the aid of improved and refined methods. The old air-condenser ‡ had consisted of ribbed brass plates, tested merely mechanically and found flat to about ·01 mm. They were held apart by ebonite posts 16 mm. long. These posts were found to change in length slightly through the absorption of oil. The new plates M and N (fig. 1) were made optically flat by polishing and then testing them with the aid of mercury fringes against a standard optical test-plate. They were nowhere in error by more than two wave-lengths of green mercury light. They were 22 cm. in diameter and were separated by three small pieces of echelon plates about 1 cm. square and 14·9174 mm. thick, placed at points 60° apart about the circumference. These echelon plates, of course, had optically perfect plane-parallel surfaces. The dimensions of the condenser, therefore, no longer introduced an error of more than about 1 part in 10,000 instead of about 1 part in 1000 as in the previous work.

The oil droplets from the atomizer A, blown by a puff of air through r (fig. 1) entered the condenser MN through five minute holes $\frac{1}{4}$ mm. in diameter in the middle of the upper plate, and were observed as in the former work by means of light from the arc, a, filtered through a trough of water, w, and one of cupric chloride, d, for the removal of heat rays. The temperature was held constant to within one or two hundredths of a degree and very close to 23° C. by the oil-bath, G. The charge on the drop, p, was changed by X-rays from the bulb X passing through the window, g. The

---

* Equations (3) and (4) are perhaps more easily visualizable if the correction term $\left(1+\dfrac{b}{pa}\right)$ is written in the form $\left(1+A\dfrac{l}{a}\right)$, in which l is the mean free path of the gas molecule. Since the exact value of l is uncertain, I have for simplicity chosen to compute it uniformly from $\eta = ·3502 nmcl$. This gives at 23° 76 cm., $l_{air} = ·000009417$ cm.

† Millikan, Phys. Rev. xxxii. p. 381 (1911).

‡ Millikan, Phys. Rev. ii. p. 122 (1913).

Determination of e, N, *and Related Constants.* 5

pressure, held absolutely constant during an experiment, was varied from 13 cm. to 76 cm. and was measured to a tenth mm. by the manometer *m*. The atomizer, A, fed from *s* with the highest grade of watch-oil, density at 23° C.

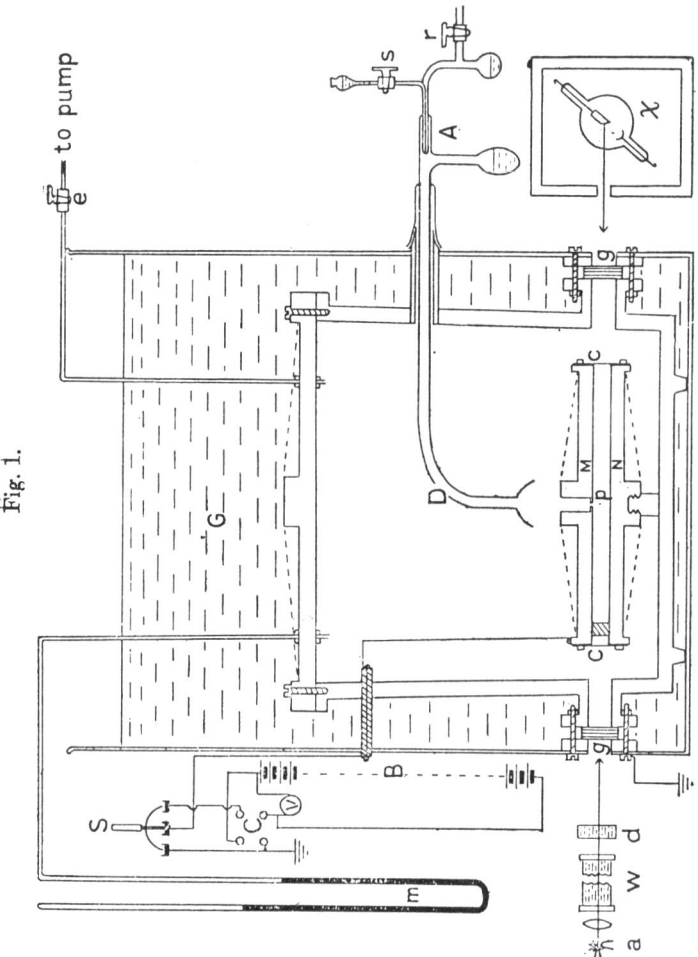

Fig. 1.

redetermined as ·9199, was blown with carefully dried and cleaned air, let in through the cock *r*, the bulbs below A being to catch excess oil. The observing optical system was a specially constructed telescope of 30 mm. objective and a magnification of 25 diameters. The objective was about

6   Prof. R. A. Millikan *on a new*

25 cm. distant from the drop, which was brought into focus by advancing or withdrawing the whole telescope system by means of a nut and screw. The distance through which the drops were timed was 1·0220 cm. It was measured precisely as in the 1912 work, and could be duplicated in successive readings to one part in a thousand. The mean value obtained from ten readings varied from 1·0218 to 1·0223—a maximum difference of one part in two thousand. This factor in the determination of *e* shares with the coefficient of viscosity of air in introducing the largest uncertainty into the final result.

The velocities of the drops both under gravity and in the field were measured with a most convenient and reliable printing chronograph made by William Gaertner & Co., of 5545 Lake Avenue, Chicago, and kindly loaned to the Laboratory for this determination. This instrument is controlled by a standard astronomical clock, and prints on a tape the hour, minute, second, and hundredth of a second at which the key is pressed as the drop crosses the cross-hair in the eyepiece—the maximum error, so far as the recording mechanism is concerned, being never more than a hundredth of a second. Some slight errors were found in the calibration of the Hipp chronoscope used in the previous determination ; but with the Gaertner Printing Chronograph the uncertainty in the time-measurements was reduced to a wholly inappreciable amount.

The electrical field strengths were determined for each drop with the aid of a 750-volt Weston Laboratory Standard voltmeter, and contain no uncertainty larger than 1 part in 3000. For this voltmeter was repeatedly calibrated in the midst of the observations against three different standard Weston cells, with results which never differed by as much as the limit indicated. The volts are then actually measured in terms of a Standard Weston cell, the above limit being merely the limit of certainty in reading the pointer in the part of the scale used. All the other elements of the problem were looked to with a care which was the outgrowth of six years of experience with measurements of this kind. This work was concluded in August 1916, and occupied the better part of two and a half years of time.

4. *The Validity of the Method.*

That portion of the investigation which has had to do with the testing of the general validity of the method and the endeavour to discover the causes of the disagreement between the results of Professor Ehrenhaft and his pupils and those obtained at the Ryerson Laboratory has been reported in detail elsewhere *. It is sufficient here to say that, although we have worked extensively with droplets of substances other than oil, particularly with mercury, upon which the irregularities are chiefly found in the Vienna work, and with particles of the same order of magnitude as those there used, we have found no indications whatever that the method, when properly used, even remotely suggests the existence of charges which are not equal to or multiples of the electron. We have studied thousands of drops of many different substances in a number of different gases, and have never found one whose charge did not fit into an arithmetic series whose greatest common divisor was the electron. We have definitely disproved Ehrenhaft's contention that this greatest common divisor is a function of the radius of the drop. Further evidence of this independence is given herewith. We have also discovered what seem to us wholly adequate reasons for the irregularities observed by Professor Ehrenhaft.

5. *The Precision of the Method.*

The consistency of the results on different drops is sufficient guarantee of the precision of the method, provided no constant error inheres in the measurement of the dimensions of the condenser, the volts, the time, or the viscosity of air, and provided the speed with which a given drop moves through the gas is strictly proportioned to the force acting upon it, as it is assumed to be. This last point was very carefully studied in the 1912 work, and considerable time has been given to studying it again. Incidentally, since for convenience most of the preceding work was done on drops charged with electrons of one sign only, and since it was thought conceivable that the electron of opposite sign might have a slightly different value and thus account for the discrepancies between different methods of getting *e*, exact demonstration has here been made that the sign of the charge is wholly immaterial. This means simply that an un-ionized gaseous molecule possesses no residual charge of either sign which is at all comparable with the electron. Some evidence upon these points is contained in Table I. (p. 8), which records experiments in which a given drop was alternately loaded with positive and negative electrons. It will be seen, not only that

* Phys. Rev. Dec. 1916, "The Existence of a Subelectron?"

Prof. R. A. Millikan *on a new*

TABLE I.

Sign of drop.	t_g sec.	t_F sec.	$n.$	$e.$
	63·118			
	63·050			
	63·186	41·728	8	
−	63·332	41·590		
	62·328			$e_1 = 6·713$
	62·728	25·740		
	62·926	25·798	11	
	62·900	25·510		
	63·214	25·806		
	Mean = 62·976			
	63·538	22·694	12	
	63·244	22·830		
	63·114	25·870		
	63·242	25·876	11	
	63·362	25·484		
+	63·136	10·830		$e_1 = 6·692$
	63·226	10·682		
	63·764	10·756		
	63·280	10·778	22	
	63·530	10·672		
	63·268	10·646		
	Mean = 63·325			
	63·642			
	63·020	71·664	6	
	62·820	71·248		
	63·514	52·668		
+	63·312	52·800	7	$e_1 = 6·702$
	63·776	52·496		
	63·300	52·860		
	63·156	71·708	6	
	63·126			
	Mean = 63·407			
	63·228	42·006	8	
	63·294	41·920		
	63·181	42·108		
	63·260	53·210		
−	63·478	52·922	7	$e_1 = 6·686$
	63·074	53·034		
	63·306	53·438		
	63·414	12·888		
	63·450	12·812	19	
	63·446	12·748		
	63·556	12·824		
	Mean = 63·335			

Duration of experiment 1 hr. 40 min.   Mean $e_1 + = 6·697$
Initial volts = 1723·5. $e_1 - = 6·700$
Final volts = 1702·1.
Pressure = 53·48 cm.

Determination of e, N, *and Related Constants.*　　　9

the mean values of the positive and negative units are the same within 3 parts in 6700, but that the time t_g under gravity is quite the same when the drop contains 22 electrons as when it contains but 6 (see column n) ; further, that the value of e_1 computed by (1) from the speeds when the drop carries from 11 to 22 electrons is the same as that found when it carries 6 or 7 electrons. These numbers show, first, that the speed produced by a given field is an exact measure of the charge ; second, that the speed under gravity and hence the apparent resistance of the medium to the motion of the drop through it is independent of the charge; and, third, that the un-ionized atom is strictly neutral, or that the positive and negative electrons are alike in charge.

6. *The Coefficient of Viscosity of Air.*

The only remaining element of uncertainty is in the coefficient of viscosity of air. In 1913, in view of measurements made in this laboratory upon this constant, in addition to measurements made elsewhere, I published * as the most probable value at 23° C. $\eta = \cdot 0001824$, and estimated that this could not contain an uncertainty of more than ·1 per cent. The correctness of this estimate was questioned by Vogel † and Gille ‡, who, while reducing the value obtained by the Halle observers by one per cent., still retained a value which was half a per cent. higher than that which I had adopted. It is to be pointed out that if this result were correct, my value of e instead of coming down would go up by three-fourths per cent., but in any case it was obviously necessary to institute further tests as to the correct value. These tests were carried out most thoroughly by Dr. E. L. Harrington, who, using the constant deflexion apparatus designed by Dr. Gilchrist and the author, succeeded, by the introduction of improvements in condition and perfections in detail, in making a determination of η which is, I think, altogether unique in its reliability and precision. I give to it alone greater weight than to all the other work of the past fifty years in this field taken together. For the individual determinations, though made with different suspensions and in such a way as to eliminate all constant sources of error save the dimensions of the cylinders, never differ among themselves by as much as ·1 per cent. and the error in the final mean can scarcely be more than one part in 2000. Indeed, the work has since been repeated by another

* *Annalen der Physik*, xli. p. 759 (1913).
† *Ibid.* xliii. p. 1235 (1914).　　　　　‡ *Ibid.* xlviii. p. 799 (1915).

10   Prof. R. A. Millikan *on a new*

observer, Mr. Stacy, and the result found dependable to within that limit of uncertainty. Dr. Harrington's * value is $\eta_{23} = \cdot 00018227$. This value is within less than a tenth per cent. of my 1913 value. The constants of the suspensions were determined by taking the periods *in vacuo*, and it is interesting that they differed from the period in air by as much as ·2 per cent., because of the moment of inertia of the air which is dragged along with the rotating cylinder.

7. *The Observations.*

The results of the final series of observations on 25 consecutive drops are given in Table II. and fig. 2. The numbers at the top of the sheet in the figure represent the approximate times of fall under gravity of the drops opposite which they stand. They are inserted to show the reader at a glance that the value of the slope and of the intercept on the $e_1^{\frac{2}{3}}$ axis, that is, the value of the electron, is not in any way a function of the radius of the drop. One can get this slope by comparing only relatively large drops at different pressures (for example, drops falling in from 14 to 20 seconds), or quite small ones (such as those falling in from 44 to 57 seconds), or by comparing drops of different sizes at the same pressure. The starred drops were those taken when the conditions of observation were considered as perfect as possible. No attempt was made to take observations on drops which fell through the fixed cross-hair distance of 1·0220 cm. in less than 14 seconds, since it was desired to keep the timing errors negligible. The value of $e^{\frac{2}{3}}$—namely, the intercept on the $e_1^{\frac{2}{3}}$ axis—was taken from the graph, as was also the slope divided by the intercept, which is the value of b in equation (4). The values thus found were

$$e^{\frac{2}{3}} = 61 \cdot 13 \times 10^{-8},$$

$$b = \cdot 000618,$$

p being measured in centimetres of mercury at 23° C. and a in centimetres. The value of A (see foregoing footnote) corresponding to this value of b is ·864, instead of ·874 as found in 1913. The difference is due, I think, to small errors which were then made in the calibration of both the Hipp chronoscope and the voltmeter, which, however, compensated each other in their effect on e, though not in that on A. The numbers given in the last column of Table II. are the values of $e^{\frac{2}{3}}$ obtained algebraically from (4), and the

* Phys. Rev. Dec. 1916.

TABLE II.

No.	Temp. °C.	P.D. (volts).	t_g (sec.).	v_1 cm./sec.	n.	$a \times 10^5$ cm.	p (cm. Hg).	$\dfrac{1}{pa}$.	$\dfrac{l}{a}$.	$e_1^{2/3} \times 10^8$.	$e^{2/3} \times 10^8$.
1	23·07	6650	16·50	·06194	7-13	23·40	74·49	57·45	·04111	63·21	61·03
2	23·00	6100	16·76	·06099	8-11	23·22	75·00	57·5	·04115	63·204	61·03
3	23·05	5308	19·73	·05180	7-15	21·34	74·49	63·0	·04509	63·54	61·16
4	23·08	4132	37·82	·02703	4-6	15·33	75·37	86·7	·06205	64·27	60·97
5	23·06	4661	40·09	·02521	3-6	14·84	75·00	90·6	·06484	64·63	61·21
6	23·12	4111	51·53	·01983	3-4	13·05	75·77	101·3	·06502	65·02	61·19
7	23·08	5299	51·48	·01985	2-5	13·05	74·98	102·4	·07329	65·07	61·20
8	23·01	6661	56·06	·01823	2-3	12·50	75·40	106·3	·07608	65·13	61·11
9	23·00	6082	59·14	·01728	1-4	12·17	75·04	109·7	·07850	65·19	61·05
10	23·10	4077	57·46	·01779	3-8	12·34	75·67	107·3	·07680	65·21	61·16
11	23·13	4663	16·58	·06165	10-12	22·72	29·26	150·6	·1078	66·70	61·01
12	23·11	4661	29·18	·03502	5-7	17·08	36·61	160·1	·1146	67·12	61·07
13	22·98	4687	18·81	·05432	8-10	21·26	30·27	155·6	·1114	67·14	61·26
14	23·12	4651	47·65	·02145	2-7	13·20	36·80	206·4	·1477	68·90	61·11
15	23·10	4648	32·72	·03129	4-6	15·92	31·35	200·7	·1437	68·97	61·39
16	23·15	3393	18·34	·05572	12-16	21·11	20·58	227·8	·1630	69·88	61·27
17	23·12	4669	46·82	·02294	2-4	13·12	29·10	262·1	·1878	70·85	60·94
18	23·12	4191	26·62	·03819	5-7	17·32	20·54	281·4	·2014	71·60	61·98
19	23·10	3339	14·10	·07249	15-19	23·00	13·24	321·4	·2297	73·34	61·20
20	23·14	4682	39·24	·02605	3-5	14·00	20·72	345·4	·2472	74··7	61·22
21	23·14	3350	18·30	·05585	10-13	20·47	13·62	359·1	·2570	74·54	60·97
22	23·00	3370	43·88	·02329	3-6	13·17	20·74	371·5	·2659	75·00	60·97
23	23·13	3381	46·90	·02179	3-6	12·69	20·74	380·6	·2724	75·62	61·24
24	23·09	3345	19·65	·05201	9-12	19·65	13·12	388·5	·2781	75·92	61·24
25	23·15	3344	26·76	·03819	6-9	16·57	13·80	483·3	·3137	77·74	61·18

Mean .... 61·126

12   Prof. R. A. Millikan *on a new*

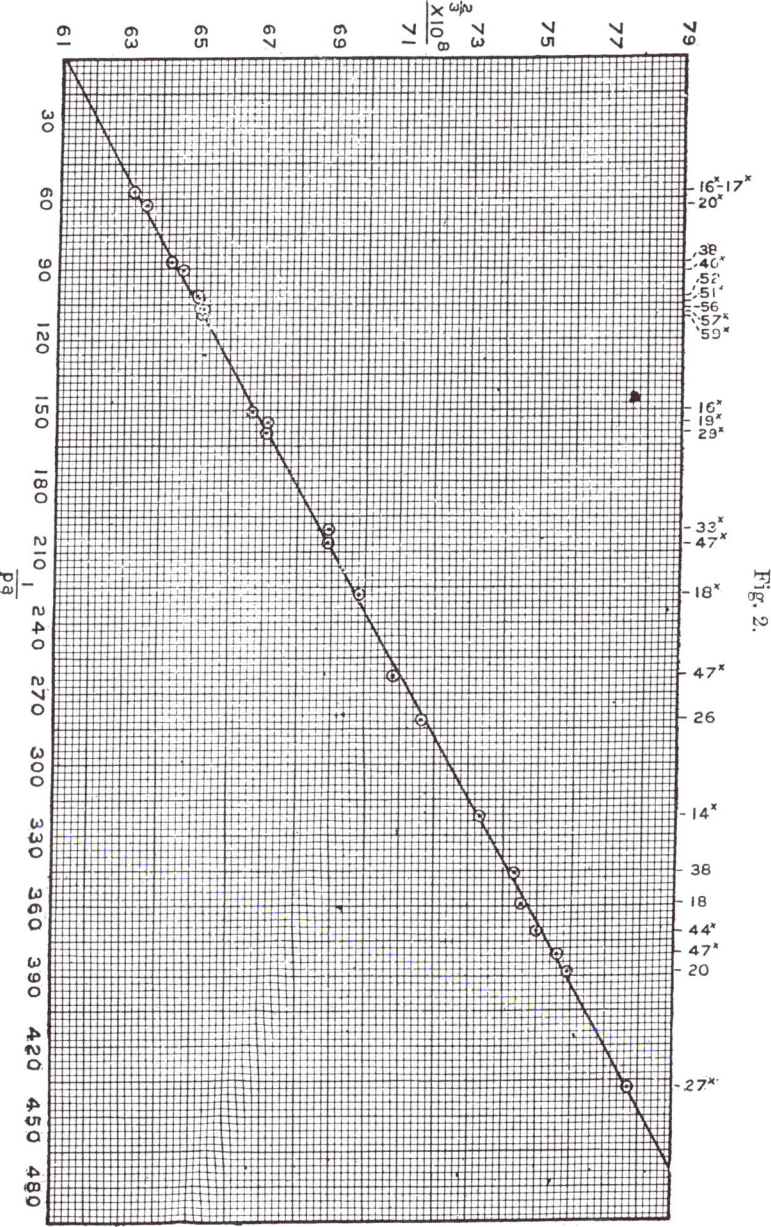

Fig. 2.

above value of b taken in connexion with each individual set of values of the observed quantities $\dfrac{1}{pa}$ and $e_1{}^{\frac{2}{3}}$. It will be seen that the final mean value of $e^{\frac{2}{3}}$ obtained by this method of analysis—a method which yields the most reliable value of $e^{\frac{2}{3}}$ obtainable from the data at hand—is

$$e^{\frac{2}{3}} = 61\cdot126 \times 10^{-8}.$$

There is but one drop in the table which yields a value of $e^{\frac{2}{3}}$ differing from this by as much as one-third of one per cent., and the probable error of the mean computed by least squares is one part in 4000.

No such precision, however, is claimed for this determination of $e^{\frac{2}{3}}$. It is ·07 per cent. higher than the value (61·086) which I published in 1913, both values being computed in terms of ·0001824, as the value of the coefficient of viscosity of air. Dr. Harrington's new value of this constant, viz. ·00018227, is, however, more reliable than the old one and is ·07 per cent. lower than that ; so that *the new value of* e *and* N *computed solely from the new data obtained in this redetermination is exactly the same as the value published in* 1913. The maximum uncertainty in this value was then estimated as one part in five hundred. This work has reduced it so that that it is no more, I think, than one part in a thousand ; for it now contains but two factors which are uncertain by as much as one part in two thousand, namely, the coefficient of viscosity of air and the cross-hair distance. The exactness of the agreement to four places with the 1913 value is, of course, accidental. This is, however, the third time that with independent determinations (one unpublished) I have come out well within one-tenth per cent. of the foregoing result.

The result of this investigation may, then, be stated as follows :—

$$e = 4\cdot774 \times 10^{-10}, \ \pm\cdot005,$$
$$N = 6\cdot062 \times 10^{23}, \ \pm\cdot006,$$

the last number being obtained from

$$N \times 4\cdot774 \times 10^{-10} = 9649\cdot4 \times 2\cdot9990 \times 10^{10}.$$

Since this is an attempt at a precise determination, and by far the most carefully carried out of any work which I have thus far done upon the evaluation of the electron, it is perhaps worth while to give more of the original data than would otherwise be justified, in order that others may better form their own estimates of its probable reliability.

14   Prof. R. A. Millikan *on a new*

Accordingly, the actual records of the observations on all of the 18 starred drops which were chiefly considered in determining the line in fig. 2 are given below. The other seven are omitted merely to save space. As the graph shows, they do not modify in any way the result. With their inclusion the graph becomes the record of 25 consecutive observations without any discards, so that the result is entirely free from the exercise of choice.

8. *The Values of some other Related Constants.*

I have already recorded in the 1913 paper the values of six fundamental but related constants which are at once known as soon as e is found, and with the same precision as that attained in its evaluation. These six are : (1) the electron, e ; (2) the Avogadro number, N ; (3) the number of molecules, n, in an ideal gas at $0°$ C. 76 cm.; (4) the kinetic energy, E_0, of molecular agitation at $0°$ C. ; (5) the constant change, , in molecular energy per degree ; (6) the entropy constant, k, or the gas constant applied to a single molecule. A seventh constant which should have been included at that time is the mass m of a hydrogen atom, given by

$$m = \frac{4·774 \times 10^{-10} \times 1·0077}{9649·4 \times 2·999 \times 10^{10}} = 1·662 \times 10^{-24} \text{ gram.}$$

This list may now be extended as follows :—
The constant of the Balmer series of hydrogen is known with the great precision attained in all wave-length determinations, and has the value $3·290 \times 10^{15}$. From Bohr's theory it is given by

$$\frac{2\pi e^4 m}{h^3} \quad \text{or} \quad \frac{2\pi e^5}{h^3 \frac{e}{m}}. \quad \ldots \ldots \quad (5)$$

I have shown that h may be determined photoelectrically[*] with an error in the case of sodium of no more than $\frac{1}{2}$ per cent., the value given by my work on sodium being $6·56 \times 10^{-27}$. The value found by Webster[†] by the X-ray method discovered by Duane and Hunt[‡] is $6·53 \times 10^{-27}$. Taking the mean of these two results, viz. $6·545 \times 10^{-27}$, obtained by wholly dissimilar methods, and substituting

[*] Phys. Rev. vii. p. 374 (1916).   [†] *Ibid.* vii. p. 599 (1916).
[‡] *Ibid.* vi. p. 166 (1915).

Determination of e, N, *and Related Constants.*　　15

in (1), after introducing Bucherer's value of $\frac{e}{m}$, viz.

$1\cdot767\times10^7$, we obtain for the Rydberg constant, $3\cdot294\times10^{15}$, *which agrees within one-tenth per cent. with the observed value.* This agreement constitutes most extraordinary justification of Bohr's equation, and warrants the use of spectroscopic data, combined with the foregoing data on *e*, for a most exact evaluation of *h*. The value of *h* computed thus from (5) with the aid of my value of *e* and the foregoing value of $\frac{e}{m}$, which is now known with a precision of one-tenth per cent., is

$$h = 6\cdot547\times15^{-27}\pm\cdot011.$$

It will be seen that the uncertainty is just $\frac{5}{3}$ the uncertainty in *e*, since *e* appears in (5) in a power $\frac{5}{3}$ that of *h*, while $\frac{e}{m}$ affects *h* by an amount which is negligible in comparison. *The foregoing value of* h *may be considered the most reliable thus far obtainable, its uncertainty being one part in six hundred.* It will be seen, too, that it agrees within just one part in five hundred with the value obtained for my sodium curves, which I estimated correct to only one part in two hundred.

Having thus fixed the value of *h* to one part in six hundred, we may obtain from Planck's equation the Wien constant, C_2, with the same precision, for it will be recalled * that

$$C_2 = \frac{hc}{k} = \frac{6\cdot547\times10^{-27}\times2\cdot999\times10^{10}}{1\cdot372\times10^{-16}}$$

$$= 1\cdot4312\pm\cdot0030 \text{ cm. degrees.}$$

The estimated error set down above is obtained from the assumption of an uncertainty of one part in six hundred for *h* and one part in one thousand for *k*. The latest experimental result on C_2 given out by the Reichsanstalt † is $C_2 = 14300$. Coblentz ‡ gives as the result of his direct experiments $C_2 = 1\cdot4369$, while his combination of total radiation experiment and theory lead him to $C_2 = 1\cdot4322$.

---

* Phys. Rev. ii. p. 142 (1913).
† *Ann. Phys.* xlviii. p. 430 (1915).
‡ Phys. Rev. vii. p. 694 (1916).

16   Prof. R. A. Millikan *on a new*

Again, from Planck's equation,

$$C_2 = \left(\frac{48\pi ak}{a}\right)^{\frac{1}{3}},$$

we can compute the Stefan-Boltzmann constant of total radiation a $\left[\text{or } \sigma = \dfrac{ac}{4}\right]$ and obtain a result which is uncertain by but six-tenths per cent. The result is

$$\sigma = 5\cdot72 \times 10^{-12} \pm \cdot034 \text{ watt cm.}^{-2} \text{ deg.}^{-4}.$$

This is exactly the value found by Coblentz from his most recent and most thorough experimental work * on σ. The exceedingly close agreement between all these values of h, C_2, and σ, computed on the one hand from the work on e, and directly observed on the other, is an indication of the exactness of the work on e.

The grating spacing in calcite computed † from the foregoing value of e is $3\cdot030$ Å.

A summary of the most important constants the values of which are fixed by this determination of e is given below, with the uncertainty attaching to each :—

The electron .... $e = 4\cdot774 \pm \cdot005 \times 10^{-10}$.

The Avogadro constant .... $N = 6\cdot062 \pm \cdot006 \times 10^{23}$.

Number of gas molecules per c. c. at 0° C. 76 cm. .... $\Big\}$ $n = 2\cdot705 \pm \cdot003 \times 10^{19}$.

Kinetic energy of translation of a molecule at 0° C. .... $\Big\}$ $E_0 = 5\cdot621 \pm \cdot006 \times 10^{-14}$.

Change of translational molecular energy per degree C. .... $\Big\}$ $\epsilon = 2\cdot058 \pm \cdot002 \times 10^{-16}$.

Mass of an atom of hydrogen in grams. $m = 1\cdot662 \pm \cdot002 \times 10^{-24}$.

Planck's element of action .... $h = 6\cdot547 \pm \cdot011 \times 10^{-27}$.

Wien's constant of spectral radiation.. $c_2 = 1\cdot4312 \pm \cdot0030$.

Stefan-Boltzmann constant of total radiation..................... $\Big\}$ $\sigma = 5\cdot72 \pm \cdot034$.

Grating spacing in calcite .... $d = 3\cdot030 \pm \cdot001$ Å.

I have to express my hearty thanks to Dr. Yoshio Ishida for invaluable aid both in observing and in computing the accompanying data.

Ryerson Physical Laboratory,
   University of Chicago,
      January 12th, 1917.

* Phys. Rev. vii. p. 694 (1916).
† Webster, Phys. Rev. vii. p. 607 (1916).

Determination of e, N, *and Related Constants.* 17

APPENDIX.—Observational Data.

t_g.	t_F.	n'.	$\dfrac{1}{n'}\left(\dfrac{1}{t_{F'}} - \dfrac{1}{t_F}\right).$	n.	$\dfrac{1}{n}\left(\dfrac{1}{t_g} + \dfrac{1}{t_F}\right).$	
			DROP No. 1.			
16·56	97·96			8	·008852	V_i=6658 volts.
16·46	97·99	1	·008831			V_f=6647 volts.
16·46	52·35			9	·008849	t=23°·07 C.
16·48	52·70	1	·008816			p=74·49 cm.
16·54	97·84			8	·008854	v_1= ·061940 $\dfrac{\text{cm.}}{\text{sec.}}$
16·58						a= ·0002340 cm.
16·50						$\dfrac{1}{pa}$=57·45
16·45	51·26	1		9	·008901	
16·46	51·30	1	·009136			$\dfrac{l}{a}$= ·04111
16·46	96·48			8	·008872	$e_1^{2/3}$=63·21
16·46	51·88	1	·008948	9	·008880	
16·50	51·68	2	·008856			
16·44	624·39			7	·008886	
			·009016			
16·55	94·18			8	·008903	
		5	·008951			
16·52	18·01			13	·008921	
16·44	18·10	5	·008930			
16·63	93·27			8	·008916	
16·44	50·90	1	·008924	9	·008917	
16·46						
Mean 16·50			·008934		·008895	$e^{2/3}$=61·03
			DROP No. 2.			
16·75	32·45			11	·008225	V_i=6107 volts.
16·72	32·49	3	·008211			V_f=6091 volts.

Part Two

Radioactivity

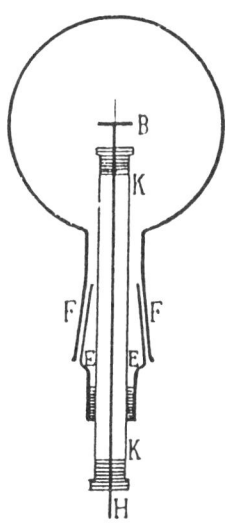

Part Two
Radioactivity

Few would argue with the selection of Michael Faraday as the greatest experimentalist of the nineteenth century. A comparable figure for the present century (at least for its first half) must undoubtedly be Ernest Rutherford. In terms of the magnitude of their achievements and the influence they had on subsequent developments in science, both men stand supreme in their respective fields. No less of an authority than Albert Einstein spoke of the two with equal reverence when he said 'I consider Rutherford to be one of the great experimental scientists of all time, and in the same class as Faraday.'

Ernest Rutherford (1871–1937), the fourth of twelve children, was born near Nelson in the South Island of New Zealand. There he completed his secondary education before winning a scholarship to Canterbury College, Christchurch, where in 1893 he obtained an MA and, one year later, a BSc after undertaking research on the magnetization of iron by high-frequency alternating currents. In 1895 he began work as a research student under J. J. Thomson at the Cavendish Laboratory, Cambridge, where his first investigations were with a novel detector of wireless waves he had devised and brought with him from New Zealand. (For a time he held the distance record of over half a mile for reception of radio signals.) After the discovery of X-rays by Röntgen in 1895, Thomson asked Rutherford to join him in a study of the effect of these rays on the ionization of gases. The collaboration produced results of considerable importance (see *Science in the Making*, Volume 2), encouraging Rutherford to carry out similar studies using ultraviolet light and subsequently the radiation from uranium, discovered by Becquerel in 1896. Intrigued by the nature of the emissions from radioactive materials themselves, he recognized that these were of two kinds with very different penetrating powers and named them, 'for convenience', alpha (α) and beta (β) rays — labels that are still used today. (A third type of radiation, gamma (γ) rays, was discovered later, in 1900, by Villard.) Studies of radioactive materials and their emissions were soon to make Rutherford famous and earn him the Nobel Prize for Chemistry.

In 1898, Rutherford, then aged 27, was offered the Macdonald Chair of Physics at McGill University in Montreal, an appointment undoubtedly aided by Thomson's glowing reference. 'I have never had a student with more enthusiasm or ability for original research than Mr Rutherford' — a testimonial written just one year after Thomson himself had secured everlasting fame by his discovery of the electron. This praise was not misplaced, for within the first years of his appointment Rutherford published at a prodigious rate on the radioactivity of uranium and also of thorium — a new radioelement discovered by Schmidt and Marie Curie. It was in fact his experiments on thorium compounds, undertaken with Frederick Soddy (then a young chemistry demonstrator at McGill), which provided the key ingredients for the transformation theory of radioactivity. A combination of Soddy's experimental skills and Rutherford's physical insight, coupled with inspired, rigorous and demanding work by both, led to the realization that one radioactive substance generated another in a series of successive transformations — each accompanied by the emission of α or β rays and with a characteristic decay constant or half-life. Two of Rutherford and Soddy's seminal papers are reproduced in this part.

Altogether Rutherford authored or co-authored 70 papers while at McGill University, 30 of which appeared in the *Philosophical Magazine*. While continuing to unravel further the complexities of radioactive series, his interests turned to the nature of the α rays themselves. In 1903 he published a classic paper on the deviation of these rays in magnetic and electric fields, experiments which confirmed they were streams of positively charged particles and also provided a measure of their charge to mass (e/m) ratio. For these difficult experiments a powerful source of radium was supplied by Marie Curie, a large electromagnet procured from the Electrical Engineering Department at McGill, and a battery of small accumulators assembled to provide 600 V for the electric field. Further details appear in the paper, which is reproduced on p 141. Suffice it to say here that the results enabled Rutherford to conclude that radioactive charge was a consequence of the projection, with high velocity, of heavy positively charged particles from the atom. Decisive proof that the particles were doubly charged helium atoms was to be obtained after Rutherford had returned to England in 1907.

Numerous honours were awarded to Rutherford for his work before he left McGill. In successive years he was made a Fellow of the Royal Society and received the Society's Rumford Medal. In 1908 he was awarded the Nobel Prize for Chemistry 'for his investigations into the disintegration of the elements and the chemistry of radioactive elements'.

At Manchester, Rutherford took the Chair of Physics vacated by Arthur Schuster (on the condition that Rutherford became his successor) and immediately began a collaboration with Hans Geiger on methods of detecting α particles. On firing these into an evacuated tube containing a central charged wire, individual particles could be detected, and by arranging for them to fall onto a zinc sulphide screen, discrete scintillations were seen. These methods of electrical and scintillation counting (the former developed principally by Geiger) became standard techniques for the detection of charged particles. By measuring the charge emitted from a radium source in a given time and dividing this by the number of observed scintillations, Rutherford and Geiger determined that the charge on the α particles was approximately twice that of the then accepted value for the electronic charge (but positive). Combining this with the known value of e/m enabled them to conclude that the α particle was indeed helium with a double charge. Direct spectroscopic verification of the proposal was obtained in 1909 by Rutherford and Royds whose classic paper is reprinted on p 152.

The only 'non-Rutherford' paper in this part is by Geiger and Nuttall. Working in Rutherford's laboratory at Manchester, they tackled one of the remaining mysteries of radioactive decay by α emission, namely the relationship between the range (and hence velocity) of the α particles and the radioactive half-life of the emitter. Geiger and Nuttall obtained an empirical relationship (the Geiger–Nuttall law) which clearly demonstrated that the longer the half-life the shorter the range, a feature which suggested that the energy of the α particles within the atom, prior to escape, was an important factor in determining the lifetime of radioactive materials. The full significance of this result was not appreciated until after the formulation of a quantum mechanical theory of α-particle tunnelling through a potential barrier, some 17 years later.

1902   4   The Cause and Nature of Radioactivity - Part I. By E. Rutherford, M.A., D.Sc., Macdonald Professor of Physics, and F. Soddy, B.A. (Oxon.), Demonstrator in Chemistry, McGill University, Montreal

It is impossible to overestimate the significance and the eventual impact of the revolutionary idea put forward in this classic paper by Rutherford and Soddy, namely that radioactivity is a manifestation of the transformation of one element into another. Throughout the nineteenth century, belief in the immutable character of elements had provided the foundation of chemical science; transmutation was no more than the dream of alchemists. Soddy warned that the chemists would 'have their heads off' if they presented the idea of transmutation without overwhelming proof.

In 1887, J. J. Thomson had startled the scientific world with his unambiguous demonstration that electrons could be torn or ejected from atoms and even more so with his speculative suggestion that these tiny charged particles were the prime constituent of matter. He further proposed that the radioactive nature of uranium, thorium and other heavy elements might be explained by the instability and eventual disintegration of atoms that contained a large number of electrons. It was certainly the case that one type of radiation emitted by some radioactive materials, the β rays, were electrons, but by 1900 it was known that two other types of radiation were sometimes emitted — namely α particles and γ rays, the nature of neither of which was known.

Becquerel's discovery in 1886 that a compound of uranium (uranium potassium sulphate) had the effect of darkening photographic plates, and without any apparent diminution in its ability to do so, aroused little excitement. His finding came too close on the heels of Röntgen's discovery of X-rays one year earlier to take centre stage. It was only after Marie Curie (the first woman in France to obtain a doctorate for scientific research and the first female anywhere to become a professor and to win the Nobel Prize) found that thorium and two new elements, polonium and radium, were intensely radioactive, that serious consideration was given to this curious phenomenon.

Ernest Rutherford's appointment as Macdonald Professor of Physics at McGill University in Montreal in 1898 was an inspired one. His interest in radioactive emissions had begun at Cambridge during work with J. J. Thomson in which he had studied the ionization of gases produced by uranium radiation. At McGill he inverted this line of research by using ionization to investigate radioactivity. Before his collaboration with Soddy began, he had developed to a high degree the combination of an ionization chamber and a quadrant electrometer for the measurement and comparison of intensities of radiation from radioactive materials. This electrical method of detection was far more sensitive than the photographic plate and permitted quantitative measurements of a kind that were indispensable to his further investigations.

The puzzling features of the radioactivity of thorium are described in the Introduction of the present paper. Referring back to earlier work published in the *Philosophical Magazine* in 1900, the authors note that 'Besides being radioactive in the same sense as the uranium compounds, the compounds of thorium continuously emit into the surrounding atmosphere a gas which possesses the property of temporary radioactivity.' Furthermore this 'emanation' had the power of exciting radioactivity on the surface of other substances in which it came into contact. The 'half-life' (i.e. the time required for the intensity of the emitted radiation to fall to one-half of its value, starting at any chosen moment) was found to be one minute for the gaseous emanation and eleven hours for the excited radioactivity. The importance of quantitative measurements for distinguishing different radio-active substances by the rate of decay of their activities is clearly evident here.

The major part of the paper, however, is concerned not with the peculiar emanation and its effects, but with the chemical separation from the original thorium compound (thorium hydroxide) of a consti-tuent which possessed an activity up to one thousand times as great as the starting material. Rutherford and Soddy named this active ingredi-ent thorium X (ThX), by analogy with a similar constituent which Crookes had succeeded in separating from uranium and had named UrX. Just as UrX was a material with chemical properties different to those of Ur, so ThX was chemically distinct from Th.

Measurement of the activity of ThX showed that, after an initial *rise* over a period of about 1 day, it decayed with a half-life of about 4 days, whereas the activity of the hydroxide from which ThX had been sepa-rated displayed, again after about 1 day during which it *fell*, an *increase* with a time constant of exactly the same value, namely 4 days. Ignoring the initial transients (shown in Figure 2 of the paper) and plotting activities in terms of percentage changes, resulted in the famous mir-ror-image curves of Figure 3. (These were later to feature in Rutherford's coat of arms when he became Lord Rutherford in 1931.) The equations governing the behaviour given below the figure are identical to those describing the charging and discharging of a capaci-tor. It took far-sighted genius and great courage to come up with an explanation for the behaviour. In the words of the paper:

> It therefore follows that if the initial irregularities of the curves are disre-garded and the residual activity of thorium is assumed to possess a *constant* value, the experimental curve obtained for the recovery of the activity will be explained if two processes are supposed to be taking place:
>
> (1) That the active constituent ThX is being produced at a constant rate;
> (2) That the activity of the ThX decays geometrically with time.
> …The primary conception is that the major part of the radioactivity of thor-ium is not due to the thorium at all, but to presence of a non-thorium sub-stance in minute amounts which is being continuously produced.

This, in essence, is the transformation theory of radioactivity. The remainder of the paper contains much of interest, although the inter-pretations of the initial rise portions of the decay and recovery curves were made without the detailed knowledge of the thorium decay series which became available only much later. For the benefit of the reader this series will be outlined below. However, the profound statements in the penultimate paragraph of the paper are first worthy of note.

Since, therefore, radioactivity is at once an atomic phenomenon and accompanied by chemical changes in which new types of matter are produced, these changes must be occurring within the atom, and the radioactive elements must be undergoing spontaneous transformation. The results that have so far been obtained, which indicate that the velocity of this reaction is unaffected by the conditions, make it clear that the changes in question are different in character from any that have been dealt with in chemistry. It is apparent that we are dealing with phenomena outside the sphere of known atomic forces. Radioactivity may therefore be considered as a manifestation of subatomic chemical change.

Current understanding of radioactivity with reference to the thorium decay series

The thorium decay series (one of three naturally occurring series, the others being known as the uranium and actinium series) is illustrated in Figure 1 below. Each member of the series is denoted by its chemical symbol with a superscript indicating its atomic weight *A*. The abscissa *Z* is the atomic number of each member (i.e. the number of protons in the

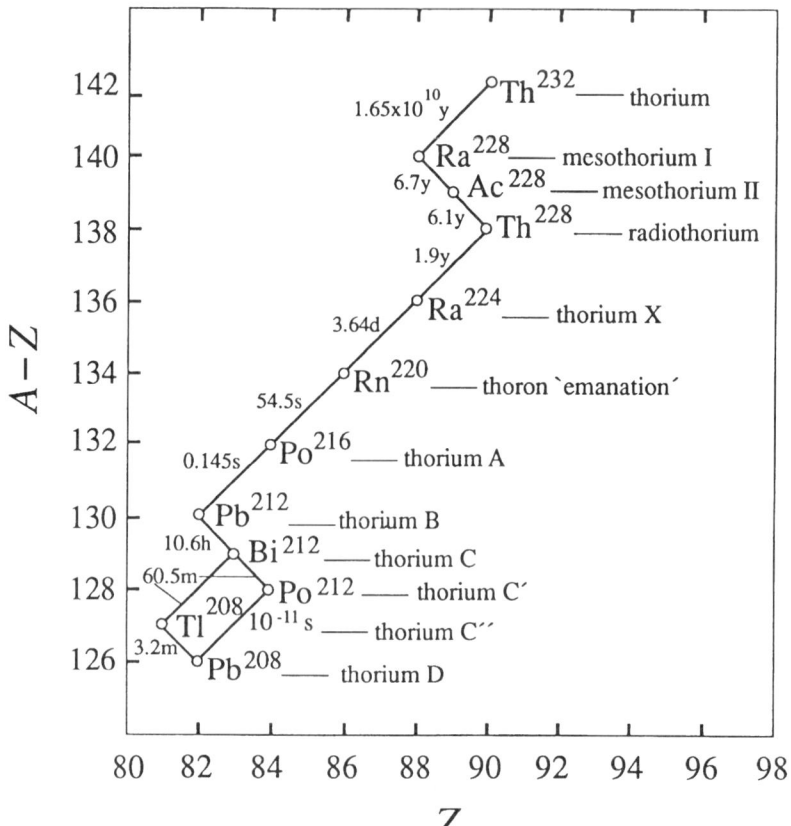

nucleus or its position in the Periodic Table) and the ordinate is $A - Z$. Radioactive decay by α-emission (non-deviable rays in the Rutherford and Soddy paper) results in the loss of two protons and two neutrons from the nucleus, i.e. 2 charges and 4 mass units, so Z decreases by 2, A by 4, and $A - Z$ by 2. The first transformation Th^{232} to Ra^{228} is an example of this, as are the transformations $Th^{228} \rightarrow Ra^{224} \rightarrow Rn^{220} \rightarrow Po^{216}$ etc. Decay by β-emission results in a loss of one electron from the nucleus (equivalent to a neutron being converted to a proton) and so Z increases by one unit while $A - Z$ decreases by one unit. Such diagonal steps to the right accompany the transformations $Ra^{228} \rightarrow Ac^{228} \rightarrow Th^{228}$ in the figure.

Also shown in the figure are the half-lives ($T = (\ln 2)/\lambda = 0.693/\lambda$ where λ is the decay constant) of each member of the series. Formerly used names for the isotopes (chemically identical materials having the same Z but different values of A) are also indicated. Thus ThX is Ra^{224}, the thorium 'emanation' Rn^{220} (radon gas), and the 'active deposit' Pb^{212}. When the members of a radioactive series are in secular equilibrium, the relative amounts (N) of each are given by

$$N_1\lambda_1 = N_2\lambda_2 = N_3\lambda_3 = \ldots$$

where the λs are the decay constants. These products are the activities (number of disintegrations per second) and so each member of the series displays the same activity. Those members with a small decay constant are present in larger amounts than those with a large decay constant, the exception being the final stable member of a series the amount of which increases steadily with time. Chemical separation of a given element disturbs the equilibrium of the series which needs time to be restored.

The thorium compound studied by Rutherford and Soddy would have contained Th^{232} and Th^{228} since they are inseparable by direct chemical means. The 'residual activity' which they could not remove must have been due to α-emission from both of these isotopes but the differences in lifetimes meant that the quantity of Ra^{224} was increasing faster than that of Ra^{228}. Chemical separation of Ra would then lead to an active constituent ThX (Ra^{224}) decaying with a half-life of about 4 days. In turn this would produce Rn^{220} (the emanation) and then, after a further fast transformation through Po^{216}, to Pb^{212} (the active deposit) with a half-life of about 11 hours. In the meantime the Th^{228} in the original compound would be transforming into Ra^{224}, increasing its activity at a rate determined by the difference between its rate of production (essentially constant on the time scale of the experiment) and its rate of decay (i.e. its half-life of about 4 days).

1903   **5**   *Radioactive Change. By E. Rutherford, M.A., D.Sc.,*
Macdonald Professor of Physics, McGill University, and
F. Soddy, M.A. (Oxon.)

No new experimental or theoretical results are presented in this paper but it does contain several very clear statements by the authors concerning their understanding of radioactivity, almost all of which hold true today.

Referring to the results described in their earlier paper, Rutherford and Soddy emphasize the *successive* nature of transformations in a radioactive series, in which one element produces (or rather transforms) into another, which in turn then creates a third element, and so on. The rays that are emitted are not a *consequence* of change that has already taken place but rather an *accompaniment* of the transformation. The authors identify three 'radio-elements' at the head of each known radioactive series and reason that the changes in naturally occurring minerals containing these elements must have been proceeding steadily over very long periods of time. So an equilibrium situation is reached in which the relative proportions of each element in the series is constant, apart of course from the ultimate stable end-product, the amount of which must grow. With regard to the α particles, they have a mass over 1000 times as great as the β particles (electrons) and of the same order as that of the hydrogen atom.

The experimental fall in activity of a member of a series which has been separated from its parents and daughters is characterized by a 'radioactive constant' λ, with a value unique to each member. The activity at any time is proportional to λ and to the amount of material that has not undergone transformation. λ cannot be changed by physical or chemical means. The authors state 'Radioactivity . . . must be regarded as the result of a process which lies wholly outside the sphere of known controllable forces, and cannot be created, altered, or destroyed.'

The three radio-elements identified by Rutherford and Soddy were uranium, thorium and radium. (The naturally-occurring radioactive series recognized today are known as the uranium, thorium and actinum series; the latter was discovered later and Rutherford's uranium and radium series are now known to be part of a single series.) In the paper the name 'metabolon' is suggested for what we now call radio-isotopes' — a name coined by Soddy in 1913.

In the final section of the paper, Rutherford and Soddy draw attention to the vast amounts of energy released in radioactive transformations. A figure of 10^8 calories per gram of radium is estimated solely from the kinetic energy of the emitted α particles. This is more than ten thousand times greater than the energy associated with chemical reactions. Furthermore the rate at which this store of energy is radiated is estimated to be 6.3×10^{11} ergs per year for radium. These figures, it is noted, may also be underestimates, because the energy of the radio-actions may be only a portion of the internal energy of the atom. It is suggested that the energy 'must be taken into account in cosmical physics' and (an inspired remark) 'the maintenance of solar energy, for example, no longer presents any fundamental difficulty if the inter-nal energy of the component elements is considered to be available, i.e. if processes of sub-atomic change are going on.'

Some of these more general aspects of radioactivity were discussed more fully a few years later by Rutherford in *The Journal of the Royal Astronomical Society of Canada* May–June (1907) 145.

1903 **5** *The Magnetic and Electric Deviation of the easily absorbed Rays from Radium. By E. Rutherford, M.A., D.Sc., Macdonald Professor of Physics, McGill University, Montreal*

In 1900, Marie Curie and others demonstrated that, of the three types of radiation emitted by radioactive materials, the β rays were easily deflected by a magnetic field whereas the α rays and γ rays were not. In the same year Becquerel found that the β rays behaved in all respects like a stream of particles with a charge to mass (e/m) ratio equal to that for electrons. The apparent absence of deflection in a magnetic field of the α rays led to their being described as 'non-devi-able' rays by Rutherford, although by 1903 he had obtained indirect evidence that these too were streams of particles. However, in contrast to the β rays, they carried a positive charge. Here he describes experi-ments which confirmed these ideas and yielded the first measurements of e/m for the α particles.

Two experimental difficulties were the smallness of the deflection of the α rays even in the strongest magnetic fields then available (less than 10000 gauss or 1 tesla) and their strong absorption in air. The small deviation required a powerful radioactive source to mitigate against loss as the particles passed through a collimating system of slits (see Figure 1 of the paper). The experiments were performed in flowing hydrogen gas which served to reduce the absorption to considerably below the level the particles would have experienced in air, while at the same time preventing the gaseous emanation (radon) from the radium source from reaching the gold-leaf electroscope used as a sen-sitive detector. Various precautions were taken to eliminate effects arising from β and γ emissions.

Application of a magnetic field perpendicular to the slits led to a fall in the rate of discharge measured by the electroscope when the α particles were deflected and absorbed by the plates. By partly capping the ends of the slits (see Figure 4), the *direction* of the deviation was determined and shown to correspond to that of positively charged particles. Application of an electric field, by applying a potential difference of 600 V between alternate plates, produced a deviation too small to be useful, although in a footnote Rutherford claims to have achieved a larger effect.

Analysis of the magnetic field data yielded values for the velocity and e/m ratio of the α particles. These are given as $2.5 \times 10^9 \text{ cm s}^{-1}$ and $6 \times 10^3 \text{ emu g}^{-1}$ respectively. Comparison of the latter value with the then accepted value for electrons ($\sim 10^7 \text{ emu g}^{-1}$) immediately suggested that, if the α particles carried a similar (but opposite) charge to electrons, they were much more massive. Indeed Rutherford proposed that a radioactive body sealed in a vessel with walls thin enough to allow escape of the α rays should show a measurable decrease in weight with time, although he considered that experiments by Heydweiler in Germany aimed at detecting such an effect were inconclusive.

With later knowledge that α particles are helium nuclei (see the following paper), their e/m ratio can be calculated to be 4.6×10^3 emu g^{-1}, i.e. somewhat less than the value deduced by Rutherford. However, these were pioneering experiments and not easy; indeed Rutherford confided to J. J. Thomson that 'it was the most difficult piece of work I have tackled for some time'.

The reason for the sudden escape of α particles with enormous velocities from radioactive materials was a mystery, not to be resolved for at least another decade.

1909 **17** *The Nature of the α Particle from Radioactive Substances. By Professor E. Rutherford, F.R.S., and T. Royds, M.Sc., 1851 Exhibition Science Scholar*

Identification of the α particle with helium was decisively confirmed in this classic and frequently referred to experimental investigation. Three years earlier Rutherford had made the first strong suggestions that α rays were streams of helium nuclei following his measurements of the e/m ratio of the particles emitted from a variety of radioactive materials, namely radium A (Po^{218}), radium C (Bi^{214}), radium F (Po^{210}) and actinium (Ac^{227}) (*Philosophical Magazine* 1906 **12** 348). These ratios all lay in the range $4.7–5.6 \times 10^3$ emu g^{-1} and were similar to those obtained from uranium and thorium radioisotopes. Rutherford concluded that the α particles had essentially the same mass in all cases studied.

Helium was known at the time to have a density about twice that of hydrogen and to be a monatomic gas. Its atomic weight was thus estimated to be close to 4. The e/m ratio for a singly charged hydrogen ion was known to be about 10^4 emu g^{-1} and so the α particle, with an e/m ratio of about half this value could be: (i) a *molecule* of singly charged hydrogen, (ii) a helium atom carrying *twice* the ionic charge of hydrogen, or (iii) *one-half* of a helium atom carrying a single charge. For several reasons given in the (cited) 1906 paper — the most interesting one of which was that helium is found in old radioactive minerals — Rutherford favoured the second of these possibilities and work with Geiger (*Philosophical Magazine* 1909 **17** 89) supported this viewpoint.

Acknowledging, however, that helium could conceivably be a *product* of a radioactive transformation involving α decay (rather than the α particle itself), Rutherford and Royds attempted the more direct method of identification as described in the present paper. The apparatus (see the figure in the paper) consisted of a glass tube containing radium A, the walls of the tube being thin enough (less than 0.01 mm) to allow the swift α particles to pass through but impervious to helium gas and other radioactive products. Surrounding this tube was a vacuum enclosure which trapped the α particles. After various intervals of time, a mercury column was used to transfer the collected gas to a capillary tube where it was subjected to a spark discharge and spectroscopically examined. A faint yellow line, characteristic of helium, became visible after a waiting time of 2 days and, after 6 days, all the strong lines of the helium spectrum were observed.

After conducting several other experiments to discount various alternative explanations of the results, the authors 'conclude with certainty ... that the α particle, after losing its charge is a helium atom'. 'Losing its charge' means of course the acquisition of two electrons.

1911   **22**   *The Ranges of the α particles from Various Radioactive Substances and a Relation between Range and Period of Transformation. By H. Geiger, Ph.D., and J. M. Nuttall, B.Sc., University of Manchester*

A relationship between the half-life of radioactive elements that transform by α decay and the range (or velocity) of the emitted α particles was first proposed by Rutherford (*Philosophical Magazine* 1907 **13** 110). In the present paper, Geiger and Nuttall describe a new method for measuring the range and obtain a simple empirical relation between the range and the half-life for several isotopes in the uranium and actinium series.

As described in the paper, the radioactive samples were placed at the centre of a silvered glass bulb, 7.95cm in radius, the pressure in which could be varied. A potential of 700 V was applied to the bulb and an electrometer connected to the central rod supporting the metal platform on which the samples sat. The ionization produced was found to be roughly constant at pressures such that the range of the α particles was less than the radius of the bulb but, as the pressure was reduced, the ionization current fell quite sharply at a critical value of pressure corresponding to that at which the particles first reached the bulb. From other experiments it was known that the range was proportional to the pressure and so, using a suitable material for calibration, Geiger and Nuttall were able to deduce the maximum ranges of α particles from different substances.

The ranges (normalized to 76 cm of mercury and 0°C) lay between 2.58 and 6.16 cm and their velocities (known, again from other experiments, to be proportional to the cube root of the maximum range) from 1.51×10^9 to 2.02×10^9 cm s^{-1}. These relatively small variations in range and velocity corresponded to half-lives lying between 5×10^9 y and 1/500 s — an enormous span. Plots of the logarithm of the transformation (decay) constant λ (defined as $(\ln 2)/T_{1/2}$ where $T_{1/2}$ is the half-life) versus the logarithm of the range R gave linear relationships for both the uranium and actinium series, suggesting the empirical relationship $\log \lambda = A + B \log R$, or alternatively $\log T_{1/2} = C - D \log v$, with A, B, C and D constants and v the velocity of emission. The authors remark that the empirical relationship explains 'why no substance has been found emitting α rays of range shorter than 2.58 cm'. Shorter ranges would imply transformation rates so slow that the corresponding activity would be beyond the limits of detection by methods existing at the time.

A physical explanation of the results was not forthcoming until 1928, after the development of quantum mechanics. Then Gamow (*Zeit. f. Phys.* 1928 **51** 204) and Gurney & Condon (*Nature* 1928 **122** 439) provided a theoretical treatment based on quantum mechanical tunnelling of an α particle through the walls of a potential barrier at the sides of a crater representing the nucleus. For an atom of atomic number Z, the outside wall of the barrier has a potential energy $V(r)$ of Coulombic form, $V(r) = 2e^2(Z - 2)/4\pi\epsilon_0 r$ (since the α particle has a charge of $2e$) and so, for $Z \sim 90$ and a nuclear radius of 3×10^{-14} m, the rim of the crater has a height of about 10 MeV. Since the energy of the α particle in the nucleus is less than half this value, classically it could never escape. According to quantum theory, however, the α particle can be represented by a standing wave inside the nuclear well, with a finite probability of tunnelling through the barrier. The important parameter is the thickness of the barrier at a height corresponding to the α particle's energy. So fast α particles with a large energy have a higher probability of escaping than slow ones, exactly as demonstrated by Geiger and Nuttall.

The full theory of α particle tunnelling actually yields a dependence $\log T_{1/2} = C + D/v$ (for nuclei of similar radii and atomic number) but, within the narrow range of observed velocities, the difference of this from the empirical relation found by Geiger and Nuttall is not great.

In their letter to Nature (loc. cit.), Gurney & Condon wrote '... disintegration is a natural consequence of the laws of quantum mechanics... Much has been written about the explosive violence with which the α particle is hurled from its place in the nucleus. But from the process pictured...one would rather say that the particle slips away almost unnoticed.'

370   Prof. E. Rutherford *and* Mr. F. Soddy *on*

be a mere point; the region of misfit (to borrow an expressive term from Prof. Osborne Reynolds) might, if necessary, have definite extension and structure.   Hypotheses of this type are most naturally (indeed, as it seems to me, unavoidably) expressed in terms of an æther which is only locally disturbed by each moving ion; so that a congeries of connected atoms like the Earth does not push it along bodily and establish any finite flow.   But there may be philosophers who prefer not to employ the term æther at all, who are satisfied with a colourless phenomenology, and who manage to escape the consideration of the possibility of an æther whose parts maintain their positions notwithstanding the motion of matter through it, by saying merely that if a certain scheme of formal relations between variables which are symbols of things unknown is altered in a certain formal way, probably originally suggested by the use of dynamical analogies such as have been referred to, the scheme will continue to group the facts under the wider conditions, and they would thus feel freed from any necessity of considering images or models, probably imperfect, of things which being outside ourselves we cannot intrinsically know.

Cambridge. August 7, 1902.

XLI. *The Cause and Nature of Radioactivity.*—Part I.   *By* E. RUTHERFORD, *M.A., D.Sc., Macdonald Professor of Physics, and* F. SODDY, *B.A.* (*Oxon.*), *Demonstrator in Chemistry, McGill University, Montreal* *.

CONTENTS.

I. *Introduction.*

THE following papers give the results of a detailed investigation of the radioactivity of thorium compounds which has thrown light on the questions connected

* Communicated by the Authors.   Accounts of these researches, during the progress of the investigation, have already been given to the London Chemical Society.

the Cause and Nature of Radioactivity. 371

with the source and maintenance of the energy dissipated by radioactive substances. Radioactivity is shown to be accompanied by chemical changes in which new types of matter are being continuously produced. These reaction products are at first radioactive, the activity diminishing regularly from the moment of formation. Their continuous production maintains the radioactivity of the matter producing them at a definite equilibrium-value. The conclusion is drawn that these chemical changes must be sub-atomic in character.

The present researches had as their starting-point the facts that had come to light with regard to thorium radioactivity (Rutherford, Phil. Mag. 1900, vol. xlix. pp. 1 & 161). Besides being radioactive in the same sense as the uranium compounds, the compounds of thorium continuously emit into the surrounding atmosphere a gas which possesses the property of temporary radioactivity. This " emanation," as it has been named, is the source of rays, which ionize gases and darken the photographic film *.

The most striking property of the thorium emanation is its power of exciting radioactivity on all surfaces with which it comes into contact. A substance after being exposed for some time in the presence of the emanation behaves as if it were covered with an invisible layer of an intensely active material. If the thoria is exposed in a strong electric field, the excited radioactivity is entirely confined to the negatively charged surface. In this way it is possible to concentrate the excited radioactivity on a very small area. The excited radioactivity can be removed by rubbing or by the action of acids, as, for example, sulphuric, hydrochloric, and hydrofluoric acids. If the acids be then evaporated, the radioactivity remains on the dish.

The emanating power of thorium compounds is independent of the surrounding atmosphere, and the excited activity it produces is independent of the nature of the substance on which it is manifested. These properties made it appear that both phenomena were caused by minute quantities of special kinds of matter in the radioactive state, produced by the thorium compound.

The next consideration in regard to these examples of radioactivity, is that the activity in each case diminishes regularly with the lapse of time, the intensity of radiation at each instant being proportional to the amount of energy remaining to be radiated. For the emanation a period of

---

* If thorium oxide be exposed to a white heat its power of giving an emanation is to a large extent destroyed. Thoria that has been so treated is referred to throughout as " de-emanated."

one minute, and for the excited activity a period of eleven hours, causes the activity to fall to half its value.

These actions—(1) the production of radioactive material, and (2) the dissipation of its available energy by radiation—which are exhibited by thorium compounds in the secondary effects of emanating power and excited radioactivity, are in reality taking place in all manifestations of radioactivity. The constant radioactivity of the radioactive elements is the result of an equilibrium between these two opposing processes.

II. *The Experimental Methods of investigating Radioactivity.*

Two methods are used for the measurement of radioactivity, the electrical and the photographic. The photographic method is of a qualitative rather than a quantitative character; its effects are cumulative with time, and as a rule long exposures are necessary when the radioactivity of a feeble agent like thoria is to be demonstrated. In addition, Russell has shown that the darkening of a photographic plate is brought about also by agents of a totally different character from those under consideration, and, moreover, under very general conditions. Sir William Crookes (Proc. Roy. Soc. (1900) lxvi. p. 409) has sounded a timely note of warning against putting too much confidence in the indications of the photographic method of measuring radioactivity. The uncertainty of an effect produced by cumulative action over long periods of time quite precludes its use for work of anything but a qualitative character.

But the most important objection to the photographic method is that certain types of rays from radioactive substances, which ionize gases strongly, produce little if any effect on the sensitive film. In the case of uranium, these protographically inactive rays form by far the greatest part of the total radiation, and much of the previous work on uranium by the photographic method must be interpreted differently (Soddy, Proc. Chem. Soc. 1902, p. 121).

On the other hand, it is possible to compare intensities of radiation by the electrical method with greater rapidity and with an error not exceeding 1 or 2 per cent. These methods are based on the property generally possessed by all radiations of the kind in question, of rendering a gas capable of discharging both positive and negative electricity. These, as will be shown, are capable of great refinement and certainty. An ordinary quadrant electrometer is capable of detecting and measuring a difference of potential of at least 10^{-2} volts. With special instruments, this sensitiveness may be increased

the Cause and Nature of Radioactivity.   373

a hundredfold. An average value for the capacity of the electrometer and connexions is 3×10^{-5} microfarads; and when this is charged up to 10^{-2} volts, a quantity of electricity corresponding to 3×10^{-13} coulombs is stored up. Now in the electrolysis of water one gram of hydrogen carries a charge of 10^5 coulombs. Assuming, for the sake of example, that the conduction of electricity in gases is analogous to that in liquids, this amount of electricity corresponds to the transport of a mass of 3×10^{-18} grams of hydrogen; that is, a quantity of the order of 10^{-12} times that detected by the balance. For a more delicate instrument, this amount would produce a large effect.

The examples of radium in pitchblende and of the thorium-excited radioactivity make it certain that comparatively large ionization effects are produced by quantities of matter beyond the range of the balance or spectroscope.

The electrometer also affords the means of recognizing and differentiating between the emanations and radiations of different chemical substances. By the rate of decay the emanation from thorium, for example, can be instantly distinguished from that produced by radium; and although a difference in the rate of decay does not of itself argue a fundamental difference of nature, the identity of the rate of decay furnishes at least strong presumption of identity of nature.

Radiations, on the other hand, can be compared by means of their penetration powers (Rutherford, Phil. Mag. 1899, vol. xlvii. p. 122). If the rays from various radioactive substances are made to pass through successive layers of aluminium-foil, each additional layer of foil cuts down the radiation to a fraction of its former value, and a curve can be plotted with the thickness of metal penetrated as abscissæ, and the intensity of the rays after penetration as ordinates, expressing at a glance the penetration power of the rays under examination. The curves so obtained are quite different for different radioactive substances. The radiations from uranium, radium, thorium, each give distinct and characteristic curves, whilst that of the last-named again is quite different from that given by the excited radioactivity produced by the thorium emanation. It has been recently found (Rutherford and Grier, *Phys. Zeit.* 1902, p. 385) that thorium compounds, in addition to a type of easily absorbed Röntgen-rays, non-deviable in the magnetic field, emit also rays of a very penetrating character deviable in the magnetic field. The latter are therefore similar to cathode-rays, which are known to consist of material particles travelling with a

374   Prof. E. Rutherford *and* Mr. F. Soddy *on*

velocity approaching that of light.   But thorium, in comparison with uranium and radium, emits a much smaller proportion of deviable radiation.   The determination of the proportion between the deviable and non-deviable rays affords a new means of investigating thorium radioactivity.

The electrometer thus supplies the study of radioactivity with methods of quantitative and qualitative investigation, and there is therefore no reason why the cause and nature of the phenomenon should not be the subject of chemical investigation.

Fig. 1 shows the general arrangement.   From 0·5 to 0·1 gram of the compound to be tested, reduced to fine powder, is uniformly sifted over a platinum plate 36 sq. cms. in area.

Fig. 1.

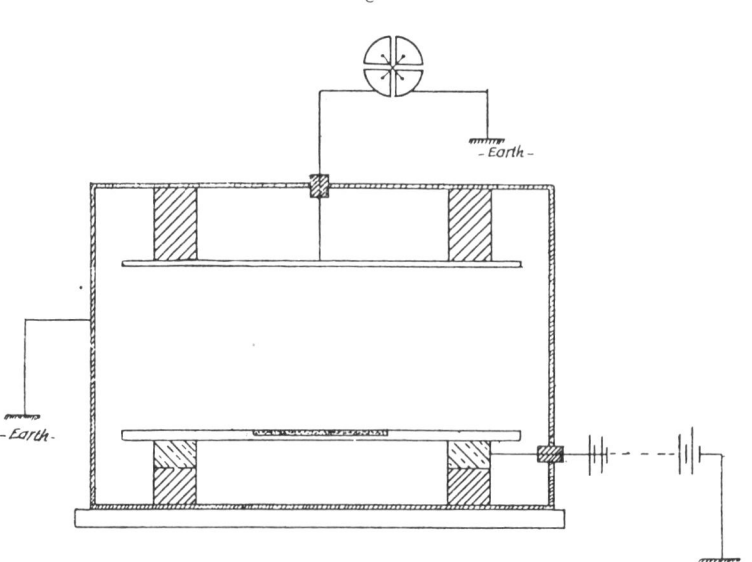

This plate was placed on a large metal plate connected to one pole of a battery of 300 volts, the other pole of which was earthed.   An insulated parallel plate was placed about 6 cm. above it, and the whole apparatus inclosed in a metal box connected to earth, to prevent electrostatic disturbance. The shaded portions in the figure represented insulators. A door was made in the apparatus so that the plate could be rapidly placed in position or removed.   Both pairs of quadrants are first connected to earth.   On connecting the one pair with the apparatus, the deflexion of the needle from zero

the Cause and Nature of Radioactivity.   375

increases uniformly with time, and the time taken to pass over 100 divisions of the scale is taken by a stop-watch. The *rate* of movement is a measure of the ionization-current between the plates. The ratio of the currents for different substances is a comparative measure of their radioactivity.

With this apparatus ·5 gr. of thorium oxide produces a current of $1·1 \times 10^{-11}$ amperes, which, with the electrometer used, working at average sensitiveness, corresponds to 100 divisions of the scale in 36 seconds. In certain cases a special modification of the Dolezalek electrometer was employed which is 100 times more sensitive. With this instrument the radioactivity of 1 milligram of thoria produces a measurable effect. If the substance gives off an emanation, the current between the plates increases with time. Under these conditions, when the thorium compound is exposed in thin layers with a maximum of radiating surface, all but one or two per cent. of the total effect is due to the straight-line radiation. Even when the effect due to the emanation has attained a maximum, this constitutes a very small fraction of the whole. This effect, however, may to a large extent be eliminated by taking the current between the electrodes immediately after the material is placed in the testing-apparatus. It may be completely eliminated by passing a current of air between the electrodes to remove the emanation as fast as it is formed.

The current between the plates observed with the electrometer at first increases with the voltage, but a stage is very soon reached when there is a very small increase for a large additional voltage. A P.D. of 300 volts was sufficient to obtain the maximum current, so that all the ions reached the electrodes before any appreciable recombination occurred.

It must, however, at once be pointed out that it is difficult to make any absolute measure of radioactivity. The radiation from thorium is half absorbed by a thickness of aluminium of ·0004 cm. ; and since thorium oxide is far denser than aluminium, it is probable that the radiation in this case is confined to a surface-layer only ·0001 cm. deep. It is obvious that different preparations, each containing the same percentage of thorium but with different densities and different states of division, will not give the same intensity of radiation. In comparing two different specimens of the same compound, it is important that the final steps in their preparation should be the same in each case. As a rule absolute measurements of this kind have been avoided. It is possible, however, to trace with great accuracy the *change* of radioactivity of any preparation with time by leaving it undisturbed on its original plate, and comparing it with a similarly undisturbed constant comparison sample. Most of the investigations have been carried out by this method.

III. *The Separation of a Radioactive Constituent from Thorium Compounds.*

During an investigation of the emanating power of thorium compounds, to be described later, evidence was obtained of the separation of an intensely radioactive constituent by chemical methods. It had been noticed that in certain cases thorium hydroxide, precipitated from dilute solutions of thorium nitrate by ammonia, possessed an abnormally low emanating power. This led naturally to an examination being made of the filtrates and washings obtained during the process. It was found that the filtrates invariably possessed emanating power, although from the nature of their production they are chemically free from thorium. If the filtrate is evaporated to dryness, and the ammonium salts removed by ignition, the small residues obtained exhibit radioactivity also, to an extent very much greater than that possessed by the same weight of thorium. As a rule these residues were of the order of one-thousandth part by weight of the thorium salt originally taken, and were many hundred, in some cases over a thousand, times more active than an equal weight of thoria. The separation of an active constituent from thorium by this method is not all dependent on the purity of the salt used. By the kindness of Dr. Knöfler, of Berlin, who, in the friendliest manner, presented us with a large specimen of his purest thorium nitrate, we were enabled to test this point. This specimen, which had been purified by a great many processes, did not contain any of the impurities found in the commercial salt before used. But its radioactivity and emanating power were at least as great, and the residues from the filtrates after precipitation by ammonia were no less active than those before obtained. These residues are free from thorium, or at most contain only the merest traces, and when redissolved in nitric acid do not appear to give any characteristic reaction.

An examination of the penetrating power of the rays from the radioactive residue, showed that the radiations emitted were in every respect identical with the ordinary thorium radiation. In another experiment the nature of the emanation from a similar intensely active thorium-free residue was submitted to examination. The rate of decay was quite indistinguishable from that of the ordinary thorium emanation; that is, substances chemically free from thorium have been

the Cause and Nature of Radioactivity.   377

prepared possessing thorium radioactivity in an intense degree.

The thorium hydroxide which had been submitted to the above process was found to be less than half as radioactive as the same weight of thorium oxide. It thus appeared that a constituent responsible for the radioactivity of thorium had been obtained, which possessed distinct chemical properties and an activity of the order of at least a thousand times as great as the material from which it had been separated.

Sir William Crookes (Proc. Roy. Soc. 1900, lxvi. p. 409) succeeded in separating a radioactive constituent of great activity and distinct chemical nature from uranium, and gave the name UrX to this substance. For the present, until more is known of its real nature, it will be convenient to name the active constituent of thorium ThX, similarly. Like UrX, however, ThX does not answer to any definite analytical reactions, but makes its appearance with precipitates formed in its solution even when no question of insolubility is involved. This accords with the view that it is present in infinitesimal quantity, and possesses correspondingly great activity. Even in the case of the most active preparations, these probably are composed of some ThX associated with accidental admixtures large in proportion.

These results receive confirmation from observations made on a different method of separating ThX. The experiment was tried of washing thoria with water repeatedly, and seeing if the radioactivity was thereby affected. In this way it was found that the filtered washings, on concentration, deposited small amounts of material with an activity often of the order of a thousand times greater than that of the original sample. In one experiment, 290 grams of thoria were shaken for a long time with nine quantities, each of 2 litres of distilled water. The first washing, containing thorium sulphate present as an impurity, was rejected, the rest concentrated to different stages and filtered at each stage. One of the residues so obtained weighed 6·4 mg., and was equivalent in radioactivity to 11·3 grams of the original thoria, and was therefore no less than 1800 times more radioactive. It was examined chemically, and gave, after conversion into sulphate, the characteristic reaction of thorium sulphate, being precipitated from its solution in cold water by warming. *No other substance than thorium could be detected by chemical analysis*, although of course the quantity was too small for a minute examination. The penetrating power of the radiation from this substance again established its identity with the ordinary thorium radiation.

Phil. Mag. S. 6. Vol. 4. No. 21. *Sept.* 1902.   2 C

In another experiment, a small quantity of thoria was shaken many times with large quantities of water. In this case, the radioactivity of the residue was examined and found to be about 20 per cent. less radioactive than the original sample.

The influence of Time on the activity of Thorium and ThX.— The preparations employed in our previous experiments were allowed to stand over during the Christmas vacation. On examining them about three weeks later it was found that the thorium hydroxide, which had originally possessed only about 36 per cent. of its normal activity, had almost completely recovered the usual value. The active residues, on the other hand, prepared by both methods, had almost completely lost their original activity. The chemical separation effected was thus not permanent in character. At this time M. Becquerel's paper (*Comptes Rendus*, cxxxiii. p. 977, Dec. 9th, 1901) came to hand, in which he shows that the same phenomena of recovery and decay are presented by uranium after it has been partially separated from its active constituent by chemical treatment.

A long series of observations was at once started to determine—

 (1) The rate of recovery of the activity of thorium rendered less active by removal of ThX ;

 (2) The rate of decay of the activity of the separated ThX ;

in order to see how the two processes were connected. The results led to the view that may at once be stated. The radioactivity of thorium at any time is the resultant of two opposing processes—

 (1) The production of fresh radioactive material at a constant rate by the thorium compound ;

 (2) The decay of the radiating power of the active material with time.

The normal or constant radioactivity possessed by thorium is an equilibrium value, where the rate of increase of radioactivity due to the production of fresh active material is balanced by the rate of decay of radioactivity of that already formed. It is the purpose of the present paper to substantiate and develope this hypothesis.

  IV. *The Rates of Recovery and Decay of Thorium Radioactivity.*

A quantity of the pure thorium nitrate was separated from ThX in the manner described by several precipitations with ammonia. The radioactivity of the hydroxide so obtained

the Cause and Nature of Radioactivity.   379

was tested at regular intervals to determine the rate of recovery of its activity. For this purpose the original specimen of ·5 gram was left undisturbed throughout the whole series of measurements on the plate over which it had been sifted, and was compared always with ·5 gram of ordinary de-emanated thorium oxide spread similarly on a second plate and also left undisturbed. The emanation from the hydroxide was prevented from interfering with the results by a special arrangement for drawing a current of air over it during the measurements.

The active filtrate from the preparation was concentrated and made up to 100 c.c. volume. One quarter was evaporated to dryness and the ammonium nitrate expelled by ignition in a platinum dish, and the radioactivity of the residue tested at the same intervals as the hydroxide to determine the rate of decay of its activity. The comparison in this case was a standard sample of uranium oxide kept undisturbed on a metal plate, which repeated work has shown to be a perfectly constant source of radiation. The remainder of the filtrate was used for other experiments.

The following table gives an example of one of a numerous series of observations made with different preparations at different times. The maximum value obtained by the hydroxide and the original value of the ThX are taken as 100 :—

Time in days.	Activity of Hydroxide.	Activity of ThX.
0	44	100
1	37	117
2	48	100
3	54	88
4	62	72
5	68	
6	71	53
8	78	
9	...	29·5
10	83	25·2
13	...	15·2
15	...	11·1
17	96·5	
21	99	
28	100	

Fig. 2 shows the curves obtained by plotting the radio-activities as ordinates, and the time in days as abscissæ. Curve II. illustrates the rate of recovery of the activity of thorium, curve I. the rate of decay of activity of ThX. It

2 C 2

:380   Prof. E. Rutherford *and* Mr. F. Soddy *on*

will be seen that neither of the curves is regular for the first two days. The activity of the hydroxide at first actually

Fig. 2.

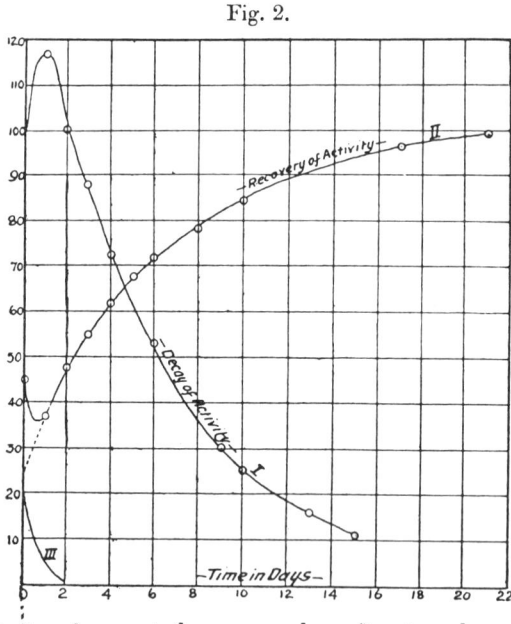

diminished and was at the same value after two days as when first prepared. The activity of the ThX, on the other hand, at first increases and does not begin to fall below the original value till after the lapse of two days (compare section IX.). These results cannot be ascribed to errors of measurement, for they have been regularly observed whenever similar preparations have been tested. The activity of the residue obtained from thorium oxide by the second method of washing decayed very similarly to that of ThX, as shown by the above curve.

If for present purposes the initial periods of the curve are disregarded and the later portions only considered, it will be seen at once that the time taken for the hydroxide to recover one half of its lost activity is about equal to the time taken by the ThX to lose half its activity, viz., in each case about 4 days, and speaking generally the percentage proportion of the lost activity regained by the hydroxide over any given interval is approximately equal to the percentage proportion of the activity lost by the ThX during the same interval. If the recovery curve is produced backwards in the normal direction to cut the vertical axis, it will be seen to do so at a

the Cause and Nature of Radioactivity.   381

minimum of about 25 per cent., and the above result holds
even more accurately if the recovery is assumed to start from
this constant minimum, as, indeed, it has been shown to do
under suitable conditions (section IX., fig. 4).

This is brought out by fig. 3, which represents the recovery

Fig. 3.

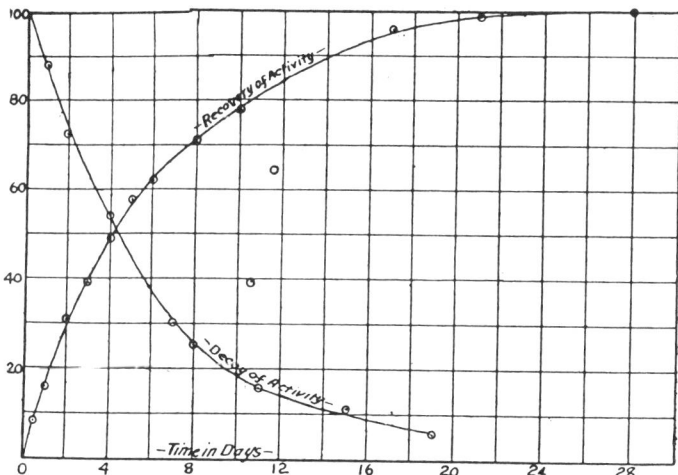

curve of thorium in which the percentage amounts of activity
recovered, reckoned from this 25 per cent. minimum, are
plotted as ordinates. In the same figure the decay curve
after the second day is shown on the same scale.

The activity of ThX decreases very approximately in a
geometrical progression with the time, *i. e.* if I_0 represent the
initial activity and I_t the activity after time t,

$$\frac{I_t}{I_0} = e^{-\lambda t}, \quad . \quad . \quad . \quad . \quad . \quad . \quad (1)$$

where λ is a constant and e the base of natural logarithms.

The experimental curve obtained with the hydroxide for
the rate of rise of its activity from a minimum to a maximum
value will therefore be approximately expressed by the
equation

$$\frac{I_t}{I_0} = 1 - e^{-\lambda t}, \quad . \quad . \quad . \quad . \quad . \quad (2)$$

where I_0 represents the amount of activity recovered when
the maximum is reached, and I_t the activity recovered after
time t, λ *being the same constant as before.*

382   Prof. E. Rutherford *and* Mr. F. Soddy *on*

Now this last equation has been theoretically developed in other places (compare Rutherford, Phil. Mag. 1900, pp. 10 and 181) to express the rise of activity to a constant maximum of a system consisting of radiating particles in which

(1) The rate of supply of fresh radiating particles is constant.

(2) The activity of each particle dies down geometrically with the time according to equation (1).

It therefore follows that if the initial irregularities of the curves are disregarded and the residual activity of thorium is assumed to possess a *constant* value, the experimental curve obtained for the recovery of activity will be explained if two processes are supposed to be taking place :

(1) That the active constituent ThX is being produced at a constant rate;

(2) That the activity of the ThX decays geometrically with time.

Without at first going into the difficult questions connected with the initial irregularities and the residual activity, the main result that follows from the curves given can be put to experimental test very simply. The primary conception is that the major part of the radioactivity of thorium is not due to the thorium at all, but to the presence of a non-thorium substance in minute amount which is being continuously produced.

V. *Chemical Properties of ThX.*

The fact that thorium on precipitation from its solutions by ammonia leaves the major part of its activity in the filtrate does not of itself prove that a material *constituent* responsible for this activity has been chemically separated. It is possible that the matter constituting the non-thorium part of the solution is rendered temporarily radioactive by its association with thorium, and this property is retained through the processes of precipitation, evaporation, and ignition, and manifests itself finally on the residue remaining.

This view, however, can be shown to be quite untenable, for upon it any precipitate capable of removing thorium completely from its solution should yield active residues similar to those obtained from ammonia. Quite the reverse, however, holds.

When thorium nitrate is precipitated by sodium or ammonium carbonate, the residue from the filtrate by evaporation and ignition is free from activity, and the thorium carbonate possesses the normal value for its activity.

The same holds true when oxalic acid is used as the

precipitant. This reagent even in strongly acid solution precipitates almost all of the thorium. When the filtrate is rendered alkaline by ammonia, filtered, evaporated, and ignited, the residue obtained is inactive.

In the case where sodium phosphate is used as the precipitant in ordinary acid solution, the part that comes down is more or less free from ThX. On making the solution alkaline with ammonia, the remainder of the thorium is precipitated as phosphate, and carries with it the whole of the active constituent, so that the residue from the filtrate is again inactive.

In fact ammonia is the only reagent of those tried capable of separating ThX from thorium.

The result of **Sir William Crookes** with uranium, which we have confirmed working with the electrical method, may be here mentioned. UrX is completely precipitated by ammonia together with uranium, and the residue obtained by the evaporation of the filtrate is quite inactive.

There can thus be no question that both ThX and UrX are distinct types of matter with definite chemical properties. Any hypothesis that attempts to account for the recovery of activity of thorium and uranium with time must of necessity start from this primary conception.

VI. *The Continuous Production of* **ThX.**

If the recovery of the activity of thorium with time is due to the production of ThX, it should be possible to obtain experimental evidence of the process. The first point to be ascertained is how far the removal of ThX by the method given reduces the total radioactivity of thorium. A preliminary trial showed that the most favourable conditions for the separation are by precipitating in hot dilute solutions by dilute ammonia. A quantity of 5 grams of thorium nitrate, as obtained from the maker, was so precipitated by ammonia, the precipitate being redissolved in nitric acid and reprecipitated under the same conditions successively *without lapse of time.*

The removal of ThX was followed by measuring the activity of the residues obtained from the successive filtrates. The activity of the ThX from the first filtrate was equivalent to 4·25 grams of thoria, from the second to 0·33 gram, and from the third to 0·07 gram. It will be seen that by two precipitations practically the whole of the ThX is removed. The radioactivity of the separated hydroxide was 48 per cent. of that of the standard de-emanated sample of thoria.

Rate of production of ThX.—A quantity of thorium nitrate solution that had been freed from ThX about a month before, was again subjected to the same process. The activity of the residue from the filtrate in an experiment in which 10 grams of this nitrate had been employed was equivalent to 8·3 grams of thorium oxide. This experiment was performed on the same day as the one recorded above, in which 5 grams of new nitrate had been employed, and it will be seen that there is no difference in the activity of the filtrate in the two cases. In one month the activity of the ThX in a thorium compound again possesses its maximum value.

If a period of 24 hours is allowed to elapse between the successive precipitations, the activity of the ThX formed during that time corresponds to about one-sixth of the maximum activity of the total thorium employed. In three hours the activity of the amount produced is about one-thirtieth. The rate of production of ThX worked out from those figures well agrees with the form of the curve obtained for the recovery of activity of thorium, if the latter is taken to express the continuous production of ThX at a constant rate and the diminution of the activity of the product in geometrical progression with the time.

By using the sensitive electrometer, the course of production of ThX can be followed after extremely short intervals. Working with 10 grams of thorium nitrate, the amount produced in the minimum time taken to carry out the successive precipitations is as much as can be conveniently measured. If any interval is allowed to lapse the effect is beyond the range of the instrument, unless the sensitiveness is reduced to a fraction of its ordinary value by the introduction of capacities into the system. Capacities of ·01 and ·02 microfarad, which reduce the sensitiveness to less than one two-hundredth of the normal, were frequently employed in dealing with these active residues.

The process of the production of ThX is continuous, and no alteration was observed in the amount produced in a given time after repeated separations. In an experiment carried out for another purpose (section IX.) after 23 successive precipitations extending over 9 days, the amount formed during the last interval was as far as could be judged no less than what occurred at the beginning of the process.

The phenomenon of radioactivity, by means of the electrometer as its measuring instrument, thus enables us to detect and measure changes occurring in matter after a few minutes interval, which have never yet been detected by the balance or suspected of taking place.

VII. *Influence of Conditions on the Changes occurring in Thorium.*

It has been shown that in thorium compounds the decay of radioactivity with time is balanced by a continuous production of fresh active material. The change which produces this material must be chemical in nature, for the products of the action are different in chemical properties from the thorium from which they are produced. The first step in the study of the nature of this change is to examine the effects of conditions upon its rate.

Effect of conditions on the rate of decay.—Since the activity of the products affords the means of measuring the amount of change, the influence of conditions on the rate of decay must be first found. It was observed that, like all other types of temporary radioactivity, the rate of decay is unaltered by any known agency. It is unaffected by ignition and chemical treatment, and the material responsible for it can be dissolved in acids and re-obtained by the evaporation of the solution, without affecting the activity. The following experiment shows that the activity decays at the same rate in solutions as in the solid state. The remainder of the solution that had been used to determine the decay curve of ThX (fig. 2) was allowed to stand, and at the end of 12 days a second quarter was evaporated to dryness and ignited, and its activity compared with that of the first which had been left since evaporation upon its original platinum dish. The activities of the two specimens so compared with each other were the same, showing that in spite of the very different conditions the two fractions had decayed at equal rates. After 19 days a third quarter was evaporated, and the activity, now very small, was indistinguishable from that of the fraction first evaporated. Re-solution of the residues after the activity had decayed does not at all regenerate it. The activity of ThX thus decays at a rate independent of the chemical and physical condition of the molecule.

Thus the rate of recovery of activity under different conditions in thorium compounds affords a direct measure of the rate of production of ThX under these conditions. The following experiments were performed :—

One part of thorium hydroxide newly separated from ThX was sealed up in a vacuum obtained by a good Töpler pump, and the other part exposed to air. On comparing the samples 12 days later no difference could be detected between them either in their radioactivity or emanating power.

In the next experiment a quantity of hydroxide freed from

ThX was divided into two equal parts; one was exposed for 20 hours to the heat of a Bunsen burner in a platinum crucible, and then compared with the other. No difference in the activities was observed. In a second experiment, one half was ignited for 20 minutes on the blast, and then compared with the other with the same result. The difference of temperature and the conversion of thorium hydroxide into oxide thus exercised no influence on the activity.

Some experiments that were designed to test in as drastic a manner as possible the effect of the chemical condition of the molecule on the rate of production of ThX brought to light small differences, but these are almost certainly to be accounted for in another way. It will be shown later (section IX.) that about 21 per cent. of the normal radioactivity of thorium oxide under ordinary conditions consists of a secondary activity excited on the mass of the material. This portion is of course a variable, and since it is divided among the total amount of matter present, the conditions of aggregation, &c., will affect the value of this part. This effect of excited radioactivity in thorium makes a certain answer to the question difficult, and on this account the conclusion that the rate of production of ThX is independent of the molecular conditions is not final. The following experiment, however, makes it extremely probable.

A quantity of thorium nitrate as obtained from the maker was converted into oxide in a platinum crucible by treatment with sulphuric acid and ignition to a white heat. The de-emanated oxide so obtained was spread on a plate, and any change in radioactivity with time, which under these circumstances could certainly be detected, was looked for during the first week from preparation. None whatever was observed, whereas if the rate of production of ThX in thorium nitrate is different from that in the oxide, the equilibrium point, at which the decay and increase of activity balance each other, will be altered in consequence. There should have therefore occurred a logarithmic rise or fall from the old to the new value. As, however, the radioactivity remained constant, it appears very probable that the changes involved are independent of the molecular condition.

It will be seen that the assumption is here made that the proportion of excited radioactivity in the two compounds is the same, and for this reason compounds were chosen which possess but low emanating power. (Compare section IX. last paragraph.)

Uranium is a far simpler example of a radioactive element than thorium, as the phenomena of excited radioactivity and emanating power are here absent. The separation of UrX and the recovery of the activity of the uranium with time appear, however, analogous to these processes in thorium, and the rate of recovery and decay of uranium activity are at present under investigation. It is proposed to test the influence of conditions on the rate of change more thoroughly in the case of uranium, as here secondary changes do not interfere.

VIII. *The Cause and Nature of Radioactivity.*

The foregoing conclusions enable a great generalization to be made in the subject of radioactivity. Energy considerations require that the intensity of radiation from any source should die down with time unless there is a constant supply of energy to replace that dissipated. This has been found to hold true in the case of all known types of radioactivity with the exception of the "naturally" radioactive elements—to take the best established cases, thorium, uranium, and radium. It will be shown later that the radioactivity of the emanation produced by thorium compounds decays geometrically with the time under all conditions, and is not affected by the most drastic chemical and physical treatment. The same has been shown by one of us (Phil. Mag. 1900, p. 161) to hold for the excited radioactivity produced by the thorium emanation. This decays at the same rate whether on the wire on which it is originally deposited, or in solution of hydrochloric or nitric acid. The excited radioactivity produced by the radium emanation appears analogous. All these examples satisfy energy considerations. In the case of the three naturally occurring radioactive elements, however, it is obvious that there must be a continuous replacement of the dissipated energy, and no satisfactory explanation has yet been put forward.

The nature of the process becomes clear in the light of the foregoing results. The material constituent responsible for the radioactivity, when it is separated from the thorium which produces it, then behaves in the same way as the other types of radioactivity cited. Its activity decays geometrically with the time, and the rate of decay is independent of the molecular conditions. The normal radioactivity is, however, maintained at a constant value by a chemical change which produces fresh radioactive material at a rate also independent of the conditions. The energy required to maintain the radiations will be accounted for if we suppose that the energy of the system after the change has occurred is less than it was before.

The work of Crookes and Becquerel on the separation of UrX and the recovery of the activity of the uranium with time, makes it appear extremely probable that the same explanation holds true for this element. The work of M. and Mme. Curie, the discoverers of radium, goes to show that this body easily suffers a temporary decrease of its activity by chemical treatment, the normal value being regained after the lapse of time, and this can be well interpreted on the new view. All known types of radioactivity can thus be brought under the same category.

IX. *The Initial Portions of the Curves of Decay and Recovery.*

The curves of the recovery and decay of the activities of thorium and ThX with time suggested the explanation that the radioactivity of thorium was being maintained by the production of ThX at a constant rate. Before this can be considered rigidly established, two outstanding points remain to be cleared up. 1. What is the meaning of the early portion of the curves? The recovery curve drops before it rises, and the decay curve rises before it drops. 2. Why does not the removal of ThX render thorium completely inactive? A large proportion of the original radioactivity is not affected by the removal of ThX.

A study of the curves (fig. 2) shows that in each case a double action is probably at work. It may be supposed that the normal decay and recovery are taking place, but are being masked by a simultaneous rise and decay from other causes. From what is known of thorium radioactivity, it was surmised that an action might be taking place similar to that effected by the emanation of exciting radioactivity on surrounding inactive matter. It will be shown later that the ThX, and not thorium, is the cause of the emanating power of thorium compounds. On this view, the residual activity of thorium might consist in whole or in part of a secondary or excited radioactivity produced on the whole mass of the thorium compound by its association with the ThX. The drop in the recovery-curve on this view would be due to the decay of this excited radioactivity proceeding simultaneously with, and at first reversing the effect of the regeneration of ThX. The rise of the decay-curve would be the increase due to the ThX exciting activity on the matter with which it is associated, the increase from this cause being greater than the decrease due to the decay of the activity of the ThX. It is easy to put this hypothesis to experimental test. If the ThX is removed from the thorium as soon as it is formed over a sufficient period, the former will be prevented from

the Cause and Nature of Radioactivity.   389

exciting activity on the latter, and that already excited will decay spontaneously. The experiment was therefore performed. A quantity of nitrate was precipitated as hydroxide in the usual way to remove ThX, the precipitate redissolved in nitric acid, and again precipitated after a certain interval. From time to time a portion of the hydroxide was removed and its radioactivity tested. In this way the thorium was precipitated in all 23 times in a period of 9 days, and the radioactivity reduced to a constant minimum. The following table shows the results :—

	Activity of Hydroxide. per cent.
After first precipitation	46
After precipitations at three intervals of 24 hours	39
At three more intervals each of 24 hours, and three more each of 8 hours	22
At three more each of 8 hours	24
At six more each of 4 hours	25

The constant minimum thus attained—about 25 per cent. of the original activity—is thus about 21 per cent. below that obtained by two successive precipitations without interval, which has been shown to remove all the ThX separable by the process. The rate of recovery of this 23 times precipitated hydroxide was then measured (fig. 4). It will be

Fig. 4.

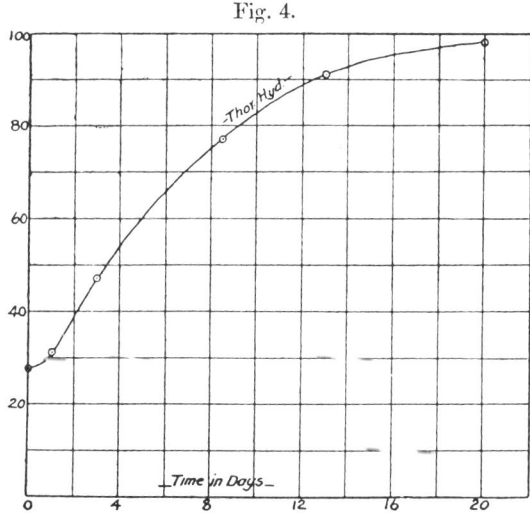

seen that it is now quite normal, and the initial drop characteristic of the ordinary curve is quite absent. It is in fact almost identical with the ordinary curve (fig. 2) that has been produced back to cut the vertical axis, and there is thus no doubt that there is a residual activity of thorium unconnected apparently with ThX, and constituting about one fourth of the whole.

The decay-curves of several of the fractions of ThX separated in this experiment after varying intervals of time were taken for the first few days. All of them showed the initial rise of about 15 per cent. at the end of 18 hours, and then a normal decay to zero. The position is thus proved that the initial irregularities are caused by the secondary radiation excited by ThX upon the surrounding matter. By suitably choosing the conditions the recovery-curve can be made to rise normally from a constant minimum, and the decay-curve be shown to consist of two curves, the first the rate of production of excited radioactivity, and the second the rate of decay of the activity as a whole.

So far nothing has been stated as to whether the excited radioactivity which contributes about 21 per cent. of the total activity of thorium is the same or different from the known type produced by the thorium emanation. All that has been assumed is that it should follow the same general law ; *i. e.* the effect will increase with the time of action of the exciting cause, and decrease with time after the cause is removed. If the rate of rise of the excited activity be worked out from the curves given (fig. 5) it will be found to agree with that of the ordinary excited activity, *i. e.* it rises to half value in about 12 hours. Curve 1 is the observed decay-curve for ThX ; curve 2 is the theoretical curve, assuming that it decreases geometrically with time and falls to half value in four days. Curve 3 is obtained by plotting the difference between these two, and therefore constitutes the curve of excited activity. Curve 4 is the experimental curve obtained for the rise of the excited radioactivity from the thorium emanation when the exciting cause is constant. But the exciting cause (ThX) in the present case is not constant, but is itself falling to half value in 4 days, and hence the difference curve, at first almost on the other, drops away from it as time goes on, and finally decays to zero. There is thus no reason to doubt that the effect is the same as that produced by the thorium emanation, which is itself a secondary effect of ThX. Curve 3 (fig. 2) represents a similar difference curve for *the decay* of excited activity, plotted from *the recovery curve* of thorium.

Since this effect of excited activity is caused by the emanation, it seemed reasonable to suppose that it will be greater, the less the emanation succeeds in escaping in the radioactive state, and therefore that de-emanated compounds should

Fig. 5.

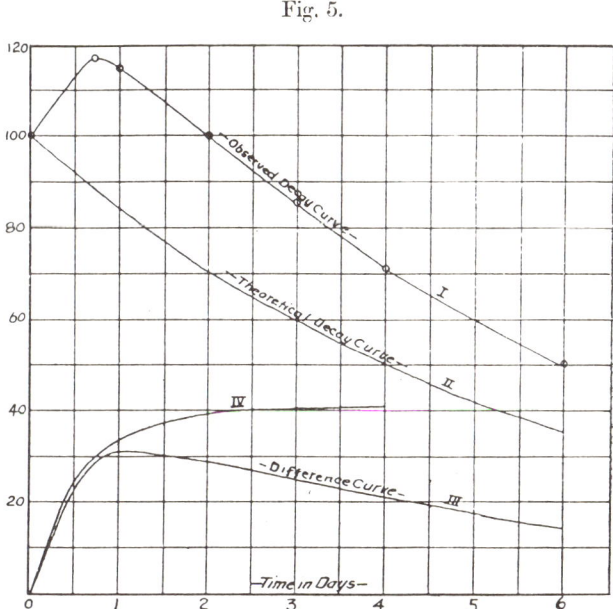

possess a greater proportion of excited radioactivity than those with high emanating power. This conclusion was tested by converting a specimen of thorium carbonate with an emanating power five times that of ordinary thoria, into oxide and de-emanating by intense ignition. The energy that before escaped in the form of emanation is now, all but a few per cent., prevented from escaping. The radioactivity of the oxide so prepared rose in the first three days about thirty per cent. of its original amount, and there thus seem to be grounds for the view that the excited radioactivity will contribute a much greater effect in a non-emanating thorium compound than in one possessing great emanating power.

Additional confirmation of this view is to be found in the nature of the radiations emitted by the two classes of compounds (Section XI.).

X. *The Non-separable Radioactivity of Thorium.*

It has not yet been found possible by any means to free thorium from its residual activity, and the place of this part in the scheme of radioactivity of thorium remains to be considered. Disregarding the view that it is a separate phenomenon, and not connected with the major part of the activity, two hypotheses can be brought forward capable of experimental test, and in accordance with the views advanced on the nature of radioactivity, to account for the existence of this part. First, if there was a second type of excited activity produced by ThX similar to that known, but with a very slow rate of decay, it would account for the existence of the non-separable activity. If this is true it will not be found possible to free thorium from this activity by chemical means, but the continuous removal of ThX over a very long period would, as in the above case, cause its spontaneous decay.

Secondly, if the change which gives rise to ThX produces a second type of matter at the same time, *i. e.* if it is of the type of a decomposition rather than a depolymerization, the second type would also in all probability be radioactive, and would cause the residual activity. On this view the second type of matter should also be amenable to separation by chemical means, although it is certain from the failure of the methods already tried that it resembles thorium much more closely than ThX. But until it is separated from the thorium producing it, its activity will not decay spontaneously. Thus what has already been shown to hold for ThX will be true for the second constituent if methods are found to remove it from the thorium.

It has been shown (Soddy, *loc. cit.*) that uranium also possesses a non-separable radioactivity extremely analogous to that possessed by thorium, and whatever view is taken of the one will in all probability hold also for the other. This consideration makes the second hypothesis, that the residual activity is caused by a second non-thorium type of matter produced in the original change, the more probable of the two.

XI. *The Nature of the Radiations from Thorium and ThX.*

From the view of radioactivity put forward it necessarily follows that the total radioactivity of thorium is altered neither in character nor amount by chemical treatment. With regard to the first, the amount of activity, it has been pointed out that the intensity of radiations *emitted* do not furnish alone a measure of the activity. The absorption in the mass of material must be considered also. The radiations of thorium oxide are derived from a very dense powder; those from ThX, on the other hand, have only to penetrate a very thin film of material. The difficulty can be overcome to some extent by taking for the comparison the radioactivity of a thin film of a soluble thorium salt produced by evaporating

the Cause and Nature of Radioactivity.   393

a solution to dryness over a large metal plate. Compared in this way, the radioactivity of ThX when first separated almost exactly equals the activity of the nitrate from which it is produced, while the hydroxide retains about two-fifths of this amount. The total activity of the products is therefore greater than that of the original salt; but this is to be expected, for it is certain that more absorption takes place in the nitrate than in the products into which it is separated.

Similar difficulties stand in the way of an answer to the second question, whether the nature of the radiations is affected by chemical treatment, for it has been experimentally observed that the penetrating power of these radiations decreases with the thickness of material traversed. The character of the radiations from ThX and thorium have, however, been compared by the method of penetration power. A large number of comparisons justifies the view that the character of thorium radioactivity is unaltered by chemical treatment and the separation of ThX, although the different types are unequally distributed among the separated products.

Determinations of the proportion of rays deviable by the magnetic field in thorium and ThX throws fresh light on the question. The general result is that ThX gives out both deviable and non-deviable rays, and the same applies to the excited activity produced by ThX. But in the experiment in which the excited radiation was allowed to spontaneously decay, by removing ThX as formed, the thorium compound obtained after 23 precipitations was found to be quite free from deviable radiation. This is one of the most striking resemblances between the non-separable radioactivities of uranium and thorium, and warrants the question whether the primary radiation of ThX is not, like that of UrX, composed entirely of cathode-rays. There is, however, no means of deciding this point owing to the excited radiation which always accompanies the primary radiation of ThX, and which itself comprises both types of rays.

Finally, it may be mentioned that the proportion of deviable and non-deviable radiation is different for different compounds of thorium. The nitrate and ignited oxide, compounds which hardly possess any emanating power, have a higher proportion of deviable radiation than compounds with great emanating power. This is indirect evidence of the correctness of the view already put forward (Section IX.), that when the emanation is prevented from escaping it augments the proportion of excited radioactivity of the compound.

Phil. Mag. S. 6. Vol. 4. No. 21. *Sept.* 1902.   2 D

XII. *Summary of Results.*

The foregoing experimental results may be briefly summarized. The major part of the radioactivity of thorium — ordinarily about 54 per cent.—is due to a non-thorium type of matter, ThX, possessing distinct chemical properties, which is temporarily radioactive, its activity falling to half value in about four days. The constant radioactivity of thorium is maintained by the production of this material at a constant rate. Both the rate of production of the new material and the rate of decay of its activity appear to be independent of the physical and chemical condition of the system.

The ThX further possesses the property of exciting radioactivity on surrounding inactive matter, and about 21 per cent. of the total activity under ordinary circumstances is derived from this source. Its rate of decay and other considerations make it appear probable that it is the same as the excited radioactivity produced by the thorium emanation, which is in turn produced by ThX. There is evidence that, if from any cause the emanation is prevented from escaping in the radioactive state, the energy of its radiation goes to augment the proportion of excited radioactivity in the compound.

Thorium can be freed by suitable means from both ThX and the excited radioactivity which the latter produces, and then possesses an activity about 25 per cent. of its original value, below which it has not been reduced. This residual radiation consists entirely of rays non-deviable by the magnetic field, whereas the other two components comprise both deviable and non-deviable radiation. Most probably this residual activity is caused by a second non-thorium type of matter produced in the same change as ThX, and it should therefore prove possible to separate it by chemical methods.

XIII. *General Theoretical Considerations.*

Turning from the experimental results to their theoretical interpretation, it is necessary to first consider the generally accepted view of the nature of radioactivity. It is well established that this property is the function of the atom and not of the molecule. Uranium and thorium, to take the most definite cases, possess the property in whatever molecular condition they occur, and the former also in the elementary state. So far as the radioactivity of different compounds of different density and states of division can be compared together, the intensity of the radiation appears to depend only on the quantity of active element present. It is not at all dependent on the source from which the element is derived, or the process of purification to which it has been subjected, provided sufficient time is allowed for the equilibrium point to be reached. It is not possible to explain the phenomena by the existence of impurities associated with the radioactive elements, even if any advantage could be derived from the assumption. For these impurities must necessarily be present always to the same extent in different specimens derived from the most widely different sources, and, moreover, they must persist *in unaltered amount* after the most refined processes of purification. This is contrary to the accepted meaning of the term impurity.

All the most prominent workers in this subject are agreed in considering radioactivity an atomic phenomenon. M. and Mme. Curie, the pioneers in the chemistry of the subject, have recently put forward their views (*Comptes Rendus*, cxxxiv. 1902, p. 85). They state that this idea underlies their whole work from the beginning and created their methods of research. M. Becquerel, the original discoverer of the property for uranium, in his announcement of the recovery of the activity of the same element after the active constituent had been removed by chemical treatment, points out the significance of the fact that uranium is giving out cathode-rays. These, according to the hypothesis of Sir William Crookes and Prof. J. J. Thomson, are *material* particles of mass one thousandth of the hydrogen atom.

Since, therefore, radioactivity is at once an atomic phenomenon and accompanied by chemical changes in which new types of matter are produced, these changes must be occurring within the atom, and the radioactive elements must be undergoing spontaneous transformation, The results that have so far been obtained, which indicate that the velocity of this reaction is unaffected by the conditions, makes it clear that the changes in question are different in character from any that have been before dealt with in chemistry. It is apparent that we are dealing with phenomena outside the sphere of known atomic forces. Radioactivity may therefore be considered as a manifestation of subatomic chemical change.

The changes brought to knowledge by radioactivity, although undeniably material and chemical in nature, are of a different order of magnitude from any that have before been dealt with in chemistry. The course of the production of new matter which can be recognized by the electrometer, by means of the property of radioactivity, after the lapse of a few hours or even minutes, might conceivably require geological epochs to attain to quantities recognized by the balance. However the well-defined chemical properties of both ThX and UrX are not in accordance with the view that the actual amounts involved are of this extreme order of minuteness. On the other hand, the existence of radioactive elements at all in the earth's crust is an *à priori* argument against the magnitude of the change being anything but small.

Radioactivity as a new property of matter capable of exact quantitative determination thus possesses an interest apart from the peculiar properties and powers which the radiations themselves exhibit. Mme. Curie, who isolated from pitch-blende a new substance, radium, which possessed distinct chemical properties and spectroscopic lines, used the property as a means of chemical analysis. An exact parallel is to be found in Bunsen's discovery and separation of cæsium and rubidium by means of the spectroscope.

The present results show that radioactivity can also be used to follow *chemical changes occurring in matter*. The properties of matter that fulfil the necessary conditions for the study of chemical change without disturbance to the reacting system are few in number. It seems not unreasonable to hope, in the light of the foregoing results, that radioactivity, being such a property, affords the means of obtaining information of the processes occurring within the chemical atom, in the same way as the rotation of the plane of polarization and other physical properties have been used in chemistry for the investigation of the course of molecular change.

Macdonald Physics Building,
Macdonald Chemistry and Mining Building,
McGill University, Montreal.

XLII. *On a Remarkable Case of Uneven Distribution of Light in a Diffraction Grating Spectrum. By* R. W. Wood, *Professor of Experimental Physics, Johns Hopkins University*[*].

I T is a well-known fact that in the spectra formed by a diffraction-grating the light is unevenly distributed, that is the total light in any one spectrum will not recombine to form white light.

I have been examining a most remarkable grating recently ruled on one of the Rowland dividing-engines in which this uneven distribution is carried to a degree almost incomprehensible. If the spectra of an incandescent lamp are viewed directly in the grating without any other optical appliance, at certain angles of incidence perfectly sharp monochromatic

[*] Communicated by the Physical Society: read June 20, 1902.

576   Prof. Rutherford *and* Mr. Soddy

practically all of the emanation comes off very suddenly at a temperature not much more than one degree above that at which only 2 per cent. has volatilized. The general indication of all the experiments, considered together, is to show that the condensed emanation possesses a true vapour-pressure, and that the emanation commences to volatilize slowly two or three degrees below the temperature of rapid volatilization even when the process occurs in a stationary atmosphere. The emanations therefore possess the usual properties possessed by ordinary gaseous matter, in so far as the phenomena of volatilization and condensation are concerned. It was shown in a recent paper that they also possess the property possessed by gases of being occluded by solids under certain conditions. These new properties, taken in conjunction with the earlier discovered diffusion phenomena, characteristic of the radioactive emanations, leave no doubt that the latter must consist of matter in the gaseous state.

McGill University, Montreal,
    March 9, 1903.

LX. *Radioactive Change. By* E. RUTHERFORD, *M.A., D.Sc., Macdonald Professor of Physics, McGill University, and* F. SODDY, *M.A. (Oxon.).*

CONTENTS.

    I. The Products of Radioactive Change, and their Specific Material
        Nature.
   II. The Synchronism between the Change and the Radiation.
  III. The Material Nature of the Radiations.
   IV. The Law of Radioactive Change.
    V. The Conservation of Radioactivity.
   VI. The Relation of Radioactive Change to Chemical Change.
  VII. The Energy of Radioactive Change and the Internal Energy of
        the Chemical Atom.

§ 1. *The Products of Radioactive Change and their
                Specific Material Nature.*

IN previous papers it has been shown that the radioactivity of the elements radium, thorium, and uranium is maintained by the continuous production of new kinds of matter which possess temporary activity. In some cases the new product exhibits well-defined chemical differences from the element producing it, and can be separated by chemical processes. Examples of this are to be found in the removal of thorium X from thorium and uranium X from uranium. In other cases the new products are gaseous in character, and

* Communicated by the Authors.

so separate themselves by the mere process of diffusion, giving rise to the radioactive emanations which are produced by compounds of thorium and radium. These emanations can be condensed by cold and again volatilized; although they do not appear to possess positive chemical affinities, they are frequently occluded by the substances producing them when in the solid state, and are liberated by solution; they diffuse rapidly into the atmosphere and through porous partitions, and in general exhibit the behaviour of inert gases of fairly high molecular weight. In other cases again the new matter is itself non-volatile, but is produced by the further change of the gaseous emanation; so that the latter acts as the intermediary in the process of its separation from the radioactive element. This is the case with the two different kinds of excited activity produced on objects in the neighbourhood of compounds of thorium and radium respectively, which in turn possess well-defined and characteristic material properties. For example, the thorium excited activity is volatilized at a definite high temperature, and redeposited in the neighbourhood, and can be dissolved in some reagents and not in others.

These various new bodies differ from ordinary matter, therefore, only in one point, namely, that their quantity is far below the limit that can be reached by the ordinary methods of chemical and spectroscopic analysis. As an example that this is no argument against their specific material existence, it may be mentioned that the same is true of radium itself as it occurs in nature. No chemical or spectroscopic test is sufficiently delicate to detect radium in pitchblende, and it is not until the quantity present is increased many times by concentration that the characteristic spectrum begins to make its appearance. Mme. Curie and also Giesel have succeeded in obtaining quite considerable quantities of pure radium compounds by working up many tons of pitchblende, and the results go to show that radium is in reality one of the best defined and most characteristic of the chemical elements. So, also, the various new bodies, whose existence has been discovered by the aid of their radioactivity, would no doubt, like radium, be brought within the range of the older methods of investigation if it were possible to increase the quantity of material employed indefinitely.

§ 2. *The Synchronism between the Change and the Radiation.*

In the present paper the nature of the changes in which these new bodies are produced remains to be considered. The experimental evidence that has been accumulated is now

578   Prof. Rutherford *and* Mr. Soddy

sufficiently complete to enable a general theory of the nature
of the process to be established with a considerable degree of
certainty and definiteness. It soon became apparent from
this evidence that a much more intimate connexion exists
between the radioactivity and the changes that maintain it
than is expressed in the idea of the production of active
matter. It will be recalled that all cases of radioactive
change that have been studied can be resolved into the
production by one substance of one other (disregarding
for the present the expelled rays). When several changes
occur together these are not simultaneous but successive.
Thus thorium produces thorium X, the thorium X produces
the thorium emanation, and the latter produces the excited
activity. Now the radioactivity of each of these substances
can be shown to be connected, not with the change in which
it was itself produced, but with the change in which it in turn
produces the next new type. Thus after thorium X has been
separated from the thorium producing it, the radiations of the
thorium X are proportional to the amount of emanation that
it produces, and both the radioactivity and the emanating
power of thorium X decay according to the same law *and at
the same rate*. In the next stage the emanation goes on to
produce the excited activity. The activity of the emanation
falls to half-value in one minute, and the amount of excited
activity produced by it on the negative electrode in an
electric field falls off in like ratio. These results are fully
borne out in the case of radium. The activity of the radium
emanation decays to half-value in four days, and so also does
its power of producing the excited activity.

Hence it is not possible to regard radioactivity as a
consequence of changes that have already taken place. The
rays emitted must be an *accompaniment* of the change of the
radiating system into the one next produced.

Non-separable activity.—This point of view at once accounts
for the existence of a constant radioactivity, non-separable by
chemical processes, in each of the three radio-elements. This
non-separable activity consists of the radiations that accompany
the primary change of the radio-element itself into the first
new product that is produced. Thus in thorium about 25
per cent. of the α radiation accompanies the first change of
the thorium into thorium X. In uranium the whole of the
α radiation is non-separable and accompanies the change of
the uranium into uranium X.

Several important consequences follow from the conclusion
that the radiations accompany the change. A body that is
radioactive must *ipso facto* be changing, and hence it is not

possible that any of the new types of radioactive matter—
e. g., uranium X, thorium X, the two emanations, &c.—can
be identical with any of the known elements. For they
remain in existence only a short time, and the decay of their
radioactivity is the expression of their continuously diminishing quantity. On the other hand, since the ultimate
products of the changes cannot be radioactive, there must
always exist at least one stage in the process beyond the
range of the methods of experiment. For this reason the
ultimate products that result from the changes remain
unknown, the quantities involved being unrecognizable,
except by the methods of radioactivity. In the naturally
occurring minerals containing the radio-elements these
changes must have been proceeding steadily over very long
periods, and, unless they succeed in escaping, the ultimate
products should have accumulated in sufficient quantity to be
detected, and therefore should appear in nature as the invariable companions of the radio-elements. We have already
suggested on these and other grounds that possibly helium
may be such an ultimate product, although, of course, the
suggestion is at present a purely speculative one. But a
closer study of the radioactive minerals would in all
probability afford further evidence on this important question.

§ 3. *The Material Nature of the Radiations.*

The view that the ray or rays from any system are
produced at the moment the system changes has received
strong confirmation by the discovery of the electric and
magnetic deviability of the α ray. The deviation is in the
opposite sense to the β or cathode-ray, and the rays thus
consist of positively charged bodies projected with great
velocity (Rutherford, Phil. Mag., Feb. 1903). The latter
was shown to be of the order of $2 \cdot 5 \ 10^9$ cms. per second. The
value of e/m, the ratio of the charge of the carrier to its mass,
is of the order $6 \ 10^3$. Now the value of e/m for the cathode-
ray is about 10^7. Assuming that the value of the charge is
the same in each case, the apparent mass of the positive
projected particle is over 1000 times as great as for the
cathode-ray. Now $e/m = 10^4$ for the hydrogen atom in the
electrolysis of water. The particle that constitutes the α ray
thus behaves as if its mass were of the same order as that of
the hydrogen atom. The α rays from all the radio-elements,
and from the various radioactive bodies which they produce,
possess analogous properties, and differ only to a slight
extent in penetrating power. There are thus strong reasons

580   Prof. Rutherford *and* Mr. Soddy

for the belief that the α rays generally are projections and that the mass of the particle is of the same order as that of the hydrogen atom, and very large compared with the mass of the projected particle which constitutes the β or easily deviable ray from the same element.

With regard to the part played in radioactivity by the two types of radiation, there can be no doubt that the α rays are by far the more important. In all cases they represent over 99 per cent. of the energy radiated *, and although the β rays on account of their penetrating power and marked photographic action have been more often studied, they are comparatively of much less significance.

It has been shown that the non-separable activity of all three radio-elements, the activity of the two emanations, and the first stage of the excited activity of radium, comprise only α rays. It is not until the processes near completion in so far as their progress can be experimentally traced that the β or cathode-ray makes its appearance †.

In light of this evidence there is every reason to suppose, not merely that the expulsion of a charged particle accompanies the change, but that this expulsion actually *is* the change.

§ 4. *The Law of Radioactive Change.*

The view that the radiation from an active substance accompanies the change gives a very definite physical meaning to the law of decay of radioactivity. In all cases where one of the radioactive products has been separated, and its activity examined independently of the active substance which gives rise to it, or which it in turn produces, it has been found that the activity under all conditions investigated falls off in a geometrical progression with the time. This is expressed by the equation

$$\frac{I_t}{I_o} = \epsilon^{-\lambda t}$$

where I_o is the initial ionization current due to the radiations, I_t that after the time t, and λ is a constant. Each ray or

---

* In the paper in which this is deduced (Phil. Mág. Sept. 1902, p. 329) there is an obvious slip of calculation. The number should be 100 instead of 1000.

† In addition to the α and β rays the radio-elements also give out a third type of radiation which is extremely penetrating. Thorium as well as radium (Rutherford, *Phys. Zeit.* 1902) gives out these penetrating rays, and it has since been found that uranium possesses the same property. These rays have not yet been sufficiently examined to make any discussion possible of the part they play in radioactive processes.

projected particle will in general produce a certain definite number of ions in its path, and the ionization current is therefore proportional to the number of such particles projected per second.   Thus

$$\frac{n_t}{n_o} = \epsilon^{-\lambda t},$$

where n_t is the number projected in unit of time for the time t and n_o the number initially.

If each changing system gives rise to one ray, the number of systems N_t which remain unchanged at the time t is given by

$$N_t = \int_t^\infty n_t \, . \, dt = \frac{n_o}{\lambda} \epsilon^{-\lambda t}.$$

The number N_o initially present is given by putting $t=0$.

$$N_o = \frac{n_o}{\lambda}$$

and

$$\frac{N_t}{N_o} = \epsilon^{-\lambda t}.$$

The same law holds if each changing system produces two or any definite number of rays.

Differentiating

$$\frac{dN}{dt} = -\lambda N_t,$$

or, the rate of change of the system at any time is always proportional to the amount remaining unchanged.

The law of radioactive change may therefore be expressed in the one statement—the proportional amount of radioactive matter that changes in unit time is a constant.   When the total amount does not vary (a condition nearly fulfilled at the equilibrium point where the rate of supply is equal to the rate of change) the proportion of the whole which changes in unit time is represented by the constant λ, which possesses for each type of active matter a fixed and characteristic value. λ may therefore be suitably called the "radioactive constant." The complexity of the phenomena of radioactivity is due to the existence as a general rule of several different types of matter changing at the same time into one another, each type possessing a different radioactive constant.

§ 5. *The Conservation of Radioactivity.*

The law of radioactive change that has been deduced holds for each stage that has been examined, and therefore holds

582   Prof. Rutherford *and* Mr. Soddy

for the phenomenon generally. The radioactive constant λ has been investigated under very widely varied conditions of temperature, and under the influence of the most powerful chemical and physical agencies, and no alteration of its value has been observed. The law forms in fact the mathematical expression of a general principle to which we have been led as the result of our investigations as a whole. Radioactivity, according to present knowledge, must be regarded as the result of a process which lies wholly outside the sphere of known controllable forces, and cannot be created, altered, or destroyed. Like gravitation, it is proportional only to the quantity of matter involved, and in this restricted sense it is therefore true to speak of the principle as the conservation of radioactivity*. Radioactivity differs of course from gravitation in being a special and not necessarily a universal property of matter, which is possessed by different kinds in widely different degree. In the processes of radioactivity these different kinds change into one another and into inactive matter, producing corresponding changes in the radioactivity. Thus the decay of radioactivity is to be ascribed to the disappearance of the active matter, and the recovery of radioactivity to its production. When the two processes balance—a condition very nearly fulfilled in the case of the radio-elements in a closed space—the activity remains constant. But here the apparent constancy is merely the expression of the slow rate of change of the radio-element itself. Over sufficiently long periods its radioactivity must also decay according to the law of radioactive change, for otherwise it would be necessary to look upon radioactive change as involving the creation of matter. In the universe therefore the total radioactivity must, according to our present knowledge, be growing less and tending to disappear.

* Apart from the considerations that follow, this nomenclature is a convenient expression of the observed facts that the total radioactivity (measured by the radiations peculiar to the radio-elements) is for any given mass of radio-element a constant under all conditions investigated. The radioactive equilibrium may be disturbed and the activity distributed among one or more active products capable of separation from the original element. But the sum total throughout these operations is at all times the same.

For practical purposes the expression "conservation," applied to the radioactivity of the three radio-elements, is justified by the extremely minute proportion that can change in any interval over which it is possible to extend actual observations. But *rigidly* the term "conservation" applies only with reference to the radioactivity of any definite quantity of radioactive matter, whereas in nature this quantity must be changing spontaneously and continually growing less. To avoid possible misunderstanding, therefore, it is necessary to use the expression only in this restricted sense.

Hence the energy liberated in radioactive processes does not disobey the law of the conservation of energy.

It is not implied in this view that radioactivity, considered with reference to the quantity of matter involved, is conserved under all conceivable conditions, or that it will not ultimately be found possible to control the processes that give rise to it. The principle enunciated applies of course only to our present state of experimental knowledge, which is satisfactorily interpreted by its aid.

The general evidence on which the principle is based embraces the whole field of radioactivity. The experiments of Becquerel and Curie have shown that the radiations from uranium and radium respectively remain constant over long intervals of time. Mme. Curie put forward the view that radioactivity was a specific property of the element in question, and the successful separation of the element radium from pitchblende was a direct result of this method of regarding the property. The possibility of separating from a radio-element an intensely active constituent, although at first sight contradictory, has afforded under closer examination nothing but confirmation of this view. In all cases only a part of the activity is removed, and this part is recovered spontaneously by the radio-element in the course of time. Mme. Curie's original position, that radioactivity is a specific property of the element, must be considered to be beyond question. Even if it should ultimately be found that uranium and thorium are admixtures of these elements with a small *constant* proportion of new radio-elements with correspondingly intense activity, the general method of regarding the subject is quite unaffected.

In the next place, throughout the course of our investigations we have not observed a single instance in which radioactivity has been created in an element not radioactive, or destroyed or altered in one that is, and there is no case at present on record in which such a creation or destruction can be considered as established. It will be shown later that radioactive change can only be of the nature of an atomic disintegration, and hence this result is to be expected, from the universal experience of chemistry in failing to transform the elements. For the same reason it is not to be expected that the rate of radioactive change would be affected by known physical or chemical influences. Lastly, the principle of the conservation of radioactivity is in agreement with the energy relations of radioactive change. These will be considered more fully in § 7, where it is shown that the energy changes involved are of a much higher order of magnitude than is the case in molecular change.

It is necessary to consider briefly some of the apparent exceptions to this principle of the conservation of radio-activity. In the first place it will be recalled that the emanating power of the various compounds of thorium and radium respectively differ widely among themselves, and are greatly influenced by alterations of physical state. It was recently proved (Phil. Mag. April 1903, p. 453) that these variations are caused by alterations in the rate at which the emanations escape into the surrounding atmosphere. The emanation is produced at the same rate both in de-emanated and in highly emanating thorium and radium compounds, but is in the former stored up or occluded in the compound. By comparing the amount stored up with the amount pro-duced per second by the same compound dissolved, it was found possible to put the matter to a very sharp experimental test which completely established the law of the conservation of radioactivity in these cases. Another exception is the apparent destruction of the thorium excited activity deposited on a platinum wire by ignition to a white heat. This has recently been examined in this laboratory by Miss Gates, and it was found that the excited activity is not destroyed, but is volatilized at a definite temperature and redeposited in unchanged amount on the neighbouring surfaces.

Radioactive " Induction."—Various workers in this subject have explained the results they have obtained on the idea of radioactive "induction," in which a radioactive substance has been attributed the power of inducing activity in bodies mixed with it, or in its neighbourhood, which are not other-wise radioactive. This theory was put forward by Becquerel to explain the fact that certain precipitates (notably barium sulphate) formed in solutions of radioactive salts are them-selves radioactive. The explanation has been of great utility in accounting for the numerous examples of the presence of radioactivity in non-active elements, without the necessity of assuming in each case the existence of a new radio-element therein, but our own results do not allow us to accept it.

In the great majority of instances that have been recorded the results seem to be due simply to the *mixture of active matter with the inactive element.* In some cases the effect is due to the presence of a small quantity of the original radio-element, in which case the "induced" activity is permanent. In other cases, one of the disintegration products, like uranium X or thorium X, has been dragged down by the precipitate, producing temporary, or, as it is sometimes termed, "false" activity. In neither case is the original character of the radiation at all affected. It is probable that a re-examination of some of the effects that have been attributed to radioactive induction would lead to new disintegration products of the known radio-elements being recognized.

Other Results.—A number of cases remain for consideration, where, by working with very large quantities of material, there have been separated from minerals possible new radio-elements, *i. e.* substances possessing apparently permanent radioactivity with chemical properties different from those of the three known radio-elements. In most of these cases, unfortunately, the real criteria that are of value, viz., the nature of the radiations and the presence or absence of distinctive emanations, have not been investigated. The chemical properties are of less service, for even if a new element were present, it is not at all necessary that it should be in sufficient quantity to be detected by chemical or spectroscopic analysis. Thus the radio-lead described by Hoffmann and Strauss and by Giesel cannot be regarded as a new element until it is shown that it has permanent activity of a distinctive character.

In this connexion the question whether polonium (radio-bismuth) is a new element is of great interest. The polonium discovered by Mme. Curie is not a permanent radioactive substance, its activity decaying slowly with the time. On the view put forward in these papers, polonium must be regarded as a disintegration product of one of the radio-elements present in pitchblende. Recently, however, Marckwald (*Ber. der D. Chem. Gesel.* 1902, pp. 2285 & 4239), by the electrolysis of pitchblende solutions, has obtained an intensely radioactive substance very analogous to the polonium of Curie. But he states that the activity of his preparation does not decay with time, and this, if confirmed, is sufficient to warrant the conclusion that he is not dealing with the same substance as Mme. Curie. On the other hand, both preparations give only α rays, and in this they are quite distinct from the other radio-elements. Marckwald has succeeded in separating his substance from bismuth, thus showing it to possess different chemical properties, and in his latest paper states that the bismuth-free product is indistinguishable chemically from tellurium. If the permanence of the radioactivity is established, the existence of a new radio-element must be inferred.

If elements heavier than uranium exist it is probable that they will be radioactive. The extreme delicacy of radio-activity as a means of chemical analysis would enable such elements to be recognized even if present in infinitesimal quantity. It is therefore to be expected that the number of

586   Prof. Rutherford and Mr. Soddy

radio-elements will be augmented in the future, and that considerably more than the three at present recognized exist in minute quantity. In the first stage of the search for such elements a purely chemical examination is of little service. The main criteria are the permanence of the radiations, their distinctive character, and the existence or absence of distinctive emanations or other disintegration products.

§ 6. *The Relation of Radioactive Change to Chemical Change.*

The law of radioactive change, that the rate of change is proportional to the quantity of changing substance, is also the law of monomolecular chemical reaction. Radioactive change, therefore, must be of such a kind as to involve one system only, for if it were anything of the nature of a combination, where the mutual action of two systems was involved, the rate of change would be dependent on the concentration, and the law would involve a volume-factor. This is not the case. Since radioactivity is a specific property of the element, the changing system must be the chemical atom, and since only one system is involved in the production of a new system and, in addition, heavy charged particles, in radioactive change the chemical atom must suffer disintegration.

The radio-elements possess of all elements the heaviest atomic weight. This is indeed their sole common chemical characteristic. The disintegration of the atom and the expulsion of heavy charged particles of the same order of mass as the hydrogen atom leaves behind a new system lighter than before, and possessing chemical and physical properties quite different from those of the original element. The disintegration process, once started, proceeds from stage to stage with definite measurable velocities in each case. At each stage one or more α " rays " are projected, until the last stages are reached, when the β " ray " or electron is expelled. It seems advisable to possess a special name for these now numerous atom-fragments, or new atoms, which result from the original atom after the ray has been expelled, and which remain in existence only a limited time, continually undergoing further change. Their instability is their chief characteristic. On the one hand, it prevents the quantity accumulating, and in consequence it is hardly likely that they can ever be investigated by the ordinary methods. On the other, the instability and consequent ray-expulsion furnishes the means whereby they can be investigated. We would therefore suggest the term *metabolon* for this purpose.

Thus in the following table the metabolons at present known to result from the disintegration of the three radio-elements have been arranged in order.

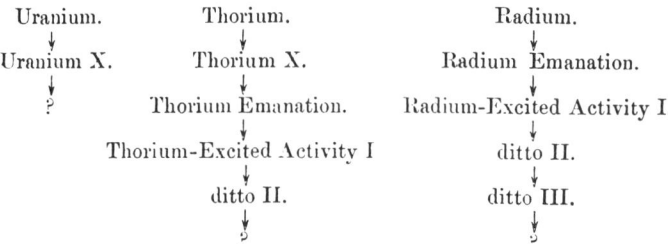

The three queries represent the three unknown ultimate products. The atoms of the radio-elements themselves form, so to speak, the common ground between metabolons and atoms, possessing the properties of both. Thus, although they are disintegrating, the rate is so slow that sufficient quantity can be accumulated to be investigated chemically. Since the rate of disintegration is probably a million times faster for radium than it is for thorium or uranium, we have an explanation of the excessively minute proportion of radium in the natural minerals. Indeed, every consideration points to the conclusion that the radium atom is also a metabolon in the full sense of having been formed by disintegration of one of the other elements present in the mineral. For example, an estimation of its "life," goes to show that the latter can hardly be more than a few thousand years (see § 7). The point is under experimental investigation by one of us, and a fuller discussion is reserved until later.

There is at present no evidence that a single atom or metabolon ever produces more than one new kind of metabolon at each change, and there are no means at present of finding, for example, either how many metabolons of thorium X, or how many projected particles, or "rays," are produced from each atom of thorium. The simplest plan therefore, since it involves no possibility of serious error if the nature of the convention is understood, is to assume that each atom or metabolon produces one new metabolon or atom and one "ray."

§ 7. *The Energy of Radioactive Change, and the Internal Energy of the Chemical Atom.*

The position of the chemical atom as a very definite stage in the complexity of matter, although not the lowest of which it is now possible to obtain experimental knowledge,

588   Prof. Rutherford *and* Mr. Soddy

is brought out most clearly by a comparison of the respective energy relations of radioactive and chemical change. It is possible to calculate the order of the quantity of energy radiated from a given quantity of radio-element during its complete change, by several independent methods, the conclusions of which agree very well among themselves. The most direct way is from the energy of the particle projected, and the total number of atoms. For each atom cannot produce less than one " ray " for each change it undergoes, and we therefore arrive in this manner at a minimum estimate of the total energy radiated. On the other hand, one atom of a radio-element, if completely resolved into projected particles, could not produce more than about 200 such particles at most, assuming that the mass of the products is equal to the mass of the atom. This consideration enables us to set a maximum limit to the estimate. The α rays represent so large a proportion of the total energy of radiation that they alone need be considered.

Let m = mass of the projected particle,
v = the velocity,
e = charge.

Now for the α ray of radium

$$v = 2 \cdot 5 \ 10^9,$$

$$\frac{e}{m} = 6 \ 10^3.$$

The kinetic energy of each particle

$$\tfrac{1}{2}mv^2 = \frac{1}{2}\frac{m}{e}v^2 e = 5 \ 10^{14} e.$$

J. J. Thomson has shown that

$e = 6 \ 10^{-10}$ E.S. Units $= 2 \ 10^{-20}$ Electromagnetic Units.

Therefore the kinetic energy of each projected particle $= 10^{-5}$ erg. Taking 10^{20} as the probable number of atoms in one gram of radium, the total energy of the rays from the latter $= 10^{15}$ ergs $= 2 \cdot 4 \ 10^7$ gram-calories, on the assumption that each atom projects one ray. Five successive stages in the disintegration are known, and each stage corresponds to the projection of at least one ray. It may therefore be stated that the total energy of radiation during the disintegration of one gram of radium cannot be less than 10^8 gram-calories, and may be between 10^9 and 10^{10} gram-calories. The energy radiated does not necessarily involve the whole of the energy of disintegration and may be only a small part of it. 10^8 gram-calories per gram may therefore be safely accepted as the least possible estimate of the energy of radioactive change in radium. The union of hydrogen and oxygen liberates approximately $4 \ 10^3$ gram-calories per gram of water produced, and this reaction sets free more energy for a given weight than any other chemical change known. The energy of radioactive change must therefore be at least twenty-thousand times, and may be a million times, as great as the energy of any molecular change.

The rate at which this store of energy is radiated, and in consequence the life of a radio-element, can now be considered. The order of the total quantity of energy liberated per second in the form of rays from 1 gram of radium may be calculated from the total number of ions produced and the energy required to produce an ion. In the solid salt a great proportion of the radiation is absorbed in the material, but the difficulty may be to a large extent avoided by determining the number of ions produced by the radiation of the emanation, and the proportionate amount of the total radiation of radium due to the emanation. In this case most of the rays are absorbed in producing ions from the air. It was experimentally found that the maximum current due to the emanation from 1 gram of radium, of activity 1000 compared with uranium, in a large cylinder filled with air, was $1\cdot65 \ 10^{-8}$ electromagnetic units. Taking $e = 2 \ 10^{-20}$, the number of ions produced per second $= 8\cdot2 \ 10^{11}$. These ions result from the collision of the projected particles with the gas in their path. Townsend (Phil. Mag. 1901, vol. i.), from experiments on the production of ions by collision, has found that the minimum energy required to produce an ion is 10^{-11} ergs. Taking the activity of pure radium as a million times that of uranium, the total energy radiated per second by the emanation from 1 gram of pure radium $= 8200$ ergs. In radium compounds in the solid state, this amount is about $\cdot4$ of the total energy of radiation, which therefore is about

$$2 \ 10^4 \text{ ergs per second,}$$
$$6\cdot3 \ 10^{11} \text{ ergs per year,}$$
$$15{,}000 \text{ gram-calories per year.}$$

This again is an under-estimate, for only the energy employed in producing ions has been considered, and this may be only a small fraction of the total energy of the rays.

Since the α radiation of all the radio-elements is extremely similar in character, it appears reasonable to assume that the feebler radiations of thorium and uranium are due to these elements disintegrating less rapidly than radium. The energy radiated in these cases is about 10^{-6} that from radium, and is therefore about ·015 gram-calorie per year. Dividing this quantity by the total energy of radiation, $2·4 \ 10^7$ gram-calories, we obtain the number $6 \ 10^{-10}$ as a maximum estimate for the proportionate amount of uranium or thorium undergoing change per year. Hence in one gram of these elements less than a milligram would change in a million years. In the case of radium, however, the same amount must be changing per gram *per year*. The " life " of the radium cannot be in consequence more than a few thousand years on this minimum estimate, based on the assumption that each particle produces one ray at each change. If more are produced the life becomes correspondingly longer, but as a maximum the estimate can hardly be increased more than 50 times. So that it appears certain that the radium present in a mineral has not been in existence as long as the mineral itself, but is being continually produced by radioactive change.

Lastly, the number of " rays " produced per second from 1 gram of a radio-element may be estimated. Since the energy of each " ray " $= 10^{-5}$ ergs $= 2·4 \ 10^{-13}$ gram-calories, $6 \ 10^{10}$ rays are projected every year from 1 gram of uranium. This is approximately 2000 per second. The α radiation of 1 milligram of uranium in one second is probably within the range of detection by the electrical method. The methods of experiment are therefore almost equal to the investigation of a single atom disintegrating, whereas not less than 10^4 atoms of uranium could be detected by the balance.

It has been pointed out that these estimates are concerned with the energy of radiation, and not with the total energy of radioactive change. The latter, in turn, can only be a portion of the internal energy of the atom, for the internal energy of the resulting products remains unknown. All these considerations point to the conclusion that the energy latent in the atom must be enormous compared with that rendered free in ordinary chemical change. Now the radio-elements differ in no way from the other elements in their chemical and physical behaviour. On the one hand they resemble chemically their inactive prototypes in the periodic system very closely, and on the other they possess no common chemical characteristic which could be associated with their radioactivity. Hence there is no reason to assume that this enormous store of energy is possessed by the radio-elements alone. It seems probable that atomic energy in general is of a similar, high order of magnitude, although the absence of change prevents its existence being manifested. The existence of this energy accounts for the stability of the chemical elements as well as for the con-

591

servation of radioactivity under the influence of the most varied conditions. It must be taken into account in cosmical physics. The maintenance of solar energy, for example, no longer presents any fundamental difficulty if the internal energy of the component elements is considered to be available, *i. e.* if processes of sub-atomic change are going on. It is interesting to note that Sir Norman Lockyer has interpreted the results of his spectroscopic researches on the latter view (Inorganic Evolution, 1900) although he regards the temperature as the cause rather than the effect of the process.

McGill University, Montreal.

THE

LONDON, EDINBURGH, and DUBLIN

PHILOSOPHICAL MAGAZINE

AND

JOURNAL OF SCIENCE.

———◆———

[SIXTH SERIES.]

FEBRUARY 1903.

XV. *The Magnetic and Electric Deviation of the easily absorbed Rays from Radium.* By E. RUTHERFORD, *M.A., D.Sc., Macdonald Professor of Physics, McGill University, Montreal*[*].

RADIUM gives out three distinct types of radiation:—
(1) The α rays, which are very easily absorbed by thin layers of matter, and which give rise to the greater portion of the ionization of the gas observed under the usual experimental conditions.

(2) The β rays, which consist of negatively charged particles projected with high velocity, and which are similar in all respects to cathode rays produced in a vacuum-tube.

(3) The γ rays, which are non-deviable by a magnetic field, and which are of a very penetrating character.

These rays differ very widely in their power of penetrating matter. The following approximate numbers, which show the thickness of aluminium traversed before the intensity is reduced to one-half, illustrate this difference.

Radiation.	Thickness of Aluminium.
α rays	·0005 cm.
β rays	·05 cm.
γ rays	8 cms.

In this paper an account will be given of some experiments which show that the α rays are deviable by a strong magnetic and electric field. The deviation is in the opposite sense to

* Communicated by the Author.

Phil. Mag. S. 6. Vol. 5. No. 26. *Feb.* 1903.   N

that of the cathode rays, so that the radiations must consist of positively charged bodies projected with great velocity. In a previous paper * I have given an account of the indirect experimental evidence in support of the view that the α rays consist of projected charged particles. Preliminary experiments undertaken to settle this question during the past two years gave negative results. The magnetic deviation, even in a strong magnetic field, is so small that very special methods are necessary to detect and measure it. The smallness of the magnetic deviation of the α rays, compared with that of the cathode rays in a vacuum-tube, may be judged from the fact that the α rays, projected at right angles to a magnetic field of strength 10,000 c.g.s. units, describe the arc of a circle of radius about 39 cms., while under the same conditions the cathode rays would describe a circle of radius about ·01 cm.

In the early experiments radium of activity 1000 was used, but this did not give out strong enough rays to push the experiment to the necessary limit. The general method employed was to pass the rays through narrow slits and to observe whether the rate of discharge, due to the issuing rays, was altered by the application of a magnetic field. When, however, the rays were sent through sufficiently narrow slits to detect a small deviation of the rays, the rate of discharge of the issuing rays became too small to measure, even with a sensitive electrometer.

I have recently obtained a sample of radium † of activity 19,000, and using an electroscope instead of an electrometer, I have been able to extend the experiments, and to show that the α rays are all deviated by a strong magnetic field.

Magnetic Deviation of the Rays.

Fig. 1 A shows the general arrangement of the experiment. The rays from a thin layer of radium passed upwards through a number of narrow slits, G, in parallel, and then through a thin layer of aluminium foil ·00034 cm. thick into the testing vessel V. The ionization produced by the rays in the testing vessel was measured by the rate of movement of the leaves of a gold-leaf electroscope B. This was arranged after the manner of C. T. R. Wilson in his experiments on

---

* Phil. Mag. Jan. 1903, p. 113. It was long ago suggested by Strutt (Phil. Trans. Roy. Soc. 1900) that the α rays consist of positively charged particles projected from the active substance. The same idea has lately been advanced by Sir Wm. Crookes (Proc. Roy. Soc. 1900).

† The sample of radium of greater activity than that usually sold was obtained from the Société Centrale de Produits Chimiques, through the kindness of M. P. Curie.

Electric Deviation of Rays from Radium.   179

the spontaneous ionization of air. The gold-leaf system was insulated inside the vessel by a sulphur bead C, and could be

Fig. 1 A.

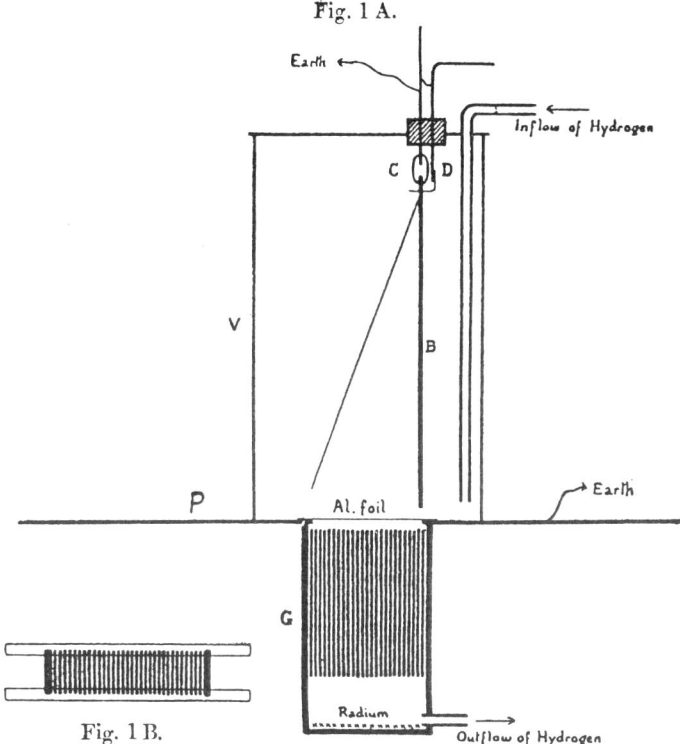

Fig. 1 B.

charged by means of a movable wire D, which was afterwards earthed. The rate of movement of the gold-leaf was observed by means of a microscope through small mica windows in the testing vessel.

In order to increase the ionization in the testing vessel, the rays passed through 20 to 25 slits of equal width, placed side by side. This was arranged by cutting grooves at regular intervals in side-plates into which brass plates were slipped. A cross section of the system of metal plates and air-spaces is shown in fig. 1 B.

The width of the slit varied in different experiments between ·042 and ·1 cm.

The magnetic field was applied perpendicular to the plane of the paper and parallel to the plane of the slits.

The testing vessel and system of plates were waxed to a lead

N 2

180   Prof. E. Rutherford *on the Magnetic and*

plate P so that the rays entered the vessel V only through the aluminium foil.

It is necessary in these experiments to have a steady stream of gas passing downwards between the plates in order to prevent the diffusion of the emanation from the radium upwards into the testing vessel. The presence in the testing vessel of a small amount of this emanation, which is always given out by radium, would produce large ionization effects and completely mask the effect to be observed.

For this purpose a steady current of dry electrolytic hydrogen of 2 c.c. per second was passed into the testing vessel, streamed through the porous aluminium foil, and passed between the plates, carrying with it the emanation from the apparatus.

The use of a stream of hydrogen instead of air greatly simplifies the experiment, for it *increases* at once the ionization current due to the α rays in the testing vessel, and (at the same time) greatly *diminishes* that due to the β and γ rays.

This follows at once from the fact that the α rays are much more readily absorbed in air than in hydrogen, while the rate of production of ions due to the β and γ rays is much less in hydrogen than in air. The intensity of the α rays after passing between the plates is consequently greater when hydrogen is used ; and since the rays pass through a sufficient distance of hydrogen in the testing vessel to be largely absorbed, the total amount of ionization produced by them in hydrogen is greater than in air.

With the largest electromagnet in the laboratory I was only able to deviate about 30 per cent. of the α rays. Through the kindness of Professor Owens, of the Electrical Engineering Department, I was, however, enabled to make use of the upper part of the field-magnet of a 30 kilowatt E lison dynamo. Suitable pole-pieces are at present being made for the purpose of obtaining a strong magnetic field over a considerable area; but with rough pole-pieces I have been enabled to obtain a sufficiently strong field to completely deviate the α rays.

The following is an example of an observation on the magnetic deviation : —

Pole-pieces 1·90 × 2·50 cms.

Strength of field between pole-pieces 8370 units.

Apparatus of 25 parallel plates of length 3·70 cms., width ·70 cm., with an average air-space between plates of ·042 cm.

Distance of radium below plates 1·4 cm.

Electric Deviation of Rays from Radium.   181

	Rate of Discharge of Electroscope in volts per minute.
(1) Without magnetic field	8·33
(2) With magnetic field	1·72
(3) Radium covered with thin layer of mica to absorb all α rays ...	0·93
(4) Radium covered with mica and magnetic field applied	0·92

The mica plate, ·01 cm. thick, was of sufficient thickness to completely absorb all the β rays, but allowed the β and γ rays to pass through without appreciable absorption. The difference between (1) and (3), 7·40 volts per minute, gives the rate of discharge due to the α rays alone; the difference between (2) and (3), 0·79 volt per minute, that due to the α rays not deviated by the magnetic field employed.

The amount of α rays not deviated by the field is thus about 11 per cent. of the total. The small difference between (2) and (4) includes the small ionization due to the β rays, for they would have been completely deviated by the magnetic field. It is probable that the ionization due to the β rays without a magnetic field was actually stronger than this; but the residual magnetic field, when the current was broken, was large enough to deviate them completely before reaching the testing vessel. (4) comprises the effect of the γ rays together with the natural leak of the electroscope in hydrogen.

In this experiment there was a good deal of stray magnetic field acting on the rays before reaching the pole-pieces. The distribution of this field at different portions of the apparatus is shown graphically in fig. 2.

Fig. 2.

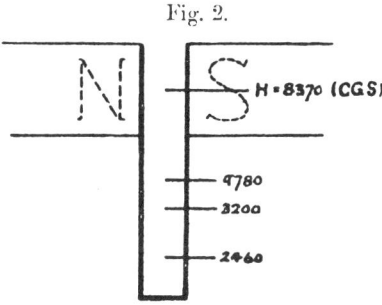

The following table shows the rate of discharge due to the α rays for different strengths of the magnetic field. The

182   Prof. E. Rutherford *on the Magnetic and*

maximum value with no magnetic field is taken as 100. These results are shown graphically in fig. 3.

Magnetic field between pole-pieces.	Rate of discharge due to *a* rays.
0	100
3720 C.G.S. units	66
4840 ,, ,,	50
6500 ,, ,,	33
7360 ,, ,,	23
8370 ,, ,,	11

The curve (fig. 3) shows that the amount deviated is approximately proportional to the magnetic field.

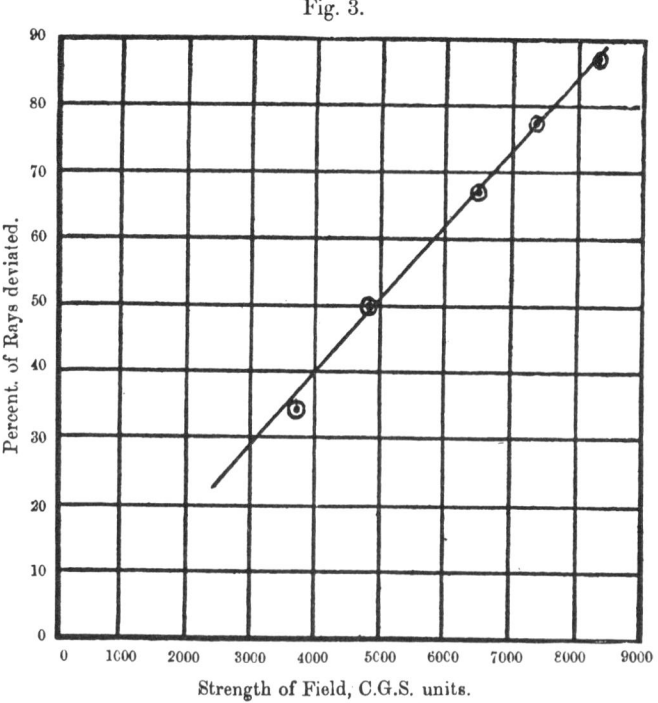

Fig. 3.

Percent. of Rays deviated.

Strength of Field, C.G.S. units.

With another apparatus, with a mean air space of ·055 cm., the rays were *completely* deviated by a uniform magnetic field of strength 8400 units extending over the length of the plates, a distance of 4·5 cms.

Electric Deviation of Rays from Radium.   1·83

Direction of the Deviation of the Rays.

In order to determine the direction of the deviation, the rays were passed through slits of 1 mm. width. Each slit was about half covered by a brass plate in which air-spaces were cut to correspond accurately with the system of parallel plates. Fig. 4 represents an enlarged section of three of the

Fig. 4.

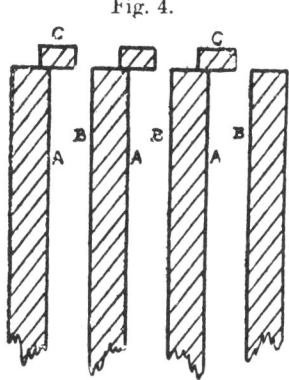

plates, with the metal plate C half covering the slit AB. If a magnetic field is applied, not sufficiently great to deviate all the rays, the rate of discharge in the testing vessel when the rays are deviated in the direction from A to B should be much greater than when the magnetic field is reversed, *i. e.* when the rays are deviated from B to A. This was found to be the case, for while the rate of discharge was not much diminished by the application of the field in one direction, it was reduced to about one quarter of its value by reversal of the field.

In this way it was found that the direction of deviation in a magnetic field was *opposite* in sense to the cathode rays, *i. e.*, the rays consisted of positively charged particles.

Electrostatic Deviation of the Rays.

The apparatus was similar to that employed for the magnetic deviation of the rays with the exception that the brass sides, which held the plates in position, were replaced by ebonite.

Twenty-five plates were used of length 4·50 cms., width 1·5 cm., and average air-space of ·055 cm. The radium was ·85 cm. below the plates. Alternate plates were connected together and charged by means of a battery of small accumulators to a potential-difference of 600 volts. A current of hydrogen was used as in the case of the magnetic experiment.

184   Prof. E. Rutherford *on the Magnetic and*

With a P.D. of 600 volts, a consistent difference * of 7 per
cent. was observed in the rate of discharge due to the α rays
with the electric field off and on.   A larger potential-
difference could not be used as a spark passed between the
plates in the presence of radium.

The amount of deviation in this experiment was too small
to determine the direction of deviation by the electric field.

Determination of the Velocity of the Rays.

It is difficult to determine with certainty the value of the
curvature of the path of the rays in a given magnetic field
from the percentage amount of rays deviated, on account of
the fact that some of the rays which strike the sides of the
parallel plates are deviated so as to pass into the testing
vessel.

From data obtained, however, by observing the value of
the magnetic field for *complete deviation* of the rays, it was
deduced that

$$H\rho = 390,000,$$

where H = value of magnetic field,
          ρ = radius of curvature of path of the rays.
This gives the higher limit of the value Hρ.

By using the usual equations of the deviation of a moving
charged body it was deduced that the velocity V of the rays
was given by

$$V = 2 \cdot 5 \times 10^9 \text{ cms. per sec.,}$$

and that the value $\dfrac{e}{m}$, the ratio of the charge of the carrier

to its mass, was given by

$$\frac{e}{m} = 6 \times 10^3.$$

These results are only rough approximations and merely
indicate the order of the values of these quantities, as the
electric deviations observed were too small for accurate
observations.   The experiments are being continued with
special apparatus, and it is hoped that much larger electrostatic deviations will be obtained, and in consequence a more
accurate determination of the constants † of the rays.

* In later experiments, which are not yet completed, I have been
able to deviate about 45 per cent. of the α rays in a strong electric field.
          † The α rays are complex, and probably consist of particles projected
with velocities lying between certain limits; for the radiations include
the α radiations from the emanation and excited activity which are distributed throughout the radium compound.

Electric Deviation of Rays from Radium.   185

The α rays from radium are thus very similar to the *Canal Strahlen* observed by Goldstein, which have been shown by Wien to be positively charged bodies moving with a high velocity. The velocity of the α rays is, however, considerably greater than that observed for the *Canal Strahlen*.

General Considerations.

The radiations from uranium, thorium, and radium, and also the radiations from the emanations and excited bodies, all include a large proportion of α rays. These rays do not differ much in penetrating power, and it is probable that *in all cases* the α radiations from them are charged particles projected with great velocities.

In a previous paper* it has been shown that the total energy radiated in the form of α rays by the permanent radioactive bodies is about 1000 times greater than the energy radiated in the form of β rays. This result was obtained on the assumption that the total number of ions produced by the two types of rays when completely absorbed in air, is a measure of the energy radiated. The α rays are thus the most important factor in the radiation of energy from active bodies, and, in consequence, any estimate of the energy radiated based on the β rays alone leads to much too small a value.

Experiments are in progress to determine the charge carried by the α rays, and from these it is hoped to deduce the rate of emission of energy in the form of α rays from the active substances.

The projection character of the α rays very readily explains some of their characteristic properties. On this view the ionization of the gas by the α rays is due to collisions of the projected masses with the gas molecules. The variation of the rate of production of the ions with the pressure of the gas and the variation of absorption of the rays in solids and gases with the density at once follows. It also offers a simple explanation of the remarkable fact that the absorption of the α rays in a given thickness of matter, when determined by the electrical method, *increases* with the thickness of matter previously traversed. It is only necessary to suppose that as the velocity of the projected particles decreases in consequence of collision with the molecules of the absorbing medium, the ionizing power of the rays decreases rapidly. This is most probably the case, for there seems to be no doubt that the positive carrier cannot ionize

* Rutherford and Grier, Phil. Mag. Sept. 1902.

the gas below a certain velocity, which is very great compared with the velocity of translation of the molecules.

It is of interest to consider the probable part that the α rays play in the radioactive bodies on the general view of radioactivity that has been put forward by Mr. Soddy and myself in the Phil. Mag. Sept. and Nov. 1902. It is there shown that radioactivity is due to a succession of chemical changes in which new types of radioactive matter are being continuously formed, and that the constant radioactivity of the well known active bodies is an equilibrium process, where the rate of production of fresh active matter is balanced by the decay of activity of that already produced. Some very interesting points arose in the course of these investigations. It was found that the residual activity of uranium and thorium when freed from UrX and ThX by chemical processes consisted entirely of α rays. On the other hand, the radiation of UrX * consisted almost entirely of β rays, while that of ThX † consisted of both α and β rays. Similar results probably hold also for radium, for the Curies have shown that radium dissolved in water and then evaporated to dryness temporarily loses to a large extent its power of emitting β rays.

It thus appears probable that the emission of α rays goes on quite independently of the emission of β rays. There seems to be no doubt that the emission of β rays by active substances is a secondary phenomenon, and that the α rays play the most prominent part in the changes occurring in radioactive matter. The results obtained so far point to the conclusion that the beginning of the succession of chemical changes taking place in radioactive bodies is due to the emission of the α rays, *i.e.* the projection of a heavy charged mass from the atom. The portion left behind is unstable, undergoing further chemical changes which are again accompanied by the emission of α rays, and in some cases also of β rays.

The power possessed by the radioactive bodies of apparently spontaneously projecting large masses with enormous velocities supports the view that the atoms of these substances are made up, in part at least, of rapidly rotating or oscillating systems of heavy charged bodies large compared with the electron. The sudden escape of these masses from their orbit may be due either to the action of internal forces or external forces of which we have at present no knowledge.

It also follows from the projection nature of the α rays that the radioactive bodies, when inclosed in sealed vessels

* Soddy, Proc. Chem. Soc. 1902.
† Rutherford and Grier, Phil. Mag. Sept. 1902.

187

sufficiently thin to allow the α rays to escape, must *decrease in weight*. Such a decrease has been recently observed by Heydweiler* for radium, but apparently under such conditions that the α rays would be largely absorbed in the glass tube containing the active matter.

In this connexion it is very important to decide whether the loss of weight observed by Heydweiler is due to a decrease of weight of the radium itself or to a decrease of weight of the glass envelope; for it is well known that radium rays produce rapid colourations throughout a glass tube, and it is possible that there may be a chemical change reaching to the surface of the glass which may account for the effects observed.

McGill University,
Montreal, Nov. 10, 1902.

[281]

XXI. *The Nature of the α Particle from Radioactive Substances. By* Professor E. RUTHERFORD, *F.R.S., and* T. ROYDS, *M.Sc.,* 1851 *Exhibition Science Scholar* *.

THE experimental evidence collected during the last few years has strongly supported the view that the α particle is a charged helium atom, but it has been found exceedingly difficult to give a decisive proof of the relation. In recent papers, Rutherford and Geiger † have supplied still further evidence of the correctness of this point of view, The number of α particles from one gram of radium have been counted, and the charge carried by each determined. The values of several radioactive quantities, calculated on the assumption that the α particle is a helium atom carrying two unit charges, have been shown to be in good agreement with the experimental numbers. In particular, the good agreement between the calculated rate of production of helium by radium and the rate experimentally determined by Sir James Dewar ‡, is strong evidence in favour of the identity of the α particle with the helium atom.

The methods of attack on this problem have been largely indirect, involving considerations of the charge carried by the helium atom and the value of e/m of the α particle. The proof of the identity of the α particle with the helium atom is incomplete until it can be shown that the α particles, accumulated quite independently of the matter from which they are expelled, consist of helium. For example, it might be argued that the appearance of helium in the radium emanation was a result of the expulsion of the α particle, in the same way that the appearance of radium A is a consequence of the expulsion of an α particle from the emanation. If one atom of helium appeared for each α particle expelled, calculation and experiment might still agree, and yet the α particle itself might be an atom of hydrogen or of some other substance.

We have recently made experiments to test whether helium appears in a vessel into which the α particles have been fired, the active matter itself being enclosed in a vessel sufficiently thin to allow the α particles to escape, but impervious to the passage of helium or other radioactive products.

* Communicated by the Authors.
† Proc. Roy. Soc. A. lxxxi. pp. 141–173 (1908).
‡ Proc. Roy. Soc. A. lxxxi. p. 280 (1908).

282   Prof. Rutherford *and* Mr. Royds *on the Nature*

The experimental arrangement is clearly seen in the figure
The equilibrium quantity of emanation from about 140 milligrams of radium was purified and compressed by means of a

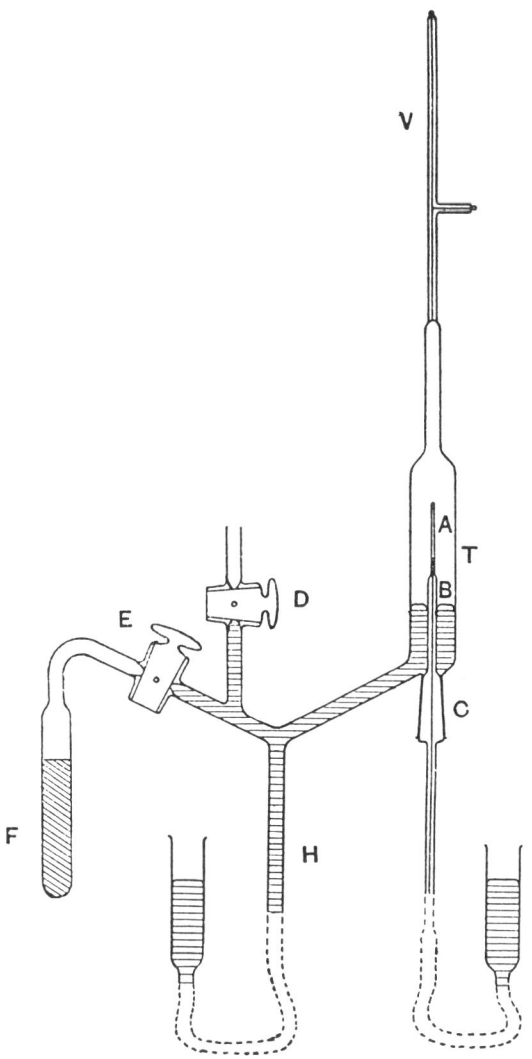

mercury-column into a fine glass tube A about 1·5 cms. long.
This fine tube, which was sealed on a larger capillary tube B,
was sufficiently thin to allow the α particles from the emanation and its products to escape, but sufficiently strong to

withstand atmospheric pressure.   After some trials, Mr.
Baumbach succeeded in blowing such fine tubes very uniform
in thickness.   The thickness of the wall of the tube employed
in most of the experiments was less than $\frac{1}{100}$ mm., and was
equivalent in stopping power of the α particle to about
2 cms. of air.   Since the ranges of the α particles from the
emanation and its products radium A and radium C are 4·3,
4·8, and 7 cms. respectively, it is seen that the great
majority* of the α particles expelled by the active matter
escape through the walls of the tube.   The ranges of the
α particles after passing through the glass were determined
with the aid of a zinc-sulphide screen.   Immediately after
the introduction of the emanation the phosphorescence showed
brilliantly when the screen was close to the tube, but practically disappeared at a distance of 3 cms.   After an hour,
bright phosphorescence was observable at a distance of
5 cms.   Such a result is to be expected.   The phosphorescence
initially observed was due mainly to the α particles of the
emanation and its product radium A (period 3 mins.).   In
the course of time the amount of radium C, initially zero,
gradually increased, and the α radiations from it of range
7 cms. were able to cause phosphorescence at a greater
distance.

The glass tube A was surrounded by a cylindrical glass
tube T, 7·5 cms. long and 1·5 cms. diameter, by means of a
ground-glass joint C.   A small vacuum-tube V was attached
to the upper end of T.   The outer glass tube T was exhausted
by a pump through the stopcock D, and the exhaustion
completed with the aid of the charcoal tube F cooled by
liquid air.   By means of a mercury column H attached to a
reservoir, mercury was forced into the tube T until it reached
the bottom of the tube A.

Part of the α particles which escaped through the walls of
the fine tube were stopped by the outer glass tube and part
by the mercury surface.   If the α particle is a helium atom,
helium should gradually diffuse from the glass and mercury
into the exhausted space, and its presence could then be
detected spectroscopically by raising the mercury and compressing the gases into the vacuum-tube.

In order to avoid any possible contamination of the
apparatus with helium, freshly distilled mercury and entirely
new glass apparatus were used.   Before introducing the
emanation into A, the absence of helium was confirmed

---

* The α particles fired at a very oblique angle to the tube would be
stopped in the glass.   The fraction stopped in this way would be small
under the experimental conditions.

284   Prof. Rutherford *and* Mr. Royds *on the Nature*

experimentally.   At intervals after the introduction of the emanation the mercury was raised, and the gases in the outer tube spectroscopically examined.   After 24 hours no trace of the helium yellow line was seen ; after 2 days the helium yellow was faintly visible; after 4 days the helium yellow and green lines were bright ; and after 6 days all the stronger lines of the helium spectrum were observed.   The absence of the neon spectrum shows that the helium present was not due to a leakage of air into the apparatus.

There is, however, one possible source of error in this experiment.   The helium may not be due to the α particles themselves, but may have *diffused* from the emanation through the thin walls of the glass tube.   In order to test this point the emanation was completely pumped out of A, and after some hours a quantity of helium, about 10 times the previous volume of the emanation, was compressed into the same tube A.

The outer tube T and the vacuum-tube were removed and a fresh apparatus substituted.   Observations to detect helium in the tube T were made at intervals, in the same way as before, but no trace of the helium spectrum was observed over a period of eight days.

The helium in the tube A was then pumped out and a fresh supply of emanation substituted.   Results similar to the first experiment were observed.   The helium yellow and green lines showed brightly after four days.

These experiments thus show conclusively that the helium could not have diffused through the glass walls, but must have been derived from the α particles which were fired through them.   In other words, the experiments give a decisive proof that the α particle after losing its charge is an atom of helium.

Other Experiments.

We have seen that in the experiments above described helium was not observed in the outer tube in sufficient quantity to show the characteristic yellow line until two days had elapsed.   Now the equilibrium amount of emanation from 100 milligrams of radium should produce helium at the rate of about ·03 c.mm. per day.   The amount produced in one day, if present in the outer tube, should produce a bright spectrum of helium under the experimental conditions.   It thus appeared probable that the helium fired into the glass must escape very slowly into the exhausted space, for if the helium escaped at once, the presence of helium should have been detected a few hours after the introduction of the emanation.

In order to examine this point more closely the experiments were repeated, with the addition that a cylinder of thin sheet lead of sufficient thickness to stop the α particles was placed over the fine emanation tube. Preliminary experiments, in the manner described later, showed that the lead-foil did not initially contain a detectable amount of helium. Twenty-four hours after the introduction into the tube A of about the same amount of emanation as before, the yellow and green lines of helium showed brightly in the vacuum-tube, and after two days the whole helium spectrum was observed. The spectrum of helium in this case after one day was of about the same intensity as that after the fourth day in the experiments without the lead screen. It was thus clear that the lead-foil gave up the helium fired into it far more readily than the glass.

In order to form an idea of the rapidity of escape of the helium from the lead some further experiments were made. The outer cylinder T was removed and a small cylinder of lead-foil placed round the thin emanation-tube surrounded the air at atmospheric pressure. After exposure for a definite time to the emanation, the lead screen was removed and gested for helium as follows. The lead-foil was placed in a glass tube between two stopcocks. In order to avoid a possible release of the helium present in the lead by pumping out the air, the air was displaced by a current of pure electrolytic oxygen*. The stopcocks were closed and the tube attached to a subsidiary apparatus similar to that employed for testing for the presence of neon and helium in the gases produced by the action of the radium emanation on water (Phil. Mag. Nov. 1908). The oxygen was absorbed by charcoal and the tube then heated beyond the melting-point of lead to allow the helium to escape. The presence of helium was then spectroscopically looked for in the usual way. Using this method, it was found possible to detect the presence of helium in the lead which had been exposed for only four hours to the α rays from the emanation. After an exposure of 24 hours the helium yellow and green lines came out brightly. These experiments were repeated several times with similar results.

A number of blank experiments were made, using samples of the lead-foil which had not been exposed to the α rays, but in no case was any helium detected. In a similar way,

* That the air was completely displaced was shown by the absence of neon in the final spectrum.

286

the presence of helium was detected in a cylinder of tinfoil exposed for a few hours over the emanation-tube.

These experiments show that the helium does not escape at once from the lead, but there is on the average a period of retardation of several hours and possibly longer.

The detection of helium in the lead and tin foil, as well as in the glass, removes a possible objection that the helium might have been in some way present in the glass initially, and was liberated as a consequence of its bombardment by the α particles.

The use of such thin glass tubes containing emanation affords a simple and convenient method of examining the effect on substances of an intense α radiation quite independently of the radioactive material contained in the tube.

We can conclude with certainty from these experiments that the α particle after losing its charge is a helium atom. Other evidence indicates that the charge is twice the unit charge carried by the hydrogen atom set free in the electrolysis of water.

University of Manchester,
    Nov. 13, 1908.

Ranges of α particles from Radioactive Substances.   613

types of rays. On the basis of this assumption the whole number of ions produced by a β particle of high speed per cm. of its path at atmospheric pressure is 67.

Our thanks are due to Professor Rutherford for his help and interest in these experiments.

Physical Laboratories.
The Victoria University, Manchester.

LVII. *The Ranges of the α particles from Various Radioactive Substances and a Relation between Range and Period of Transformation.   By* H. GEIGER, *Ph.D., and* J. M. NUTTALL, *B Sc., University of Manchester* *.

IT is well known that the α particles from different radioactive substances are characterized by their ranges, *i. e.* by the distance through which they can travel in air at atmospheric pressure. This was first pointed out by Bragg, and the ranges of a number of products have been determined by him and his co-workers. The method applied by Bragg to determine the ranges is well known. By means of a set of parallel tubes placed directly above the active plate α rays with practically parallel paths were obtained. The ionization produced by these rays was measured at different distances in a shallow ionization vessel, and the distance in air at which the ionization just disappeared was taken as the range of the α particles.

It was shown by Rutherford that at the same distance from the source at which the α particles fail to produce ionization they also lose their power of producing scintillations. The observation of the scintillations at different distances from the source therefore presents another way of determining the ranges of the α particles, and this method has frequently been made use of by Hahn and other observers. It appears, however, that the scintillations method gives somewhat smaller values for the ranges than the ionization method.

Great difficulty has been experienced in the determination of the ranges of the very inactive substances uranium and thorium.   In these cases, the methods mentioned above are not applicable. Estimates of the ranges of these products were, however, made by Bragg†, but more accurate values

* Communicated by Prof. E. Rutherford, F.R.S.
† W. H. Bragg, Phil. Mag. xi. p. 754 (1906).

614   Dr. H. Geiger *and* Mr. J. M. Nuttall *on the Ranges*

—in the case of uranium—have recently been given by Geiger and Rutherford*, and by Foch†.

It would be of great importance if a method could be devised which would be equally suitable for the accurate determination of the ranges of all the known α ray products, but at present no such method has been found. The main difficulties result from the fact that some of the products are gases, whilst some are very feebly active or only available in the presence of other α ray products.

In the present investigation we have employed a method which appears specially suitable for the determination of the ranges of the substances whose activities are small. The arrangement is indicated in fig. 1. The inside of a large glass bulb is silvered and connected to a battery of about 700 volts. The active film is placed in the centre of the bulb on a small metal disk B which is connected to the electrometer by means of the wire H. The brass tube KK surrounding the wire, and insulated from it by ebonite plugs, serves as electrostatic protection, and prevents any electrical leak from the glass bulb to the wire. The tube passes air-tight through the ground-glass joint E, which fits into the corresponding part F, sealed to the glass bulb. The length of the wire H is carefully adjusted so that the plate B is exactly in the centre of the bulb. If the pressure in the bulb is reduced the ionization produced by the α particles will remain practically constant so long as the range of the α particles at the particular pressure does not exceed the radius of the bulb, viz. 7·95 cm. But as soon as the pressure is decreased below that value the ionization current will also decrease. From this critical pressure and the radius of the bulb the range of the α particles can easily be deduced. It adds greatly to the accuracy of the determination of the critical pressure if the active layer is very thin, and if the area over which it is spread is small. Experiments with polonium showed that plates with diameters up to 2·6 cm. could be used without introducing an appreciable error.

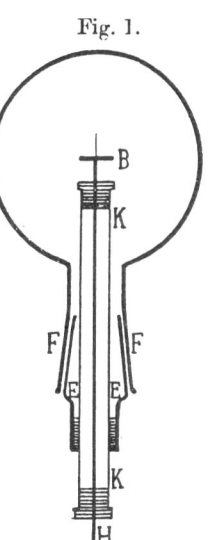

Fig. 1.

   * H. Geiger and E. Rutherford, *Le Radium*, vii. p. 225 (1910), and Phil. Mag. **xx**. p. 691 (1910).
   † A. Foch, *Le Radium*, viii. p. 101 (1911).

of α particles from various Radioactive Substances.   615

In fig. 2 the curves for air obtained with different substances are given. The abscissæ give the pressure in

Fig. 2.

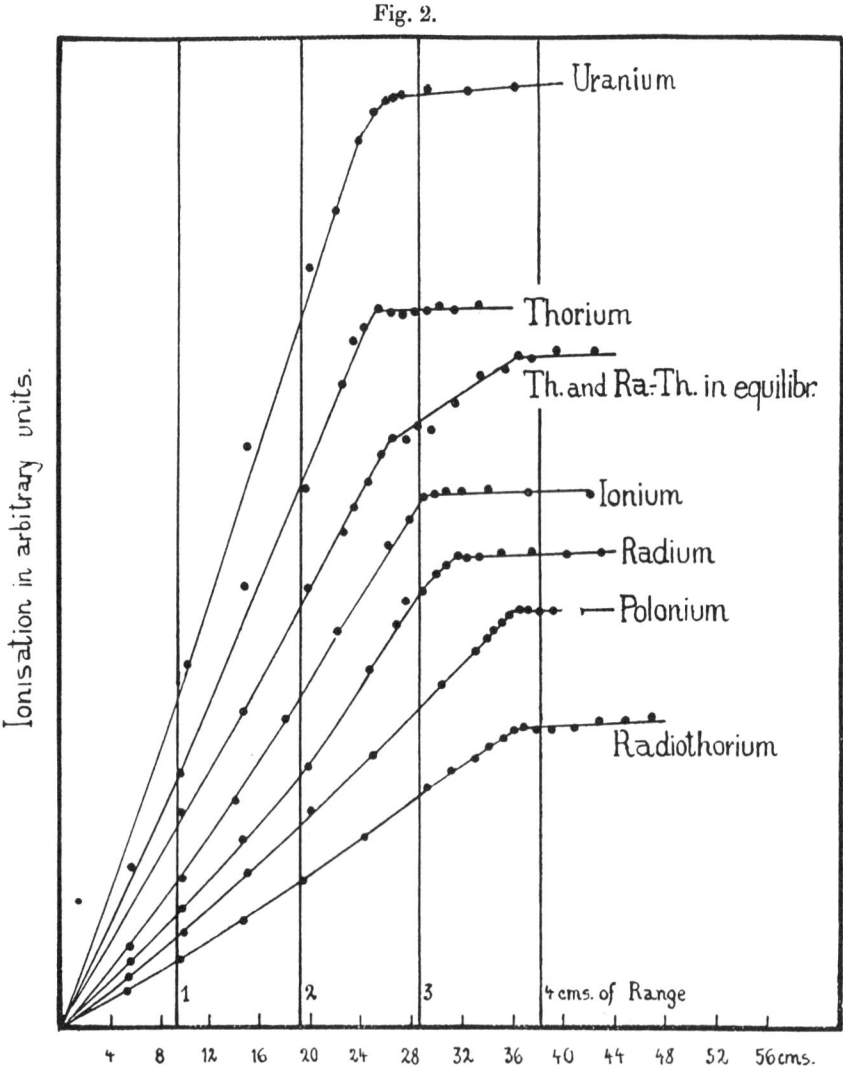

centimetres of mercury, and for convenience also the corresponding ranges reduced to 76 cm. pressure and 15° centigrade are marked. The ordinates give the ionization current measured by the electrometer for each particular

2 S 2

616   Dr. H. Geiger *and* Mr. J. M. Nuttall *on the Ranges*

pressure.   It will be seen that the critical pressure is sharply defined in each case.

A few details regarding the radioactive substances used in these experiments may be given.

1. *Uranium.*—3 mgr. of uranium oxide were spread uniformly over an area of 2·5 cm. diameter.   Experiments made with thicker films gave practically the same result.

2. *Thorium.*—It is well known that thorium cannot be separated from radiothorium.   Through the kindness of Prof. Boltwood we have, however, obtained some thorium which, on account of frequent precipitations over a period of many years, was repeatedly freed from mesothorium, and thus the radiothorium had practically died away.   Immediately before the measurements were taken the thorium was precipitated several times in order to get rid of the thorium X and the subsequent products which might have been present.

3. *Thorium and Radiothorium in equilibrium.*— Some thorium with the radiothorium in equilibrium was separated from thorite mineral, and by several precipitations freed from thorium X and the following products.   The curve shows two distinct breaks corresponding to the α rays from thorium and radiothorium.

4. *Ionium.*—The source consisted of a small and hardly visible spot of ionium mixed with thorium on a thin aluminium-foil.   The activity due to the thorium was too small to be detected.

5. *Radium.*—A radium solution was first freed from polonium, emanation, and active deposit.   A drop of the solution was evaporated on a platinum plate and the measurements taken immediately.

6. *Polonium.*— The polonium films were prepared by bringing the polonium solution in contact with a carefully cleaned copper-foil.   By electrolytic action the polonium is then deposited on the copper-foil.

7. *Radiothorium.*—Radiothorium was separated from a mesothorium preparation and freed from thorium X and the following products.   A thin film was prepared and measurements taken at once.

The results for the different products employed in these measurements are collected in the following table.   In each case the average value of all the measurements is given and reduced to a pressure of 76 cm. and a temperature of 15° and 0° centigrade respectively.   But it must be remembered that recent investigations have shown that actually only a small fraction of the α particles traverses the whole range,

of α particles from various Radioactive Substances.   617

and that on account of scattering and possibly other causes many are stopped at a somewhat shorter distance. One of us [*] has shown that the velocity of expulsion of the α particles is proportional to the cube root of its maximum range. Thus, knowing the velocity of one product the velocities of the others can be deduced. Comparison is most suitably made with Ra C, for which the initial velocity has been measured by Rutherford[†] and was found to be $2·06 \times 10^9$ cm./sec. The initial velocities of expulsion calculated in this way are added in the table.

Substance.	Range 15° C.	Range 0° C.	Initial Velocity of Expulsion.
Uranium	2·72 cm.	2·58 cm.	$1·51 \times 10^9$ cm./sec.
Ionium	3·00 ,,	2·84 ,,	$1·56 \times 10^9$,,
Radium	3·30 ,,	3·13 ,,	$1·61 \times 10^9$,,
Polonium	3·77 ,,	3·58 ,,	$1·68 \times 10^9$,,
Thorium	2·72 ,,	2·58 ,,	$1·51 \times 10^9$,,
Radiothorium ...	3·87 ,,	3·67 ,,	$1·70 \times 10^9$,,

We have already mentioned the earlier estimates of the ranges of uranium and thorium. The range of ionium has been measured by Boltwood[‡], that of radium by Bragg and Kleeman[§], that of polonium by Kucera and Masek[||], and by Levin[¶], and that of radiothorium by Hahn[**]. Nearly all the figures given in the table above are in fair agreement with those obtained by previous observers. In some cases, however, pressure and temperature of the air are not mentioned, so that it is difficult to make a comparison of value between the present and older determinations.

In connexion with the redetermination of the ranges of the α particles from a number of products, given above and in a previous paper by one of us[††], we have reconsidered the possibility of a relation between the period of the active substances and the ranges of the α particles emitted by them.

[*] H. Geiger, Proc. Roy. Soc. A. lxxxiii. p. 505 (1910).
[†] E. Rutherford, Phil. Mag. xii. p. 348 (1906).
[‡] B. B. Boltwood, Amer. Journ. Sc. xxv. p. 365 (1908).
[§] W. H. Bragg and R. D. Kleeman, Phil. Mag. x. p. 318 (1905).
[||] Kucera and Masek, *Phys. Zeitschr.* vii. p. 337 (1906).
[¶] M. Levin, *Phys. Zeitschr.* vii. p. 519 (1906).
[**] O. Hahn, *Phys. Zeitschr.* vii. p. 456 (1906).
[††] H. Geiger, Phil. Mag. xxii. p. 201 (1911).

618   Dr. H. Geiger *and* Mr. J. M. Nuttall *on the Ranges*

It has been already pointed out by Rutherford * in 1907 that possibly a relation exists between these quantities. It appeared that the range of the α particles was greater the smaller the period of transformation of the substance. There were, however, several products for which this relation did not seem to hold.

In order to find any possible relation between the range and the period we have plotted in fig. 3 the logarithms of

Fig. 3.

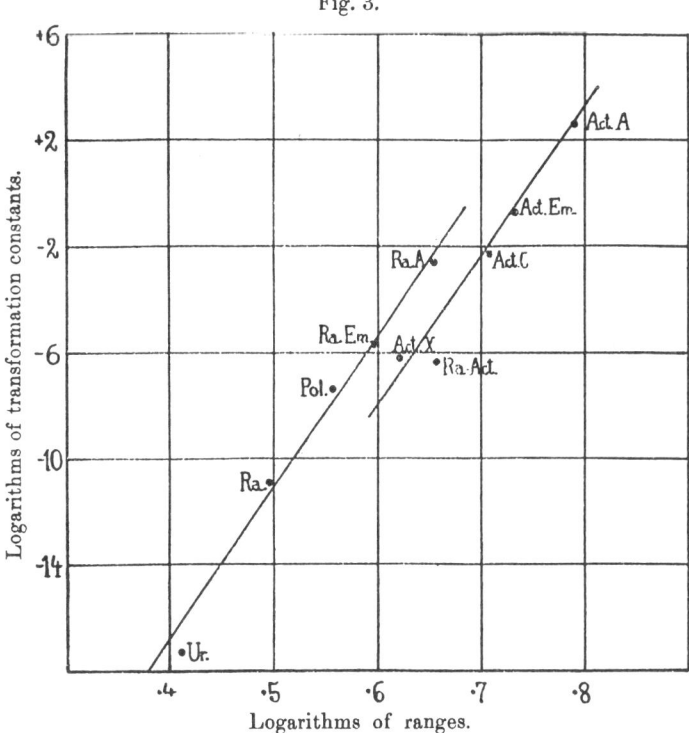

the transformation constants of the different products against the logarithms of the corresponding ranges for the products in the uranium-radium series and in the actinium series. The data from which these curves are plotted are given in the following table. For convenience the initial velocity of expulsion is added as well as the half-value period and the transformation constant. The ranges are reduced to 0° centigrade.

* E. Rutherford, Phil. Mag. xiii. p. 110 (1907).

of α particles from various Radioactive Substances. 619

Substance.	Range at 0°.	Initial Velocity.	Transformation Constant.	Half-value Period.
Uranium	2·58 cms.	$1·51 \times 10^9 \frac{cm.}{sec.}$	$4·6 \times 10^{-18}$	5×10^9 years
Ionium	2·84 ,,	1·56 ,, ,,	————	
Radium	3·13 ,,	1·61 ,, ,,	$1·1 \times 10^{-11}$	2000 years
Ra Emanation ...	3·94 ,,	1·74 ,, ,,	$2·085 \times 10^{-6}$	3·85 days
Radium A	4·50 ,,	1·81 ,, ,,	$3·85 \times 10^{3}$	3·0 minutes
Radium C	6·57 ,,	2·06 ,, ,,	————	
Polonium	3·58 ,,	1·68 ,, ,,	$5·60 \times 10^{-8}$	143 days
Radioactinium ...	4·55 ,,	1·82 ,, ,,	$4·1 \times 10^{-7}$	19·5 days
Actinium X	4·17 ,,	1·77 ,, ,,	$7·6 \times 10^{-7}$	10·5 days
Act Emanation ...	5·40 ,,	1·93 ,, ,,	$1·8 \times 10^{-1}$	3·9 seconds
Actinium A	6·16 ,,	2·02 ,, ,,	350	$\frac{1}{500}$ second
Actinium C	5·12 ,,	1·90 ,, ,,	$5·4 \times 10^{-3}$	2·15 minutes

All the products in the uranium-radium series are marked on the curve except ionium and radium C. In the case of ionium the period has not yet been determined, but according to Soddy[*] it is between 5×10^4 and 10^6 years. It will be seen that the numbers of the uranium-radium series lie very closely on a straight line. Assuming that ionium is no exception to the rule, and taking the range of its α particles to be 2·84 cm., an inspection of the curve shows that its half-value period should be nearly as high as one million years.

The discussion of radium C presents some difficulties, for from its range the half-value period of the product should be exceedingly short, about 10^{-6} second. The recent work of Hahn and Meitner[†] and of Fajans[‡] has shown that the substance ordinarily called radium C is undoubtedly complex, and that the changes occurring in these substances are perhaps irregular. There is certainly no definite evidence yet available which would contradict the possibility that the period of transformation of the product which gives rise to the long-range α particles is very short.

It should be pointed out that a certain difficulty exists with regard to uranium. Boltwood[§] has shown that the change in uranium gives rise to two α particles for one in each of the successive products, and this has been confirmed

[*] F. Soddy, *Le Radium*, vii. p. 295 (1910).
[†] O. Hahn and L. Meitner, *Phys. Zeitschr.* x. p. 697 (1909).
[‡] K. Fajans, *Phys. Zeitschr.* xii. p. 369 (1911).
[§] B. B. Boltwood, Amer. Journ. of Sci. xxv. p. 269 (1908).

620   *Ranges of α particles from Radioactive Substances.*

by the scintillation method by Geiger and Rutherford*. This may be ascribed to the existence of two successive α-ray products, or to the emission of two α particles in the disintegration of each atom.   This latter hypothesis is excluded by the experiments of Marsden and Barratt†, who found no evidence that two α particles were emitted simultaneously. In regard to the first supposition Marsden and Barratt's work also indicates that the period of the second product is more than a few seconds.   It seems not impossible that there may exist two successive products, each of long period, which cannot be separated by ordinary chemical methods.   For example, if the periods of these two substances were of the same order of magnitude the ranges of the α particles would differ very little and therefore be difficult to distinguish.

In regard to actinium, it will be observed that the previous notation has been changed in consequence of the discovery of a new α-ray product following the emanation.   The new product is called actinium A, the second product actinium B, &c.   The reason for the change of nomenclature is given in the following paper by Rutherford and Geiger.   It will be seen from the figure that the relation between range and period is again represented by a straight line falling below the corresponding line of the uranium-radium series.   Radioactinium does not lie exactly on the line ; this may be due to a slight error in the range, which is in this case difficult to determine.

The nomenclature for the active deposit of the thorium emanation has also been changed in a similar manner to that of actinium in consequence of the discovery of another quickly decaying α-ray product, but the ranges of this product and of the emanation have not yet been determined with accuracy. Some preliminary experiments, however, made by one of us indicated that the numbers referring to the products of the thorium series also lie on a straight line when plotted in the same way as has been done for the uranium-radium and actinium series.   The details will be discussed later when some experiments now in progress have been completed.

The connexion indicated above between the period and range is at present only empirical, but it may depend on some simple relation which may ultimately be brought to light.   Similar straight lines to those above would be obtained by plotting period against initial velocity of expulsion, since the range is proportional to the third power of the velocity.

It is of interest that the relation discussed above offers a

* H. Geiger and E. Rutherford, Phil. Mag. xx. p. 691 (1910).
† E. Marsden and J. Barratt, Proc. Phys. Soc. xxiii. p. 367 (1911).

621

possible explanation of the reason why no substance has been found emitting α rays of range shorter than 2·58 cm. For example, the life of a substance which emitted α rays of range 1 cm. would be so long, and consequently its transformation so slow, that its activity would be beyond the limits of detection by present methods.

Experiments are at present in progress with the view of determining with accuracy the ranges of the products which are yet uncertain. The result of such a complete investigation may be expected to show definitely whether the relation given holds generally for all the substances emittting α rays.

We are indebted to Prof. Rutherford for his help and his interest in these experiments.

The Victoria University, Manchester,
Physical Laboratories.

Part Three

Structure of the Atom

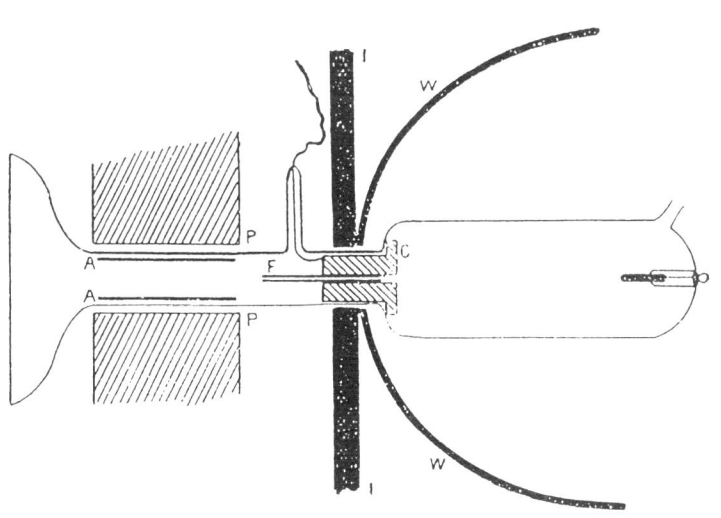

Part Three
Structure of the Atom

The papers in this part describe the birth of atomic physics and with it the dawn of a new epoch in science. They describe the unveiling of the innermost secrets of the atom via discoveries which had an influence on subsequent developments in physics and an importance for mankind generally which have scarcely been equalled in any sphere of human endeavour.

By the early 1900s, sufficient time had elapsed and enough evidence accumulated since J. J. Thomson's discovery of the electron in 1897 to convince all but the most sceptical that this minute negatively charged particle was a basic constituent of all matter. Electrons had been found to be present not only in gaseous discharges, but to be emitted from metals when heated or irradiated with ultraviolet light. They had been identified with the β rays of radioactivity, and their motion within atoms had been associated with the oscillators required for the absorption and emission of light. As carriers of electricity, the degree of freedom of electrons to move in solids served to distinguish good from bad conductors, i.e. metals from insulators.

But did matter consist of anything else but electrons? For a while J. J. Thomson, the acknowledged leader in the field, thought not. In a sense he revived Benjamin Franklin's one-fluid theory of electricity; positive charge was associated with nothing material — it signified simply the absence of electrons. In a letter to Oliver Lodge in 1904 he wrote 'I have always had hopes (not yet realized) of being able to do without positive electrification as a separate entity and to replace it by some properties of the corpuscles [electrons]'. In an early model for the atom, Thomson had the electrons moving in circular planetary orbits within a massless sphere of uniform positive electricity. In order to account for the mass of atoms, his model required thousands of electrons — even in the lightest element, hydrogen. The problem of the stability of such a system of charges, particularly that associated with the loss of energy of the electrons by radiation (a difficulty to be encountered later in Rutherford's nuclear model of the atom), was ingeniously tackled by Thomson who postulated that electromagnetic waves radiated by one electron were neutralized by interference with those from another in a diametrically opposite orbit. Any breakdown in this balance could, he argued, account for radioactive disintegration.

Eventually, however, Thomson was forced to abandon his many-electron atom, principally on the basis of results obtained by one of his own students, Barkla, who deduced from X-ray scattering experiments that the number of electrons in a given atom was comparable to half its atomic weight, except in the case of hydrogen where it was one. Since the mass of the electron was about 1/2000 that of the hydrogen atom, Thomson conceded, somewhat reluctantly, that atoms must contain something other than electrons to account for their mass.

Thomson's experiments on 'positive rays', one account of which appears in the first paper of this Part, were driven by his attempts to find the positive analogue of negative electricity. His first results appeared fruitful and exciting, in that whatever gas was used for the electric discharge, positively charged particles with a mass (actually an e/m ratio) comparable to that of the hydrogen atom were identified. In addition, he found heavier particles which he identified with Rutherford's α particles. Thomson felt sure he had identified *two* natural units of positive electricity. His conclusions, however, were premature; the signals he observed were later found to be a result of residual hydrogen arising from the dissociation of water vapour trapped in the glass walls of the discharge tubes. The most important consequence of Thomson's work on positive rays turned out to be the development of his apparatus, by Francis Aston, into a mass spectrograph and the use of this in identifying isotopes — an advance described in the last article of this Part.

Rutherford's studies of radioactivity and the nature of the α particle have been covered in Part Two. In the present Part, one of his papers for which he is perhaps even better known is reproduced. It describes what must be considered the greatest of several outstanding achievements made by Rutherford during his remarkable career. The seeds for his nuclear model of the atom — in which Thomson's sphere of positive charge was replaced by a small but massive (relative to the electron) nucleus — were sown back in 1905 when Rutherford himself drew attention to the scattering of α particles as they passed through air or thin sheets of mica. He commented then that 'It can easily be calculated that the change of direction of $2°$ ... in passing through 0.03 cm of mica would require ... an average transverse field of 100 million V/cm. Such a result brings out clearly the fact that the atoms of materials must be the seat of very intense electrical forces.'

After his move from McGill to Manchester in 1907, Rutherford's pioneering work with Geiger on methods of detecting and counting single α particles provided the techniques for an experiment which he proposed for a research student. The suggestion he made to Marsden (later Sir Ernest) was to study the scattering of α particles as they passed through thin metallic foils, with a view to seeing if any were deflected through large angles. Whether Rutherford expected a positive result is not clear but within days Geiger reported excitedly that he and Marsden had found that about 1 in 20 000 of α particles incident on a gold foil only 0.00004 cm thick were completely reversed in their direction, producing scintillations on the same side of the foil as the source. Such events, although rare, could not be explained by a succession of multiple deflections each of small angle, as Rutherford showed in his famous paper of 1911, published two years after the discovery. Instead they required a *single* encounter at close range with a source of intense electric field. Rutherford's paper contains a 'textbook' theoretical treatment of such an encounter. Assuming a Coulomb inverse square law of repulsion between the α particle and a much more massive, and hence stationary, point positive charge, he shows that the path follows a hyperbola. An attractive force, such as would occur if the 'nucleus' (a term not used by Rutherford in this paper) were negatively charged, leads to the same trajectory — in this case exactly analogous to that followed by a comet approaching and receding from the sun. It was some time later before Rutherford reached the conclusion that the nucleus was positive and that circulating electrons provided the compensating charge. However the triumph of the nuclear model of the atom was already his.

Detailed experiments by Geiger and Marsden confirmed all the essential theoretical predictions of Rutherford's model, as can be read in the facsimile of their paper (p 224) which they published two years later. By then C. T. R. Wilson had obtained beautiful cloud-chamber pictures of tracks of α particles, most of which displayed abrupt bends at their ends when the particles' velocity fell and they suffered single large deflections. These photographs (see Plate 5) provided direct verification of Rutherford's model.

Of those near to Rutherford just after his formulation of the nuclear model of the atom, Niels Bohr seems to have been the one who was the most impressed. Whilst others pondered over its significance, Bohr eagerly accepted the new ideas and within two years was to develop a theoretical treatment which incorporated proposals so radical that even Rutherford had difficulty in accepting them.

Niels Bohr (1855–1962) was born and raised in Copenhagen by intellectual parents. His doctoral dissertation on the electron theory of metals was submitted to the University of Copenhagen in 1911 — the year in which Rutherford's classic paper appeared. After a short time in Cambridge, where he failed to interest Thomson in his work, he joined Rutherford in Manchester for 4 months in 1912. During this short stay, Bohr — a shy and unassuming young man — struck up what was to be a lasting friendship with the jovial and outgoing Rutherford, each recognizing the other's strengths and abilities. It was also then that Bohr had his brilliant insights which were to result in a famous trilogy of papers which he wrote after his return to Denmark and published in the *Philosophical Magazine*. The first of these is reproduced in this Part.

The problem of the electromagnetic radiation emitted by circulating electrons, mentioned earlier in connection with Thomson's model of the atom, was just as severe in Rutherford's model. Any loss of energy by the electrons ensured they would eventually but inevitably spiral into the nucleus, just as a satellite encountering air resistance returns to Earth. Bohr's answer to this seemingly insurmountable difficulty was to assert that, on the very small scale of the atom, classical electro-dynamics broke down and that an electron revolving in a well defined planetary orbit simply did not radiate energy. Furthermore he proposed that, unlike orbits in the corresponding gravitational analogy, there were only a limited number of allowed orbits for the electrons. Then in a strange and puzzling mixture of classical and quantum theories (a true understanding of which did not occur until much later), Bohr postulated that an electron could jump from one orbit to another, losing or gaining energy in units of $h\nu$ where h is Planck's constant and ν is the frequency of electromagnetic radiation emitted or absorbed. This condition, as he shows in the paper, is equivalent to the electron in each orbit having an angular momentum quantized in units of $h/2$. An immediate result from this analysis was a formula for the frequencies of the spectral lines of hydrogen in excellent quantitative agreement with experimental data.

The initial reaction to Bohr's theory varied from scepticism to out-right hostility! There was admiration for the manner in which it accounted for spectral series but bewilderment as to the underlying meaning of it all. In an account of Bohr's work at a meeting of the British Association at Birmingham in September 1913, James Jeans hailed it 'most ingenious and suggestive' but also pointed out that the only justification of Bohr's postulates 'is the very weighty one of success'. Rutherford was nervous about so much being deduced on the basis of his model but nevertheless gave Bohr every encouragement. J. J. Thomson was anything but supportive and, in the same volume of the *Philosophical Magazine* as Bohr's paper appeared, presented a new model of his own which had little merit and was not heard of again.

Bohr was keenly aware of the 'delicate question of the simultaneous use of the old mechanics and of the new assumptions introduced by Planck's theory of radiation' and did not flinch from attempting a dee-per understanding of the meaning of his own postulates. He returned to Manchester for two years as a lecturer and would have joined Rutherford on his move to Cambridge in 1919, had not the Danish authorities offered him a professorship and a new laboratory — the Institute for Theoretical Physics in Copenhagen, where he held the directorship for the rest of his life. There followed a period of intense theoretical activity, not only in Bohr's Institute but at rival schools led by Sommerfeld in Munich and Max Born in Göttingen. At first effort was directed towards extending the Bohr model to multi-electron atoms for which it was found necessary to introduce orbits much more complicated than the circular ones which had sufficed for hydrogen and ionized helium. Sommerfeld, in particular, proposed elliptical orbits inclined at various angles to each other (space quantization). Eventually the orbits were replaced by 'shells' of electrons, each specified by four quantum numbers.

The theoretical developments were, in the meantime, being complemented by experimental work of considerable importance, not least that undertaken by Moseley, two of whose classic papers on X-ray spectroscopy published in 1913 and 1914 are reproduced next. Henry Moseley (1887–1915) had a privileged education at Summer Fields preparatory school, at Eton and at Trinity College, Oxford, before he joined Rutherford as a demonstrator in 1910. After an 'apprenticeship' of two years, during which he mastered the techniques used in the Manchester laboratory at that time, he became intrigued by the recently reported results of W. H. Bragg and his son W. L. Bragg on the reflection of X-rays by the atomic planes of crystals (see Part Six for a paper by W. L. Bragg). Moseley's interest in X-rays was not shared by Rutherford at the time and he was forced to visit W. L. Bragg in Leeds who instructed him in the necessary techniques.

Work by Barkla, extended by Whiddington in 1911, had shown that targets bombarded by cathode rays emitted X-rays which could be separated in terms of their penetrating powers into two groups, known as K and L. Heavier elements produced more penetrating rays than lighter ones, and it was deduced that their frequencies were roughly proportional to the atomic weight of the material from which the target was made. Moseley went further, arguing that the rays might be characteristic of excited vibrations of the innermost electrons of the atoms. Armed with Bragg's method for determining the wavelength of X-rays, he set about measuring the frequencies of the K and L characteristic rays from a series of elements. He quickly established that these followed a regular progression in terms of the placement of the element in the Periodic Table. For the K series (which in fact he established contained two characteristic lines, K_α and K_β), he found frequencies which were proportional to $(N - 1)^2$ where N is a whole number which increased by one in proceeding through a succession of elements. For the L series (which again contained more than one line), the variation was as $(N - 7.4)^2$. The results suggested that N was a fundamental quantity for each element and Moseley demonstrated convincingly that this was not the atomic weight, but rather the atomic number, i.e. the number of electrons in the atom, or equivalently, the number of elementary positive charges on the nucleus. In a few celebrated cases he was able to show that the then-accepted order of elements in the Periodic Table was incorrect, and furthermore to predict the existence of several as then undiscovered elements.

Moseley's work was completed in the space of a few years' which included a move to Oxford and the rebuilding of his equipment. He was tragically killed during war service in the Dardanelles in 1915. His legacy to science in establishing the classification of elements by atomic numbers rather than atomic weight is a fitting memorial to his brilliant work.

The penultimate paper in this Part is a masterful summary by Rutherford of the situation, as he saw it, of the structure of the atom. It was published just three years after his first proposal of the nuclear atom, but during this time a significant amount of theoretical and experimental work had been done in supporting and extending his basic idea. The paper is notable in many respects (see the commentary on p 184), but is particularly relevant here with regard to his clear insistence on the existence of isotopes — a concept which was to be verified principally by Aston in his work at Cambridge, as described in the last paper in this Part.

1907 **13** *On Rays of Positive Electricity. By J. J. Thomson, M.A., F.R.S.*

The motivation for this study by J. J. Thomson of 'rays of positive electricity' had its origin in the discovery by Goldstein in 1886 that when small holes were made in the cathode of a gas discharge tube, luminous rays streamed from its reverse side. Goldstein called these rays *Kanalstrahlen* or Canal Rays. From 1898 onwards, Wien studied the deflection of these rays in magnetic and electric fields and deduced, firstly, that they carried a positive charge and, secondly, that if they were regarded as a stream of particles, their charge to mass ratio (e/m) had a maximum value which corresponded to that of an ionized hydrogen atom. Thomson, buoyed up no doubt by his success in identifying negatively charged corpuscles (electrons) with cathode rays in 1897, saw in experiments of this type the possibility of unravelling the mysteries of *positive* electricity.

Thomson's apparatus (Figure 2 of the paper) was similar to that used by Wien; in particular he followed Wien's arrangement for the alignment of the electric and magnetic fields which was such that the deflections each produced were at right angles. With this geometry all particles with the same e/m ratio but differing velocities create a phosphorescent patch on the flat end of the tube which takes the form of a parabola, whereas particles with different e/m ratios and the same velocity produce a straight line pattern. In the paper Thomson outlines the theory giving rise to these shapes and describes a neat method which he devised for taking into account any variation in the magnetic field strength along the path of the rays.

The experimental results described in this first paper by Thomson on positive rays are confined principally to air, hydrogen, argon and helium. In a later publication (*Philosophical Magazine* 1908 **16** 657–691) he extended the investigations to other gases. At relatively high pressures of the four gases studied here, the patterns of the phosphorescence bands indicated the presence of particles with a range of e/m values. In the units used (emu g^{-1}) the maximum value of this ratio was 10^4, corresponding to that expected for a singly charged positive hydrogen atom. Lower values of e/m were attributed by Thomson to particles of the same mass which had carried a charge during only part of their journey from the cathode to the screen. In the case of hydrogen, a second band suggested the presence of particles with an e/m ratio of 5×10^3 in addition to those with $e/m = 10^4$. For helium in the tube, a third type of particle with an e/m ratio half as much again was detected under certain conditions. When the pressure in the discharge tube was reduced, the bands split into patches, but e/m ratios of 10^4 and half this value were still found for air, hydrogen and helium and, in addition, for two other gases — carbonic acid and neon. Although helium at intermediate pressures exhibited the smaller value of e/m (2.5×10^3), even this signature disappeared at very low pressures. These results suggested to Thomson the predominance of *two* kinds of particles for all gases.

Although he was later to modify his views, he concludes here that '... under very intense electric fields, different substances give out particles charged with positive electricity, and these particles are independent of the nature of the gas from which they originate. These particles are, as far as we know at present, of two kinds, for one kind e/m has the value of 10^4, that of an atom of hydrogen; for the other kind e/m has half this value, i.e. it has the same value as for the α particles [emitted] from radioactive substances'. His wish to restrict to two the number of fundamental particles carrying positive electricity appears to have been influenced by a similar independence on the nature of the gas he had found for negative particles in 1897, an observation which had led him to propose the 'electron' as a universal constituent of matter.

Thomson's identification of the particles having $e/m = 5 \times 10^3$ with an α particle was plausible but not correct. Two years later Rutherford and Royds (see p. 152) established that α particles were doubly ionized helium atoms (actually helium nuclei). Their mass is thus four times that of hydrogen and their (nuclear) charge twice that of hydrogen; their e/m ratio is therefore indeed about 5×10^3. But so is that of the singly charged hydrogen molecule. It would seem that the ubiquitous presence of two particles with e/m ratios of 10^4 and 5×10^3 arose respectively from hydrogen atoms and hydrogen molecules which, in spite of precautions taken, were always present in the discharge tube. However, the signal Thomson observed for helium gas (with $e/m = 2.5 \times 10^3$) was undoubtedly correctly identified as singly charged ionized helium.

The absence of signatures for the nuclei of the other gases studied was very mystifying. Later, it was established that the pressure in the tube needed to be lower and also that the willemite (zinc sulphide) screen he used was insensitive to heavy nuclei. When signals characteristic of other nuclei were found, they presented mysteries of their own. In the hands of Aston, the technique and apparatus devised by Thomson was developed into a mass spectrograph capable of determining atomic masses with great precision — an advance which led to the first identification of isotopes of non-radioactive elements (see p 303).

1911   **21**   *The Scattering of α and β Particles by Matter and the Structure of the Atom. By Professor E. Rutherford, F.R.S., University of Manchester*

Publication of this paper marked no less than the 'discovery' by Rutherford of the atomic nucleus — the small positively charged central core at the heart of all elements. Its importance, historically, can hardly be overstated and yet it took two years before the scientific world accepted the full significance of Rutherford's proposal.

The experimental result which convinced Rutherford that atoms were essentially devoid of material — that minute negative charges (electrons) circulated like planetary objects around a more massive, but nevertheless still small, central positive core — was the observation that α particles in their passage through matter occasionally suffered large angular deflections. Indeed some of the particles were found to rebound backwards! He was later to recall that the effect 'was almost as incredible as if you fired a 15-inch shell at a piece of tissue paper and it came back and hit you'.

The theoretical part of the paper is directed at showing that a succession of scattering events, each of which deflects the α particle by a small amount but which in principle might combine to give a single large deflection, cannot in fact account for the angular distribution found experimentally. Here Rutherford had in mind J. J. Thomson's 'plum-pudding' model of the atom in which the electrons were assumed to move in a uniform sphere of positive charge and for which single scattering events were incapable of causing large deflections. On the other hand, single scattering from a highly concentrated positive charge *can* produce such deflections.

Figure 1 of the paper, illustrating the encounter of a positively charged particle with a stationary charge of the same sign (situated at S), is very familiar to students of physics, as indeed is the theoretical analysis of the event which yields equation 5.

In the final section, General Considerations, Rutherford writes 'Considering the evidence as a whole, it seems simplest to suppose that the atom contains a central charge distributed through a very small volume, and that the large single deflexions are due to the central charge as a whole, and not to its constituents.' But he follows this with caveats. There is 'the possibility that a small fraction of the positive charge may be carried by satellites extending some distance from the centre' and, more seriously, 'The deductions from the theory so far considered are independent of the sign of the central charge, and it has not so far been found possible to obtain definite evidence to determine whether it be positive or negative.' (This latter possibility can be visualized by considering an encounter in which the α particle is 'swung around' the nucleus by an attractive rather than a repulsive force, in the manner that space probes use encounters with planets to alter their trajectory.) Definitive evidence that the nucleus of atoms is indeed positive was to come later.

1913 **25** *The Laws of Deflexion of α Particles through Large Angles. By Dr H. Geiger and E. Marsden*

During the two years between publication of Rutherford's classic work on the nuclear atom and the present paper, Geiger and Marsden were undertaking experiments in the same laboratory to test the predictions of the formula which Rutherford had derived for the scattering of α particles by single encounters with charged nuclei. The theoretical predictions were that the number of particles scattered through an angle ϕ should be proportional to (1) $\operatorname{cosec}^4\phi/2$, (2) the thickness of the foil used to scatter the particles, (3) the square of the nuclear charge, and (4) the inverse fourth power of the particle's velocity.

A cross-sectional view of the simple apparatus used to determine the angular dependence of the scattering is shown in Figure 1 of the paper. An intense radioactive source of radium was placed behind a thin scattering foil and the scattered α particles detected on a zinc sulphide scintillation screen attached to a microscope. The latter could be rotated to permit observations at scattering angles of up to 150°. Somewhat different arrangements were used to test the other predictions of the theory.

The angular dependence was measured for both silver and gold, and the thickness dependence for silver, copper and aluminium. After various corrections for the decay in activity of the source, etc. the experimental results verified the first two theoretical predictions with good accuracy. The third prediction, namely that the scattering should be proportional to the square of the nuclear charge, could not be tested directly because, at the time of the experiments, this charge was not known for most elements. The dependence on the atomic weight, A, of the scattering material was therefore measured. On the assumption that the central charge was proportional to A (in fact it is proportional to the atomic number Z), and allowing for the variation in the aereal density of atoms in different foils, an $A^{3/2}$ dependence of the scattering was expected. This was experimentally verified for six different elements.

Geiger and Marsden also confirmed the fourth prediction of the theory concerning the dependence of the scattering on the incident velocity. Sheets of mica were interposed to vary the range of the α particles and a previously obtained formula, relating the velocity to the range, used to obtain relative velocities.

A final experiment was undertaken to determine an absolute value for the fraction of particles scattered through a given angle and hence to obtain the nuclear charge directly. The results here were somewhat uncertain but they did allow Geiger and Marsden to claim at the end of the paper 'that the number of elementary charges composing the centre of the atom is equal to half the atomic weight'. It was later found that this is only approximately true for light elements and even less valid for heavier ones, but the finding did alert scientists to the fact that the nucleus might contain particles other than charged ones.

1913 **26** *On the Constitution of Atoms and Molecules. By N. Bohr, Dr. phil. Copenhagen*

Niels Bohr was just 26 when he began to lay the foundations of his remarkable theoretical treatment of Rutherford's nuclear model of the atom. Working in Rutherford's laboratory in Manchester, he recognized the far-reaching implications of the new model and was the first to recognize that the nuclear charge on the atoms in a given element was a quantity more fundamental than the atomic weight. It is the charge which determines the number of peripheral electrons and the chemical properties, and hence the position of an element in the Periodic Table.

The single outstanding problem of Rutherford's atom concerned its stability. If the planetary electrons circulate around a central positive core, what prevents them from spiralling into the nucleus as their energy diminishes due to loss by radiation?[1]

Bohr's 'answer' to this question was formulated on the basis of an assumption that stable orbits do in fact exist and that energy is radiated only when the electron makes a transition between such 'stationary' orbits. Furthermore he proposed that the radiation is emitted in units of Planck's quantum energy $h\nu$. Such bold assertions were not for the faint-hearted and would have been dismissed out of hand were it not for the startling agreement that Bohr found between the observed frequencies of the line spectra of hydrogen and the predictions of his model.

The paper starts with classical expressions for the frequency of revolution and size of orbit of an electron describing a closed elliptical path around a heavy positively charged nucleus. These are given (equation (1) of the paper) in terms of the energy, W, to remove the electron from the orbit to rest at an infinite distance away.[2]

Bohr's next step was to assume that 'during the binding of the electron, a homogenous radiation is emitted of frequency ν, equal to half the frequency of revolution of the electron in its final orbit'. Thus $W = \tau h \omega / 2$ where τ is a whole number. The smallest and most stable orbit then corresponds to $\tau = 1$ and leads, after substitution in equation (1) and with values of the constants known at the time, to values of $2a = 1.1 \times 10^{-8}$ cm, $\omega = 6.2 \times 10^{15}$ s^{-1} and $W = 13$ eV, for hydrogen with a nuclear charge $E = e$. The order of magnitude agreement with estimates of the atomic size and with the known ionization energy was truly impressive.

Calculation of the frequencies of lines of the hydrogen spectra follows immediately. Passage from one stationary state to another involves emission of radiation of energy equal to the difference in energy between the orbits, i.e. with frequencies given by

$$\nu = \frac{2\pi^2 me^4}{h^3} \left\{ \frac{1}{\tau_2^2} - \frac{1}{\tau_1^2} \right\}$$

The visible Balmer series is obtained with $\tau_2 = 2$ and integral values of τ_1 (3, 4, 5, etc.); likewise the infrared Paschen series is found with $\tau_2 = 3$ and varying τ_1. Other series in the infrared and ultraviolet (as then unobserved) were predicted. (In 1916 Lyman found an ultraviolet series with $\tau_2 = 1$, in 1922 Brackett discovered one in the infrared with $\tau_2 = 4$, and in 1924 Pfund identified one with $\tau_2 = 5$ — five series in all).

The value of the constant in front of the term in brackets is Bohr's true claim to fame. With values of m, e and h as then known, it had the value 3.1×10^{15} s^{-1}, in excellent agreement with the experimental value of 3.29×10^{15} s^{-1}. (In the MKS system of units, the constant is $me^4/8\varepsilon_0^2h^3$ which, with today's values of the fundamental constants, yields 3.29×10^{15} s^{-1} or, in terms of energy, 13.6 eV — the Rydberg constant.)

Bohr then extends his theory to account for the spectral lines observed for helium and other elements containing more than one electron, before returning to the general considerations of his theoretical treatment. In this section, he finds another useful way of interpreting stationary orbits, namely that the angular momentum of the electron is quantized in such orbits. In his own words at the end of the paper: 'In any molecular system consisting of positive nuclei and electrons in which the nuclei are at rest relative to each other and the electrons move in circular orbits, the angular momentum of every electron round the centre of its orbit will in the permanent state of the system be equal to $h/2\pi$, where h is Planck's constant'.

The paper is the first of a trilogy by Bohr. The other two appeared in *Philosophical Magazine* 1913 **26** 476 and 857.

1913 **26** *The High-Frequency Spectra of the Elements. By H. G. J. Moseley, M.A.*

The practical use of X-rays was recognized by the medical profession almost immediately after their discovery by Wilhelm Konrad Röntgen in 1895, but some 18 years passed before W. H. Bragg and his son W. L. Bragg analysed the frequency distribution of the radiation emitted by X-ray tubes. This was made possible by using the closely spaced atomic planes of a crystal to diffract the X-rays, in an analogous manner to the diffraction of the much larger *optical* wavelengths by the rulings of a diffraction grating. It was discovered soon thereafter that, superimposed on a broad continuous background, the emission spectra contain lines characteristic of the material from which the anode target is made.

Harry Moseley secured a prominent place in the history of science by analysing the characteristic frequencies emitted from an X-ray tube in which different materials were used as targets. He found that the frequencies were intimately related to the atomic number (nuclear charge) of the elements, a discovery which brought definitive order to the Periodic Table and provided unambiguous support for the Rutherford model of atomic structure.

In this first of the two Moseley papers reproduced here, the results from twelve elements with atomic weights varying from about 40 to 65 are presented. From the frequencies ν of the principal spectral lines, Moseley calculated the quantity $Q = \sqrt{\nu / \frac{3}{4}\nu_0}$ and found that this increased by a constant amount in going from one element to the next when these were ordered according to their then-accepted place in the Periodic Table. The constant ν_0 was taken to be the Rydberg frequency, which had been known since 1890 (see *Science in the Making*, Volume 2) to describe the frequencies of the lines in the *optical* spectrum of hydrogen via the equation

$$\nu = \nu_0 \left\{ \frac{1}{n_1^2} - \frac{1}{n_2^2} \right\} \qquad \text{with } \nu_0 = 109\,720c = 3.29 \times 10^{15} \text{ s}^{-1}$$

where n_1 and n_2 are integers (denoted by τ in the previous paper by Bohr). For example, the Balmer series of hydrogen is accounted for by putting $n_1 = 1$ and $n_2 = 2, 3, 4$, etc. The factor $\frac{3}{4}\nu_0$ in Moseley's expression for Q clearly arises when $n_1 = 1$ and $n_2 = 2$ but, before discussing this further, it will be advantageous to follow Moseley in regarding Q simply as a parameter suggested by the experimental results.

In order to make the steps in Q equal to unity, Moseley found it necessary to multiply $\nu^{\frac{1}{2}}$ by a constant factor, so that, in effect, Q became $N - k$ with N a whole number and k a constant, equal to one for the series here considered. With this adjustment, each element had a unique identifying number N approximately equal to half its atomic weight A. This atomic number (N in the paper, today denoted by Z) is identified with the nuclear charge of Rutherford's atom, which Moseley concludes is a more significant quantity than A in ordering the elements in the Periodic Table. For example cobalt ($N = 27$) precedes nickel ($N = 28$) even though the order of A is the opposite.

Moseley now makes a connection with the Bohr theory of stationary electronic states in atoms, in which each ring of electrons has a well-defined angular momentum and ν_0 is given in terms of fundamental constants and the square of the nuclear charge. Using this theory the frequency of radiation emitted when an electron passes between stationary states with $n_2 = 2$ and $n_1 = 1$ is

$$\nu = \frac{3}{4} \frac{2\pi^2 e^4 m}{h^3} (N - \sigma_n)^2$$

The small correction factor σ_n to N is associated with there being more than one electron in each ring. In the case of hydrogen with one electron, and alkali halides with an outer shell containing one electron, σ_n is zero, but for electronic transitions between inner shells (such as is the case for X-ray spectra), the factor must be included.

As Moseley states, 'The numerical agreement between the experimental values and those calculated from a theory designed to explain the ordinary hydrogen spectrum is remarkable, as the wave-length dealt with in the two cases differs by a factor of about 2000.'

1914 **27** *The High-Frequency Spectra of the Elements. Part II. By H. G. J. Moseley, M.A.*

In this well-written and important publication, Moseley extends his measurements of X-ray spectra, described in the previous paper, to over thirty more materials, thereby covering in total almost half the Periodic Table of the elements.

The apparatus is nicely illustrated in two diagrams. Figure 1 shows the ingenious method Moseley used for bringing different targets into the stream of cathode rays without breaking the vacuum of the X-ray tube. A vertical section of this tube, showing its connection to the spectrometer, is illustrated in Figure 2. The diffracting crystal is not shown but would have sat on the central movable table B.

Details of the measurement techniques and the nature of the samples are described at some length. In some cases, only alloys of the elements were available. The wavelengths of several emission lines in both the so-called K and L series are reported in the tables, together with the corresponding Q values (defined in the paper) and the deduced atomic numbers N of the elements. For the principal α-line of both series, the frequency (c/wavelength) could be expressed as

$$\nu = A(N - b)^2$$

with $A = \left(\dfrac{1}{1^2} - \dfrac{1}{2^2}\right)\nu_0 = \dfrac{3}{4}\nu_0$ and $b = 1$ for the K lines

and $A = \left(\dfrac{1}{2^2} - \dfrac{1}{3^2}\right)\nu_0 = \dfrac{5}{36}\nu_0$ and $b = 7.4$ for the L lines

Energy

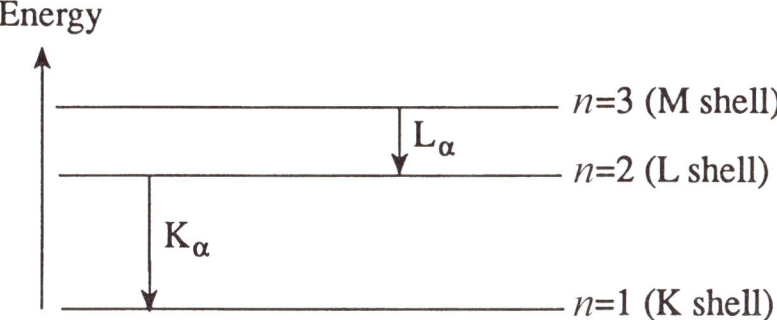

The electronic transitions responsible for the α-line emissions are sketched on the electron energy level diagram drawn (with the benefit of hindsight) below.

Moseley proposes that the larger 'correction factor' b for the L lines 'suggests that the L system is situated further from the nucleus'. We would now interpret this in terms of the screening of the nuclear charge being larger for electrons in the $n = 3$ level, owing to the presence of electrons in the $n = 2$ and $n = 1$ levels.

In the remarkable graphical representation of the data in Figure 3 of the paper, Moseley assigns an atomic number to the elements listed and furthermore predicts the existence of three 'undiscovered' elements with $N = 43$, 61 and 75. The first and last of these correspond to spaces long recognized by chemists; they are now known to be technetium (Tc), which does not exist naturally because it has no stable isotope, and rhenium (Re). The other missing element ($N = 61$) is promethiem (Pm), which also does not exist naturally in any abundance.

A few other 'mistakes' can be identified by comparison with a modern Periodic Table. The elements holmium (Ho) ($N = 67$) and dysprosium (Dy) ($N = 66$) should be interchanged. Also the sequence of rare-earth elements ends at lutetium (Lu) ($N = 71$); Moseley assigned a value of N one too high for this element and also for ytterbium (Yb) ($N = 70$), so thulium II (Tm II) has disappeared. The element keltium, proposed by Urbain and mentioned towards the end of the paper, is not now recognized; the missing element with $N = 72$ is hafnium (Hf), discovered in 1922.

The above, relatively minor, discrepancies should not detract from the significance of Moseley's work in unravelling some of the innermost secrets of the electronic structure of atoms and in supporting Rutherford's model of a central nucleus, the charge on which was of more significance than the mass. It was not until 1932 that the *neutron* was discovered, although the existence of neutral particles in the nucleus which, together with the protons, accounted for the atomic weight, had been anticipated by Rutherford and Chadwick in 1924.

1914 **27** *The Structure of the Atom. By Sir Ernest Rutherford, F.R.S., Professor of Physics, University of Manchester*

In this paper Rutherford surveys the nuclear model of the atom which he had proposed in 1911 (see p 204), in the light of relevant experimental results and theories published in the intervening years. These enabled him to expound much more confidently than previously his ideas on atomic structure and to extend them to considerations of the nature of the nucleus itself.

The first point of significance occurs on the first page where he writes that the atom contains a *positively* charged nucleus of small dimensions in which practically all the mass is concentrated. The uncertainty concerning the sign of the charge, expressed in his 1911 paper, has now disappeared. In reviewing the evidence in favour of the model, Rutherford cites the α-particle scattering experiments of Geiger and Marsden, and the first photographs by C. T. R. Wilson of cloud tracks produced by α particles undergoing large-deflection scattering events (see Plate 5) which had been published two years earlier. Later on in the paper he also cites the evidence from Moseley's studies of X-ray spectra and, in the final paragraph (almost, it appears, as an afterthought) he refers to Bohr's model for hydrogen and helium, albeit with reservations about the physical meaning of some of Bohr's assumptions.

After a brief discussion of the possibility of radioactive β particles suffering large deflections in a manner similar to α particles, Rutherford turns his attention to α-particle scattering on their passage through light gases. An accompanying paper by Darwin (*Philosophical Magazine* 1914 **27** 499) contains a prediction that a close encounter with a hydrogen atom should lead to its acquiring a velocity about 1.6 times as great as that of the α particle itself, and Rutherford reports (as then) unpublished results of Marsden providing evidence for this knock-on effect. Subsequent experiments of this type were to lead to the first artificial disintegration of a nucleus (actually nitrogen, see p 326) but here Rutherford uses the data to estimate the size of the hydrogen nucleus. This, he argues, must be less than 1.7×10^{-13} cm. The name *proton* was not introduced until 1919; hence usage of the term *positive electron* throughout the paper. (After Dirac's prediction of the existence of antiparticles, the true positive counterpart of the electron eventually became known as the *positron*.)

With the knowledge that the helium atom had a mass nearly four times that of hydrogen, Rutherford proposes that the nucleus of this element contains four 'positive electrons' and two negative electrons. It was of course established later that the helium nucleus contains two protons and two neutrons.

Today's interpretation of the fact that the helium atom does not have quite four times the mass of hydrogen differs considerably from that proposed by Rutherford, although his suggestion is interesting in the context of views prevailing at the time. It was based on the proposal by J. J. Thomson that the mass of the electron is entirely electromagnetic in origin. Assuming the same applies to protons, Rutherford recognized that the 'electrical mass' of a tightly packed system of positive and negative charges would depend on the way their fields interact. Differences between nuclear masses and the sum of the masses of their component particles later became known as *mass defects* Δm, related to nuclear binding energies ΔE via the Einstein relation $\Delta E = \Delta mc^2$. However, theories of nuclear binding could not advance until the true constituents of the nucleus had been identified. Rutherford's insistence that nuclei must contain (negative) electrons found its strongest argument in the phenomenon of radioactive β decay. He guessed (correctly) that the high energy associated with β rays could not arise if the electrons were ejected from the external electron distribution but could do so if they were liberated from the nucleus. The actual mechanism of β emission was subsequently shown to be the decay of a neutron into a proton and an electron, with the subsequent emission of the electron.

In the final section of the paper Rutherford considers the relation between the charge and mass of nuclei. The greater importance of the former in determining chemical properties had become clear from the work of Moseley and others, but there was still the mystery of why the mass or atomic weight was equal to or greater than twice that calculated from the number of positively charged particles in the nucleus. The existence of atoms with the same nuclear charge but different masses is beautifully illustrated here by Rutherford with reference to the radioactive series of uranium. Transformation of the first five members of this series occurs by α or β emissions in the order: α, β, β, α, α. So, starting with uranium, which Rutherford took to have an atomic weight of 238.5 and a nuclear charge of 100, the fourth member of the series has a mass that is lower by 4 units but carries the same charge.[3] The same applies to the second and fifth elements in the sequence. Confirmation of the existence of *isotopes* of the same element was to be decisively demonstrated some years later by Aston (see next paper).

1920 **39**   *The Mass-Spectra of Chemical Elements. By F. W. Aston, M.A., D.Sc., Clerk Maxwell Student of the University of Cambridge*

As a research student under Poynting at Birmingham University, Francis Aston discovered a narrow non-luminous region adjacent to the cathode in a gas-discharge tube, a 'dark space' which was to bear his name. He is, however, better known for his development of the Aston mass spectrograph which led to a clear identification of isotopes and an accurate determination of their atomic weights. For this work he was awarded the Nobel Prize for Chemistry in 1922.

In 1910 Aston was appointed as J. J. Thomson's assistant at the Cavendish Laboratory, Cambridge, where he was given the task of improving the apparatus used by Thomson in his study of positive rays (see p 189). Of the several modifications he made, the most significant was the introduction of a very narrow collimating tube between the vessel supporting the discharge and the chamber in which the positive ions were deflected and detected. This allowed for a sufficient pressure for the discharge on the one side whilst permitting the deflection part of the apparatus on the other side to be highly evacuated. Another improvement made by Aston was the replacement of the previously used willemite screen by photographic plates placed inside the deflection tube, which considerably improved the sensitivity of detection. With the redesigned equipment, parabolic traces corresponding to e/m ratios characteristic of various gases — patterns which Thomson had predicted but failed to find in 1907 — began to appear. Thomson wrote '...this effect may furnish a valuable means of analysing the gases in the tubes and determining their atomic weights' (*Philosophical Magazine* 1910 **20** 752).

In 1912 Thomson reported on two parabolic traces which appeared when neon was used for the discharge gas. When analysed, these gave masses, relative to that of hydrogen, of 20 and 22. The atomic weight of neon was known at the time to be 20.2 and so Thomson identified the lower mass with pure neon and the higher mass with an unknown hydride of neon (NeH_2^+) or perhaps with a new isotope. The problem was left for Aston to solve, but the First World War intervened and it was not until 1919 that investigations could be continued.

After the war, Aston built a new positive ray spectrograph with a resolving power of 1 in 1000. Its design (*Philosophical Magazine* 1919 **38** 707) was a considerable advance on earlier forms of the apparatus. Instead of the electric and magnetic fields being applied over the same region of space as previously, the positive rays were first dispersed by passing them through an 'electrostatic lens' and then deflected in the opposite direction by a magnetic field. With this arrangement, particles with the same e/m ratio were focused to the same point, irrespective of their velocities. With the aid of reference lines from elements of known atomic weight, Aston's new instrument had sufficient precision to allow unambiguous determination of the masses associated with different lines (see Plate 13). Within a year he had conclusively demonstrated that neon contained two isotopes with atomic weights of 20.00 and 22.00 (to an accuracy of about one-tenth of a percent) and that their proportions (as estimated from relative intensities) were 90% and 10% by volume in the natural gas. The possibility of a third isotope of mass 21 was also indicated (*Philosophical Magazine* 1920 **39** 449).

In the present paper, the apparatus — now called a mass-spectro-graph — is illustrated and described in detail. Measurements are reported for oxygen, carbon, neon, chlorine, argon, nitrogen, hydro-gen, helium and mercury, which, apart from hydrogen, produced lines associated with atomic or molecular masses that are whole numbers. This 'whole-number rule' resolved the puzzle of fractional atomic weights which Aston here explains as 'merely fortuitous statistical effects due to the relative quantities of the isotopic constituents'. On the scale chosen, with oxygen ($O = 16$) as the standard, hydrogen has a mass of 1.008. The mass of heavier elements is not exactly the sum of the masses of the constituent particles because of the need to subtract the mass equivalent of their binding energies.

Notes

1. Classical electrodynamics predicts that energy is radiated whenever a charged particle is accelerated. In the case of circulating electrons the acceleration is centripetal.

2. The notation used by Bohr differs from that used today. The symbol ω refers to a frequency, *not* an angular frequency, of revolution; the nuclear charge E is normally written Ze where Z is the atomic number: and the formulae are in cgs units. If we preserve Bohr's notation, the equations can be derived for a circular orbit of radius r using the following expressions:

Force of attraction $=$ centripetal force

$$eE/a^2 = mv^2/a$$

Energy required to remove electron (ionization energy)

$$W = -\{\text{potential energy} + \text{kinetic energy}\}$$
$$= eE/a - mv^2/2$$

Orbital velocity $v = 2\pi a/T = 2\pi a\omega$ where $T =$ period of revolution.

Algebraic manipulation of these expressions yields the two equations labelled (1) in the paper. They also apply for elliptical orbits of major axis 2*a*.

3. The fact that Rutherford took the atomic weight and atomic number of uranium to be 238.5 and 100, instead of 238 and 92, does not alter his conclusions in any way. The series he considered is written in modern notation below.

$$^{238}_{92}U \xrightarrow{\alpha} {}^{234}_{90}Th \xrightarrow{\beta} {}^{234}_{91}Po \xrightarrow{\beta} {}^{234}_{92}U \xrightarrow{\alpha} {}^{230}_{90}Th \xrightarrow{\alpha} {}^{226}_{88}Ra$$

THE

LONDON, EDINBURGH, AND DUBLIN

PHILOSOPHICAL MAGAZINE

AND

JOURNAL OF SCIENCE.

[SIXTH SERIES.]

MAY 1907.

XLVII. *On Rays of Positive Electricity.*
By J. J. THOMSON, *M.A., F.R.S.*[*]

Fig. 1.

IN 1886 Goldstein discovered that when the cathode in a discharge-tube is perforated, rays pass through the openings and produce luminosity in the gas behind the cathode ; the colour of the light depends on the gas with which the tube is filled and coincides with the colour of the velvety glow which occurs immediately in front of the cathode. The appearance of these rays is indicated in fig. 1, the anode being to the left of the cathode KK. Since the rays appeared through narrow channels in the cathode, Goldstein called them " Kanalstrahlen " : now that we know more about their nature, "positive rays" would, I think, be a more appropriate name. Goldstein showed that a magnetic force which would deflect cathode rays to a very considerable extent was quite without effect on the " Kanalstrahlen." By using intense magnetic fields, W. Wien showed that these rays could be deflected, and that the deflexion was in the opposite direction to that of the cathode rays, indicating that these rays carry a positive charge of electricity. This was confirmed by measuring the electrical charge received by a vessel into which the rays passed through a small hole, and also by observing the direction in which

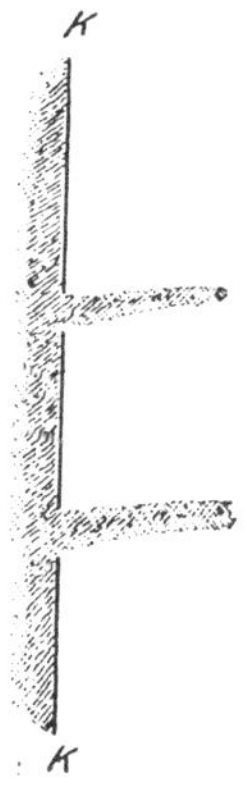

[*] Communicated by the Author.

Phil. Mag. S. 6. Vol. 13. No. 77. *May* 1907.   2 Q

562   Prof. J. J. Thomson *on*

they are deflected by an electric force.   By measuring
the deflexions under magnetic and electric forces, Wien
found by the usual methods the value of e/m and the velocity
of the rays.   He found for the maximum value of e/m the
value of 10^4, which is the same as that for an atom of hydrogen
in the electrolysis of solutions.   A valuable summary of the
properties of these rays is contained in a paper by Ewers *.

As these rays seem the most promising subjects for investigating the nature of positive electricity, I have made a
series of determinations of the values of e/m for positive rays
under different conditions.   The results of these I will now
proceed to describe.

Apparatus.

Screen used to detect the rays.—The rays were detected
and their position determined by the phosphorescence they
produced on a screen at the end of the discharge-tube.   A
considerable number of substances were examined to find the
one which would fluoresce most brightly under the action of
the rays.   As the result of these trials, Willemite was selected.
This was ground to a very fine powder and dusted uniformly
over a flat plate of glass.   Considerable trouble was found in
obtaining a suitable substance to make the powder adhere to
the glass.   All gums &c. when bombarded by the rays are
liable to give off gas ; this renders them useless for work in
vacuum-tubes.   The method finally adopted was to smear a
thin layer of " water-glass " (sodium-silicate) over the glass
plate, and then dust the powdered Willemite over this layer
and allow the water-glass to dry slowly before fastening the
plate to the end of the tube.

The form of tube adopted is shown in fig. 2.   A hole is
bored through the cathode, and this hole leads to a very fine
tube F.   The bore of this tube is made as fine as possible so
as to get a small well-defined fluorescent patch on the screen.
These tubes were either carefully made glass tubes, or else
the hollow thin needles used for hypodermic injections, which
I find answer excellently for this purpose.   After getting
through the needle, the positive rays on their way down the
tube pass between two parallel aluminium plates A, A.
These plates are vertical, so that when they are maintained
at different potentials the rays are subject to a horizontal
electric force, which produces a horizontal deflexion of the
patch of light on the screen.   The part of the tube containing the parallel aluminium plates is narrowed as much as
possible, and passes between the poles P, P of a powerful
electromagnet of the Du Bois type.   The poles of this magnet

---

* *Jahrbuch der Radioaktivität,* iii. p. 291 (1906).

Rays of Positive Electricity.   563

are as close together as the glass tube will permit, and
are arranged so that the lines of magnetic force are horizontal and at right angles to the path of the rays. The
magnetic force produces a vertical deflexion of the patch of
phosphorescence on the screen. To bend the positive rays it
is necessary to use strong magnetic fields, and if any of the
lines of force were to stray into the discharge-tube in front

Fig. 2.

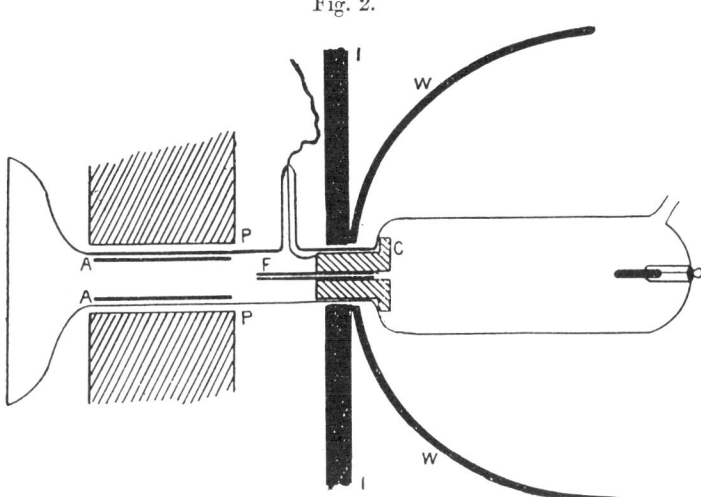

of the cathode, they would distort the discharge in that part
of the tube. This distortion might affect the position of the
phosphorescent patch on the screen, so that unless we shield
the discharge-tube we cannot be sure that the displacement
of the phosphorescence is entirely due to the electric and
magnetic fields acting on the positive rays after they have
emerged from behind the cathode.

To screen off the magnetic field, the tube was placed in a
soft iron vessel W with a hole knocked in the bottom, through
which the part of the tube behind the cathode was pushed.
Behind the vessel a thick plate of soft iron with a hole bored
through it was placed, and behind this again as many thin
plates of soft iron, such as are used for transformers, as there
was room for were packed. When this was done it was
found that the magnet produced no perceptible effect on the
discharge in front of the cathode.

The object of the experiments was to determine the value
of e/m by observing the deflexion produced by magnetic and
electric fields. When the rays were undeflected they produced a bright spot on the screen ; when the rays passed

2 Q 2

through electric and magnetic fields the spot was not simply deflected to another place, but was drawn out into bands or patches, sometimes covering a considerable area. To determine the velocity of the rays and the value of *e/m*, it was necessary to have a record of the shape of these patches. This might have been done by substituting a photographic plate for the Willemite screen. This, however, was not the method adopted, as, in addition to other inconveniences, it involves opening the tube and repumping for each observation, a procedure which would have involved a great expenditure of time. The method actually adopted was as follows :—The tube was placed in a dark room from which all light was carefully excluded, the tube itself being painted over so that no light escaped from it. Under these circumstances the phosphorescence on the screen appeared bright and its boundaries well defined. The observer traced in Indian ink on the outside of the thin flat screen the outline of the phosphorescence. When this had been satisfactorily accomplished the discharge was stopped, the light admitted into the room, and the pattern on the screen transferred to tracing-paper; the deviations were then measured on these tracings.

Calculation of the Magnetic and Electric Deviation of the Rays.

If we assume the electric field to be uniform between the plates and zero outside them, then we can easily show that *x*, the horizontal deflexion of a ray whose charge is *e*, mass *m*, and velocity *v*, is given by the equation

$$x = \tfrac{1}{2} X \frac{e}{mv^2} l(l+2d),$$

where X is the force between the plates, *l* the length of path of the rays between the plates, and *d* the distance of the screen from the nearer end of the parallel plates.

To find the deflexion due to the magnetic field, we have, if ρ is the radius of curvature of the path at a point where the magnetic force is H,

$$\frac{mv^2}{\rho} = Hev,$$

or

$$\frac{1}{\rho} = \frac{e}{mv} H.$$

Rays of Positive Electricity. 565

If y is the vertical displacement of the particle, we have

$$\frac{1}{\rho} = \frac{d^2y}{dz^2} \text{ approximately,}$$

where z is measured along the path of the ray   Hence

$$\frac{d^2y}{dz^2} = \frac{e}{mv}H \; ;$$

$$y = \frac{e}{mv}\left[\int_0^{l+d}\int_0^z H dz\right]. \quad \cdot \quad \cdot \quad \cdot \quad \cdot \quad (1)$$

In these strong fields there are considerable variations of H along the path, so that to calculate the integrals we should have to map out the value of H along the path of the ray. This would be a very laborious process, and it was rendered unnecessary by the following simple method, which, while not involving anything like the labour of the direct method, gives much more accurate results.   The method is shown in fig. 3.

Fig. 3.

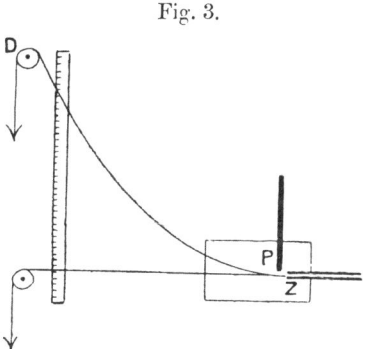

The part of the tube through which the rays pass was cut off, and a metal rod placed so that its tip Z coincided with the aperture of the narrow tube through which the positive rays had emerged. A very fine wire soldered to the end of this tube passed over a light pulley, and carried a weight at the free end. The pulley was supported by a screw by means of which it could be raised or lowered ; a known current passed through the wire, entering it at Z and leaving it through the pulley. The pulley was first placed so that the path of the stretched wire when undeflected by a magnetic field coincided with the path of the undeflected rays. A vertical scale whose edge was at the same distance from the opening through which the rays emerge as the screen on which the phosphoresence had been observed, was placed just behind the wire, and was read by a reading microscope with a micrometer eyepiece. When the magnetic field was put on, the wire was deflected; and if T is the tension of the wire, ρ the radius of curvature into which it is bent, i the current through the wire,

$$\frac{T}{\rho} = Hi \; ;$$

or, if y_1 is the vertical displacement of the wire,

$$\frac{d^2 y_1}{dz^2} = \frac{i}{T} \cdot H.$$

Now if $\frac{dy_1}{dz} = 0$ when $z = 0$ we have, if y_1 is the displacement of the wire at the scale,

$$y_1 = \frac{i}{T} \int_0^l \cdot \int_0^z H dz. \quad \ldots \ldots \quad (2)$$

Hence, comparing (1) and (2) we have

$$\frac{y}{y_1} = \frac{\frac{e}{mv}}{\frac{i}{T}}, \quad \ldots \ldots \ldots \quad (3)$$

a relation from which the magnetic force is eliminated. To ensure that the tangent to the wire is horizontal when $z = 0$, the following method is used. P is a chisel-edge carried by a screw and placed about 1 mm. in front of the fixed end of the wire; this is adjusted so that when the magnetic field is not on the wire just touches the edge: this can be ascertained by making the contact with the wire complete an electric circuit in which a bell is placed. When the magnetic field is put on the wire is pulled off from the edge, and the tangent at $z = 0$ is no longer horizontal; it can, however, be brought horizontal by raising or lowering the pulley D until the wire is again in contact with P, which can be ascertained again by the ringing of the bell. Then y_1 is the vertical distance between the point where the wire now crosses the edge of the scale and the point where it crossed it before the magnetic field was put on. Since y, y_1, i, and T can easily be measured, equation (3) gives us the value of e/mv, while the deflexion under the electric force gives the value of e/mv^2.

Rays of Positive Electricity.   567

If y is the vertical displacement of the patch of phosphorescent light on the screen produced by the magnetic field, x the horizontal displacement due to the electrostatic field, we see that

$$y = \frac{y_1}{(i/\mathrm{T})}\frac{e}{mv} = \mathrm{B}\frac{e}{mv},$$

$$x = \mathrm{A}\frac{e}{mv^2},$$

where A and B are constants depending on the position of the screen and the magnitudes of the electric and magnetic forces. These quantities can be calculated by means of the equations just given.

Since

$$\frac{y}{x} = \frac{\mathrm{B}}{\mathrm{A}}v,$$

$$\frac{y^2}{x} = \frac{\mathrm{B}^2}{\mathrm{A}}\frac{e}{m}.$$

We see that if the pencil is made up of rays having a constant velocity but having all values of e/m up to a maximum value, the spot of light will be spread out by the magnetic and electric fields into a straight line extending a finite distance from the origin. While if it is made up of two sets of

Fig. 4.   Fig. 5.

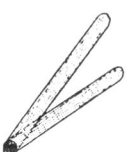

rays, one having the velocity v_1 the other the velocity v_2, the spot will be drawn out into two straight lines as in fig. 4.

If e/m is constant and the velocities have all values up to a maximum, the spot of light will be spread out into a portion of a parabola, as indicated in fig. 5.

We shall later on give examples of each of these cases.

The discharge was produced by means of a large induction-coil, giving a spark of about 50 cm. in air, with a vibrating make and break apparatus. Many tubes were used in the course of the investigation, the dimensions of these varied slightly. The distance of the screen from the hole from which the rays emerged was about 9 cm., the length of the parallel plates about 3 cm., and the distance between them ·3 cm.

568   Prof. J. J. Thomson *on*

Properties of the Positive Rays when the Pressure is not exceedingly low.

The appearance of the phosphorescent patch after deflexion in the electric and magnetic fields depends greatly upon the pressure of the gas. I will begin by considering the case when the pressure is comparatively high, say of the order of 1/50 of a millimetre. At these pressures, though the walls of the tube in front of the cathode were covered with bright phosphorescence and the dark space extended right up to the walls of the tube and was several centimetres thick, traces of the positive column could be detected in the neighbourhood of the anode. I will first take the case where the tube was filled with air. Special precautions were taken to free the air from hydrogen; it was carefully dried, and a subsidiary discharge-tube having a cathode made of the liquid alloy of sodium and potassium was fused on to the main tube. When the discharge passes from such a cathode it absorbs hydrogen. The discharge was sent through this tube at the lowest pressure at which enough light was produced in the gas to give a visible spectrum, until the hydrogen lines disappeared and the only lines visible were those of nitrogen and mercury vapour. This pressure was a little higher than that used for the investigation of the positive rays, but a pump or two was sufficient to bring the pressure down to this value. The appearance of the phosphorescence on the screen when the rays were deflected by magnetic and electric forces separately and conjointly is shown in fig. 6.

Fig. 6.

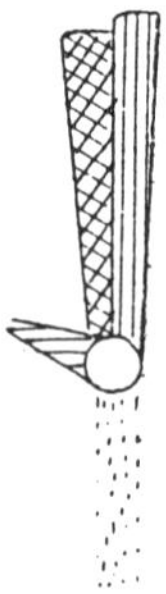

The deflexion under magnetic force alone is indicated by vertical shading, under electric force alone by horizontal shading, and under the two combined by cross shading.

The spot of phosphorescence is drawn out into a band on either side of its original position. The upper portion, which is very much the brighter, is deflected in the direction which indicates that the phosphorescence is produced by rays having a positive charge; the lower portion (indicated by dots in the figure), which though faint is quite perceptible on the Willemite screen, is deflected as if *the rays carried a negative charge*. The length of the lower portion is somewhat shorter than that of the upper one, but is quite comparable with it. The intensity of the luminosity in the upper portion is at these pressures quite continuous; no abrupt variations such as would show themselves as bright patches could be detected, although, as will be seen later on, these make their appearance at lower pressures. Considering for the present the upper portion, the straightness of the edges shows that the velocity of the rays is approxi-

Rays of Positive Electricity.   569

mately constant, while the values of e/m range from zero at the undeflected portion to the value approximately equal to 10^4 at the top of the deflected band. This value of e/m is equal to that for a charged hydrogen atom, and moreover there was no specially great luminosity in the positions corresponding to $e/m = 10^4/14$ and $10^4/16$, the values for rays carried by nitrogen or oxygen atoms, though these places were carefully scrutinised. As hydrogen when present as an impurity in the tube has a tendency to accumulate near the cathode, the following experiment was tried to see whether the Kanalstrahlen were produced from traces of hydrogen in the tube. The discharge was sent through the tube in the opposite direction, *i.e.*, so that the perforated electrode was the anode, the electric and magnetic fields being kept on. When the discharge passed in this way there was of course no luminosity on the screen; on reversing the coil again so that the perforated electrode was the cathode, the luminosity flashed out instantly, presenting exactly the same appearance as it had done when the tube had been running for some time with the perforated electrode as cathode.

The fact that a spot of light produced by the undeflected positive rays is under the action of electric and magnetic forces drawn out into a continuous band was observed by W. Wien, who was the first to measure the deflexion of the positive rays under electric and magnetic forces. The values of e/m obtained from the deflexions of various parts of this band range continuously from zero, the value corresponding to the undeflected portion, to 10^4, the value corresponding to those most deflected. Wien explained this by the hypothesis that the charged particles which make up the positive rays act as nuclei round which molecules of the gas through which the rays pass condense, so that very complex systems made up of a very large number of molecules get mixed up with the particles forming the positive rays, and that it is these heavy and cumbrous systems which give rise to that part of the luminosity which is only slightly deflected. I think that the constancy of the velocity of the rays, indicated by the straight edges of the deflected band, is a strong argument against this explanation, and that the existence of the negative rays is conclusive against it. These negatively electrified rays, which form the faintly luminous portion of the phosphorescence indicated in fig. 6, are not cathode rays. The magnitude of their deflexion shows that the ratio of e/m for these rays, instead of being as great as $1 \cdot 7 \times 10^7$, the value for cathode rays, is less than 10^4. The particles forming these rays are thus comparable in size with those which form the positive rays. The existence of these negatively electrified rays suggests at once an explanation, which I think is the true one, of the continuous band into which the spot of phosphorescence is drawn out by the electric and magnetic fields. The values of e/m which are determined by this method are really the mean values of e/m, while the particle is in the electric and magnetic fields. If the particles are for a part of their course through these fields without charge, they will not during this part of their course be deflected, and in consequence the deflexions observed on the screen, and consequently the values of e/m, will be smaller than if the particle had retained its charge during the whole of its career. Thus, suppose that some of the particles constituting the positive rays, after starting with a positive charge, get this charge neutralized by attracting to them a negatively electrified corpuscle : the mass of the corpuscle is so small in comparison with that of the particle constituting the positive ray, that the addition of the particle will not appreciably diminish the velocity of the positive particle. Some of these neutralized particles may get positively ionized again by collision, while others may get a negative charge by the adhesion to them of another corpuscle, and this process might be repeated during the course of the particle. Thus there would be among the rays some which were for part of their course unelectrified, at other parts positively electrified, and at other parts negatively electrified. Thus the mean value of e/m might have all values ranging from α, its initial value, to $-\alpha'$, where α' might be only a little less than α. This is just what we observe, and when we remember that the gas through which the rays are passing is ionized, and contains a large number of corpuscles, it is, I think, what we should expect.

At very low pressures, when there are very few ions in the gas, this continuous band stretching from the origin is replaced by discontinuous patches.

Positive Rays in Hydrogen.

In hydrogen, when the pressure is not too low, the brightness of the phosphorescent patch is greater than in air at the same pressure ; the shape of the deflected phosphorescence is markedly different from that in air. In air, the deflected phosphorescence is usually a straight band, whereas in hydrogen the boundary of the most deflected side is distinctly curved and is concave to the undeflected position. The appearance of the deflected phosphorescence is indicated in fig. 7.

The result indicated in fig. 8, which was also obtained with hydrogen, shows that we have here a mixture of two bands, as indicated in fig. 4, the two bands being produced by carriers having different maximum values of e/m. The greatest value

Rays of Positive Electricity.   571

of e/m obtained with hydrogen was the same as in air, $1\cdot2\times10^4$, the velocity was $1\cdot8\times10^8$ cm./sec. The presence

Fig. 7.   Fig. 8.

of the second band indicates that mixed with these we have another set of carriers, for which the maximum value e/m is half that in the other band, *i. e.* 5×10^3. The curvature of the boundary generally observed is due to the admixture of these two rays.

Positive Rays in Helium.

In helium the phosphorescence is bright and the deflected patch has in general the curved outline observed in hydrogen. I was fortunate enough, however, to find a stage in which  the deflected patch was split up into two distinct bands, as shown in fig. 9. The maximum value of e/m in the band *a* was $1\cdot2\times10^4$, the same as in air and hydrogen, and the velocity was $1\cdot8\times10^8$; while the maximum value of e/m in band *b* was almost exactly one quarter of that in *a* (*i. e.* $2\cdot9\times10^3$). As the atomic weight of helium is four times that of hydrogen, this result indicates that the carriers which produce the band *b* are atoms of helium. This result is interesting because it is the only case (apart from hydrogen) in which I have found values of e/m corresponding to the atomic weight of the gas ; and even in the case of helium, when the pressure in the discharge-tube is very low and the electric field very intense, the characteristic rays with $e/m=2\cdot9\times10^3$ sometimes disappear and, as in all the gases I have tried, we get two sets of rays, for one set of which $e/m=10^4$ and for the other 5×10^3.

Although the helium had been carefully purified from hydrogen, the band *a* (for which $e/m=10^4$) was generally the brighter of the two. The case of helium is an interesting one; for the class of positive rays, known as the α rays, which are given off by radioactive substances, would *a priori* seem to consist most probably of helium, since helium is one of the products of disintegration of these substances. The value of e/m for these substances is 5×10^3, where we have seen that in helium it is possible to obtain rays for which $e/m=2\cdot9\times10^3$. It is true that, at very low pressures and with strong electric

Fig. 9.
a
b

fields, we get rays for which $e/m = 5 \times 10^3$; but this is not a peculiarity of helium : all the gases which I have tried show exactly the same effect.

Argon.

When the discharge passed through argon. the effects observed were very similar to those occurring in air. The sides were perhaps a little more curved, and there was a tendency for bright spots to develop. The measurements of the electric and magnetic deflexion of these spots gave $e/m = 10^4$, the value obtained for other cases. There was no appreciable increase of luminosity in the positions corresponding to $e/m = 10^4/40$, as there would have been if an appreciable number of the carriers had been argon atoms.

Positive Rays in Gases at very low pressures.

As the pressure of the gas in the discharge-tube is gradually reduced, the appearance of the deflected phosphorescence changes : instead of forming a continuous band, the phosphorescence breaks up into two isolated patches ; that part of the phosphorescence in which the deflexion was very small disappears, as also does the phosphorescence produced by the negatively electrified portion of the rays.

In the earlier experiments considerable difficulty was experienced in working at these very low pressures ; for when the pressure was reduced sufficiently to get the effects just described, the discharge passed through the tube with such difficulty, that in a very few seconds after this stage was reached sparks passed from the inside to the outside of the tube, perforating the glass and destroying the vacuum. In spite of all precautions, such as earthing the cathode and all conductors in its neighbourhood, perforation took place too quickly to permit measurements of the deflexion of the phosphorescence.

This difficulty was overcome by taking advantage of the fact that, when the cathode is made of a very electropositive metal, the discharge passes with much greater ease than when the cathode is made of aluminium or platinum. The electropositive metals used for the cathode were (1) the liquid alloy of sodium and potassium which was smeared over the cathode, and (2) calcium, a thin plate of which was affixed to the front of the cathode. With these cathodes the pressure in the tube could be reduced to very low values without making the discharge so difficult as to lead to perforation of the tube by sparking, and accurate measurements of the position of the patches of phosphorescence could be obtained at leisure.

The results obtained at these low pressures are very interesting. Whatever kind of gas may be used to fill the

Rays of Positive Electricity.   573

tube, or whatever the nature of the electrode, the deflected
phosphorescence splits up into two patches. For one of these
patches the maximum value of e/m is about 10^4, the value for
the hydrogen atom ; while the value for the other patch is

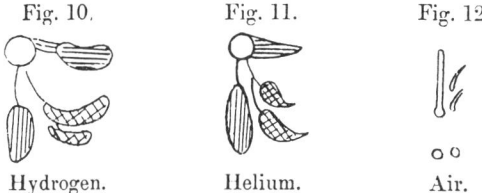

Fig. 10.   Fig. 11.   Fig. 12.

Hydrogen.   Helium.   Air.

about 5×10^3, the value for α particles or the hydrogen
molecule. Examples of the appearance of this phosphorescence are given in figs. 10, 11, 12 ; in fig. 12 the magnetic
force was reversed.

The differences in the appearance are due to differences in
the pressure rather than to differences in the gas ; for at
slightly higher pressures than that corresponding to fig. 12,
the appearance shown in figs. 10 and 11 can be obtained in
air. In all these cases the more deflected patch corresponds
to a value of about 10^4 for e/m, while e/m for the less deflected
patch is about 5×10^3.

It will be noticed that in fig. 11 there is no trace in the
helium tube of rays for which $e/m = 2 \cdot 5 \times 10^3$, which were
found in helium tubes at higher pressures ; at intermediate
pressures there are *three* distinct patches in helium, for
the first of which $e/m = 10^4$, for the second $e/m = 5 \times 10^3$,
and for the third $e/m = 2 \cdot 5 \times 10^3$ approximately. Helium
is a case where there are characteristic rays—*i. e.*, rays
for which $e/m = 10^4/M$, where M is the atomic weight of
the gas, when the discharge potential is comparatively
small, and not when, as at very low pressures, the discharge
potential is very large. I think it very probable that if we
could produce the positive rays with much smaller potential
differences than those used in these experiments, we might
get the characteristic rays for other gases. I am at present
investigating with this object the positive rays produced when
the perforated cathode is, as in Wehnelt's method, coated with
lime, when a potential difference of 100 volts or less is able to
produce positive rays. The interest of the experiments at very
low pressures lies in the fact that in this case the rays are the
same whatever gas may be used to fill the tube ; the characteristic rays of the gas disappear, and we get the same kind
of carriers for all substances.

I would especially call attention to the simplicity of the
effects produced at these low pressures : only two patches of

phosphorescence are visible.   This is, I think, an important
matter in connexion with the interpretation of these results ;
for at these low pressures we have to deal not only with the
gas with which the tube was originally filled, but also with
the gas which is given off by the electrodes and the walls of
the tube during the discharge : and it might be urged that at
these low pressures the tube contained nothing but hydrogen
given out by the electrodes.   I do not think this explanation
is feasible, for the following reasons :—

(1) The gas developed during the discharge is not wholly
hydrogen : if the discharge is kept passing long enough to
develop so much gas that the discharge through the gas
is sufficiently luminous to be observed by a spectroscope, the
spectrum always showed, in addition to the hydrogen lines,
the nitrogen bands ; indeed, the latter were generally the
most conspicuous part of the spectrum.   If the phosphorescent
screen on which the positive rays impinge is observed during
the time this gas is being given off, the changes which take
place in the appearance of the screen are as follows :—If, to
begin with, the pressure is so slow that the phosphorescent
patches are reduced to two bright spots, then, as the pressure
begins to go up owing to the evolution of the gas, the deflexion
of the spots increases.   This is owing to the reduction in the
velocity of the rays consequent upon the reduction of the
potential difference between the terminals of the tube, as
at this stage an increase in the pressure facilitates the passage
of the discharge.   In addition to the increase in the displacement, there is an increase in the area of the spots giving a
greater range of values of e/m ; this is owing to the increase
in the number of collisions made by the particles in the rays
on their way to the screen.   As more and more gas is
evolved, the patches get larger and finally overlap ; the
existence of the second patch being indicated by a diminution
in the brightness of the phosphorescence at places outside its
boundary.   As the pressure increases the luminosity gets
more and more continuous, and we finally get to the continuous band as shown in fig. 6.   At this stage it is
probable that there may be enough luminosity to give a
spectrum showing the nitrogen lines, indicating that a considerable part of the gas in the tube is air.   It is especially
to be noted that during this process, when gas was coming
into the tube, there has been no development of patches
in the phosphorescence indicating the presence of new rays ;
on the contrary, one type of carrier—that corresponding to
$e/m = 5 \times 10^3$—has disappeared.   The presence of the nitrogen
bands in the spectrum shows that nitrogen is carrying part
of the discharge, and yet there are no rays characteristic of

Rays of Positive Electricity.   575

nitrogen to be observed on the screen ; a proof, it seems to me, that different gases may be made by strong electric fields to give off the same kind of carriers of positive electricity.

Another result which shows that the positive rays are the same even although the gases are different is the following. The tube was pumped until the pressure was much too low for the discharge to pass, then small quantities of the following gases were put into the tube : air, carbonic oxide, hydrogen, helium, neon (for which I am indebted to the kindness of Sir James Dewar); the quantity admitted was adjusted so that it was sufficient to cause the discharge to pass and yet did not raise the pressure beyond the point where the phosphorescence is discontinuous. In every case there were patches corresponding to $e/m = 10^4$, $e/m = 5 \times 10^3$, and except with helium these were the only patches ; in helium, in addition to the two already mentioned, there was a third patch for which $e/m = 2 \cdot 5 \times 10^3$.

I also tried another method of ensuring that at these low pressures there were other gases besides hydrogen in the tube. I filled the tube with helium, and after exhausting to a fairly low pressure by means of the mercury pump, I performed the last stages of the exhaustion by means of charcoal cooled with liquid air. This charcoal absorbs very little helium in comparison with other gases ; so that it is certain that there was helium in the tube. The appearance of the phosphorescent screen of tubes exhausted in this way did not differ from those exhausted solely by the pump.

The most obvious explanation of these effects seems to me to be that under very intense electric fields different substances give out particles charged with positive electricity, and that these particles are independent of the nature of the gas from which they originate. These particles are, as far as we know at present, of two kinds : for one kind e/m has the value of 10^4, that of an atom of hydrogen; for the other kind e/m has half this value, *i. e.* it has the same value as for the α particles from radioactive substances.

This agreement in the maximum value of e/m at different pressures is a proof that this is a true maximum, and that there are not other more deflected rays not strong enough to produce visible phosphorescence ; for if this were the case— *i. e.*, if the value of e/m for a particle that had never lost its charge temporarily by collision were greater than 10^4—we should expect to get larger values for e/m at low pressures than at high.

I have much pleasure in thanking my assistant Mr. E. Everett for the assistance he has given me in these experiments.

[669]

LXXIX. *The Scattering of α and β Particles by Matter and the Structure of the Atom.* By Professor E. RUTHERFORD, F.R.S., University of Manchester* .

§ 1. IT is well known that the α and β particles suffer deflexions from their rectilinear paths by encounters with atoms of matter. This scattering is far more marked for the β than for the α particle on account of the much smaller momentum and energy of the former particle. There seems to be no doubt that such swiftly moving particles pass through the atoms in their path, and that the deflexions observed are due to the strong electric field traversed within the atomic system. It has generally been supposed that the scattering of a pencil of α or β rays in passing through a thin plate of matter is the result of a multitude of small scatterings by the atoms of matter traversed. The observations, however, of Geiger and Marsden † on the scattering of α rays indicate that some of the α particles must suffer a deflexion of more than a right angle at a single encounter. They found, for example, that a small fraction of the incident α particles, about 1 in 20,000, were turned through an average angle of 90° in passing through a layer of gold-foil about ·00004 cm. thick, which was equivalent in stopping-power of the α particle to 1·6 millimetres of air. Geiger ‡ showed later that the most probable angle of deflexion for a pencil of α particles traversing a gold-foil of this thickness was about 0°·87. A simple calculation based on the theory of probability shows that the chance of an α particle being deflected through 90° is vanishingly small. In addition, it will be seen later that the distribution of the α particles for various angles of large deflexion does not follow the probability law to be expected if such large deflexions are made up of a large number of small deviations. It seems reasonable to suppose that the deflexion through a large angle is due to a single atomic encounter, for the chance of a second encounter of a kind to produce a large deflexion must in most cases be exceedingly small. A simple calculation shows that the atom must be a seat of an intense electric field in order to produce such a large deflexion at a single encounter.

Recently Sir J. J. Thomson § has put forward a theory to

* Communicated by the Author. A brief account of this paper was communicated to the Manchester Literary and Philosophical Society in February, 1911.

† Proc. Roy. Soc. lxxxii. p. 495 (1909).

‡ Proc. Roy. Soc. lxxxiii. p. 492 (1910).

§ Camb. Lit. & Phil. Soc. xv. pt. 5 (1910).

670   Prof. E. Rutherford *on the*

explain the scattering of electrified particles in passing through small thicknesses of matter. The atom is supposed to consist of a number N of negatively charged corpuscles, accompanied by an equal quantity of positive electricity uniformly distributed throughout a sphere. The deflexion of a negatively electrified particle in passing through the atom is ascribed to two causes—(1) the repulsion of the corpuscles distributed through the atom, and (2) the attraction of the positive electricity in the atom. The deflexion of the particle in passing through the atom is supposed to be small, while the average deflexion after a large number m of encounters was taken as $\sqrt{m} \cdot \theta$, where θ is the average deflexion due to a single atom. It was shown that the number N of the electrons within the atom could be deduced from observations of the scattering of electrified particles. The accuracy of this theory of compound scattering was examined experimentally by Crowther * in a later paper. His results apparently confirmed the main conclusions of the theory, and he deduced, on the assumption that the positive electricity was continuous, that the number of electrons in an atom was about three times its atomic weight.

The theory of Sir J. J. Thomson is based on the assumption that the scattering due to a single atomic encounter is small, and the particular structure assumed for the atom does not admit of a very large deflexion of an α particle in traversing a single atom, unless it be supposed that the diameter of the sphere of positive electricity is minute compared with the diameter of the sphere of influence of the atom.

Since the α and β particles traverse the atom, it should be possible from a close study of the nature of the deflexion to form some idea of the constitution of the atom to produce the effects observed. In fact, the scattering of high-speed charged particles by the atoms of matter is one of the most promising methods of attack of this problem. The development of the scintillation method of counting single α particles affords unusual advantages of investigation, and the researches of H. Geiger by this method have already added much to our knowledge of the scattering of α rays by matter.

§ 2. We shall first examine theoretically the single encounters † with an atom of simple structure, which is able to

---

* Crowther, Proc. Roy. Soc. lxxxiv. p. 226 (1910).
† The deviation of a particle throughout a considerable angle from an encounter with a single atom will in this paper be called "single" scattering. The deviation of a particle resulting from a multitude of small deviations will be termed "compound" scattering.

produce large deflexions of an α particle, and then compare the deductions from the theory with the experimental data available.

Consider an atom which contains a charge ±Ne at its centre surrounded by a sphere of electrification containing a charge ∓Ne supposed uniformly distributed throughout a sphere of radius R. e is the fundamental unit of charge, which in this paper is taken as $4·65 \times 10^{-10}$ E.S. unit. We shall suppose that for distances less than 10^{-12} cm. the central charge and also the charge on the α particle may be supposed to be concentrated at a point. It will be shown that the main deductions from the theory are independent of whether the central charge is supposed to be positive or negative. For convenience, the sign will be assumed to be positive. The question of the stability of the atom proposed need not be considered at this stage, for this will obviously depend upon the minute structure of the atom, and on the motion of the constituent charged parts.

In order to form some idea of the forces required to deflect an α particle through a large angle, consider an atom containing a positive charge Ne at its centre, and surrounded by a distribution of negative electricity Ne uniformly distributed within a sphere of radius R. The electric force X and the potential V at a distance r from the centre of an atom for a point inside the atom, are given by

$$X = Ne\left(\frac{1}{r^2} - \frac{r}{R^3}\right)$$

$$V = Ne\left(\frac{1}{r} - \frac{3}{2R} + \frac{r^2}{2R^3}\right).$$

Suppose an α particle of mass m and velocity u and charge E shot directly towards the centre of the atom. It will be brought to rest at a distance b from the centre given by

$$\tfrac{1}{2}mu^2 = NeE\left(\frac{1}{b} - \frac{3}{2R} + \frac{b^2}{2R^3}\right).$$

It will be seen that b is an important quantity in later calculations. Assuming that the central charge is 100 e, it can be calculated that the value of b for an α particle of velocity $2·09 \times 10^9$ cms. per second is about $3·4 \times 10^{-12}$ cm. In this calculation b is supposed to be very small compared with R. Since R is supposed to be of the order of the radius of the atom, viz. 10^{-8} cm., it is obvious that the α particle before being turned back penetrates so close to

the central charge, that the field due to the uniform distribution of negative electricity may be neglected. In
general, a simple calculation shows that for all deflexions
greater than a degree, we may without sensible error suppose
the deflexion due to the field of the central charge alone.
Possible single deviations due to the negative electricity, if
distributed in the form of corpuscles, are not taken into
account at this stage of the theory. It will be shown later
that its effect is in general small compared with that due to
the central field.

Consider the passage of a positive electrified particle close
to the centre of an atom. Supposing that the velocity of
the particle is not appreciably changed by its passage through
the atom, the path of the particle under the influence of a
repulsive force varying inversely as the square of the distance
will be an hyperbola with the centre of the atom S as the
external focus. Suppose the particle to enter the atom in
the direction PO (fig. 1), and that the direction of motion

Fig. 1.

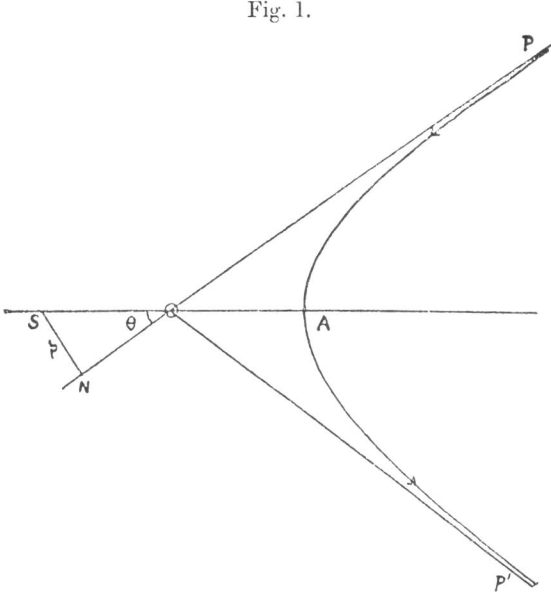

on escaping the atom is OP'. OP and OP' make equal angles
with the line SA, where A is the apse of the hyperbola.
$p=SN=$ perpendicular distance from centre on direction of
initial motion of particle.

Scattering of α and β Particles by Matter.   673

Let angle POA = θ.

Let V = velocity of particle on entering the atom, v its velocity at A, then from consideration of angular momentum

$$pV = SA \cdot v.$$

From conservation of energy

$$\tfrac{1}{2}mV^2 = \tfrac{1}{2}mv^2 - \frac{NeE}{SA},$$

$$v^2 = V^2\left(1 - \frac{b}{SA}\right).$$

Since the eccentricity is sec θ,

$$SA = SO + OA = p \operatorname{cosec}\theta(1 + \cos\theta)$$

$$= p \cot \theta/2,$$

$$p^2 = SA(SA - b) = p\cot\theta/2(p\cot\theta/2 - b),$$

$$\therefore \quad b = 2p\cot\theta.$$

The angle of deviation ϕ of the particle is $\pi - 2\theta$ and

$$\cot\phi/2 = \frac{2p}{b} \overset{*}{} \quad .. \quad . \quad . \quad . \quad . \quad . \quad (1)$$

This gives the angle of deviation of the particle in terms of b, and the perpendicular distance of the direction of projection from the centre of the atom.

For illustration, the angle of deviation ϕ for different values of p/b are shown in the following table :—

p/b	10	5	2	1	·5	·25	·125
ϕ	5°·7	11°·4	28°	53°	90°	127°	152°

§ 3. *Probability of single deflexion through any angle.*

Suppose a pencil of electrified particles to fall normally on a thin screen of matter of thickness t. With the exception of the few particles which are scattered through a large angle, the particles are supposed to pass nearly normally through the plate with only a small change of velocity. Let n = number of atoms in unit volume of material. Then the number of collisions of the particle with the atom of radius R is $\pi R^2 n t$ in the thickness t.

---

* A simple consideration shows that the deflexion is unaltered if the forces are attractive instead of repulsive.

674   Prof. E. Rutherford *on the*

The probabilty m of entering an atom within a distance p of its centre is given by

$$m = \pi p^2 nt.$$

Chance dm of striking within radii p and $p + dp$ is given by

$$dm = 2\pi p nt \cdot dp = \frac{\pi}{4} ntb^2 \cot \phi/2 \operatorname{cosec}^2 \phi/2 \, d\phi, \quad (2)$$

since $$\cot \phi/2 = 2p/b.$$

The value of dm gives the *fraction* of the total number of particles which are deviated between the angles ϕ and $\phi + d\phi$.

The fraction ρ of the total number of particles which are deflected through an angle greater than ϕ is given by

$$\rho = \frac{\pi}{4} ntb^2 \cot^2 \phi/2. \quad \cdot \quad \cdot \quad \cdot \quad \cdot \quad \cdot \quad (3)$$

The fraction ρ which is deflected between the angles ϕ_1 and ϕ_2 is given by

$$\rho = \frac{\pi}{4} ntb^2 \left(\cot^2 \frac{\phi_1}{2} - \cot^2 \frac{\phi_2}{2} \right). \quad \cdot \quad \cdot \quad \cdot \quad (4)$$

It is convenient to express the equation (2) in another form for comparison with experiment. In the case of the α rays, the number of scintillations appearing on a *constant* area of a zinc sulphide screen are counted for different angles with the direction of incidence of the particles. Let r = distance from point of incidence of α rays on scattering material, then if Q be the total number of particles falling on the scattering material, the number y of α particles falling on unit area which are deflected through an angle ϕ is given by

$$y = \frac{Q dm}{2\pi r^2 \sin \phi \cdot d\phi} = \frac{ntb^2 \cdot Q \cdot \operatorname{cosec}^4 \phi/2}{16 r^2} \cdot \quad \cdot \quad \cdot \quad (5)$$

Since $b = \dfrac{2NeE}{mu^2}$, we see from this equation that the number of α particles (scintillations) per unit area of zinc sulphide screen at a given distance r from the point of

Scattering of α and β Particles by Matter.   675

incidence of the rays is proportional to

(1) cosec⁴ $\phi/2$ or $1/\phi^4$ if ϕ be small ;
(2) thickness of scattering material t provided this is small ;
(3) magnitude of central charge Ne ;
(4) and is inversely proportional to $(mu^2)^2$, or to the fourth power of the velocity if m be constant.

In these calculations, it is assumed that the α particles scattered through a large angle suffer only one large deflexion. For this to hold, it is essential that the thickness of the scattering material should be so small that the chance of a second encounter involving another large deflexion is very small. If, for example, the probability of a single deflexion ϕ in passing through a thickness t is $1/1000$, the probability of two successive deflexions each of value ϕ is $1/10^6$, and is negligibly small.

The angular distribution of the α particles scattered from a thin metal sheet affords one of the simplest methods of testing the general correctness of this theory of single scattering. This has been done recently for α rays by Dr. Geiger *, who found that the distribution for particles deflected between 30° and 150° from a thin gold-foil was in substantial agreement with the theory. A more detailed account of these and other experiments to test the validity of the theory will be published later.

§ 4. *Alteration of velocity in an atomic encounter.*

It has so far been assumed that an α or β particle does not suffer an appreciable change of velocity as the result of a single atomic encounter resulting in a large deflexion of the particle. The effect of such an encounter in altering the velocity of the particle can be calculated on certain assumptions. It is supposed that only two systems are involved, viz., the swiftly moving particle and the atom which it traverses supposed initially at rest. It is supposed that the principle of conservation of momentum and of energy applies, and that there is no appreciable loss of energy or momentum by radiation.

* Manch. Lit. & Phil. Soc. 1910.

676   Prof. E. Rutherford *on the*

Let m be mass of the particle,

v_1 = velocity of approach,

v_2 = velocity of recession,

M = mass of atom,

V = velocity communicated to atom as result of encounter.

Let OA (fig. 2) represent in magnitude and direction the momentum mv_1 of the entering particle, and OB the momentum of the receding particle which has been turned through an angle AOB = ϕ. Then BA represents in magnitude and direction the momentum MV of the recoiling atom.

Fig. 2.

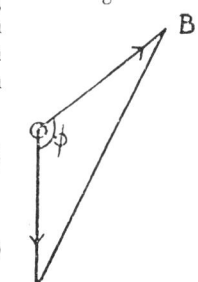

$$(MV)^2 = (mv_1)^2 + (mv_2)^2 - 2m^2 v_1 v_2 \cos \phi. \quad (1)$$

By the conservation of energy

$$MV^2 = mv_1^2 - mv_2^2. \quad . \quad . \quad (2)$$

Suppose $M/m = K$ and $v_2 = \rho v_1$, where ρ is < 1.

From (1) and (2),

$$(K+1)\rho^2 - 2\rho \cos \phi = K - 1,$$

or $\quad \rho = \dfrac{\cos \phi}{K+1} + \dfrac{1}{K+1} \sqrt{K^2 - \sin^2 \phi}.$

Consider the case of an α particle of atomic weight 4, deflected through an angle of 90° by an encounter with an atom of gold of atomic weight 197.

Since K = 49 nearly,

$$\rho = \sqrt{\frac{K-1}{K+1}} = \cdot 979,$$

or the velocity of the particle is reduced only about 2 per cent. by the encounter.

In the case of aluminium K = 27/4 and for $\phi = 90°$ $\rho = \cdot 86$.

It is seen that the reduction of velocity of the α particle becomes marked on this theory for encounters with the lighter atoms. Since the range of an α particle in air or other matter is approximately proportional to the cube of the velocity, it follows that an α particle of range 7 cms. has its range reduced to 4·5 cms. after incurring a single

deviation of 90° in traversing an aluminium atom. This is
of a magnitude to be easily detected experimentally. Since
the value of K is very large for an encounter of a β particle
with an atom, the reduction of velocity on this formula is
very small.

Some very interesting cases of the theory arise in considering the changes of velocity and the distribution of
scattered particles when the α particle encounters a light
atom, for example a hydrogen or helium atom. A discussion
of these and similar cases is reserved until the question has
been examined experimentally.

§ 5. *Comparison of single and compound scattering.*

Before comparing the results of theory with experiment, it
is desirable to consider the relative importance of single and
compound scattering in determining the distribution of the
scattered particles. Since the atom is supposed to consist of
a central charge surrounded by a uniform distribution of the
opposite sign through a sphere of radius R, the chance of
encounters with the atom involving small deflexions is very
great compared with the chance of a single large deflexion.

This question of compound scattering has been examined
by Sir J. J. Thomson in the paper previously discussed (§ 1).
In the notation of this paper, the average deflexion ϕ_1 due to
the field of the sphere of positive electricity of radius R and
quantity Ne was found by him to be

$$\phi_1 = \frac{\pi}{4} \cdot \frac{NeE}{mu^2} \cdot \frac{1}{R}.$$

The average deflexion ϕ_2 due to the N negative corpuscles
supposed distributed uniformly throughout the sphere was
found to be

$$\phi_2 = \frac{16}{5} \frac{eE}{mu^2} \cdot \frac{1}{R} \sqrt{\frac{3N}{2}}.$$

The mean deflexion due to both positive and negative electricity
was taken as

$$(\phi_1^2 + \phi_2^2)^{1/2}.$$

In a similar way, it is not difficult to calculate the average
deflexion due to the atom with a central charge discussed in
this paper.

Since the radial electric field X at any distance r from the centre is given by

$$X = Ne\left(\frac{1}{r^2} - \frac{r}{R^3}\right),$$

it is not difficult to show that the deflexion (supposed small) of an electrified particle due to this field is given by

$$\theta = \frac{b}{p}\left(1 - \frac{p^2}{R^2}\right)^{3/2},$$

where p is the perpendicular from the centre on the path of the particle and b has the same value as before. It is seen that the value of θ increases with diminution of p and becomes great for small values of ϕ.

Since we have already seen that the deflexions become very large for a particle passing near the centre of the atom, it is obviously not correct to find the average value by assuming θ is small.

Taking R of the order 10^{-8} cm., the value of p for a large deflexion is for α and β particles of the order 10^{-11} cm. Since the chance of an encounter involving a large deflexion is small compared with the chance of small deflexions, a simple consideration shows that the average small deflexion is practically unaltered if the large deflexions are omitted. This is equivalent to integrating over that part of the cross section of the atom where the deflexions are small and neglecting the small central area. It can in this way be simply shown that the average small deflexion is given by

$$\phi_1 = \frac{3\pi}{8}\frac{b}{R}.$$

This value of ϕ_1 for the atom with a concentrated central charge is three times the magnitude of the average deflexion for the same value of Ne in the type of atom examined by Sir J. J. Thomson. Combining the deflexions due to the electric field and to the corpuscles, the average deflexion is

$$(\phi_1^2 + \phi_2^2)^2 \quad \text{or} \quad \frac{b}{2R}\left(5{\cdot}54 + \frac{15{\cdot}4}{N}\right)^{1/2}.$$

It will be seen later that the value of N is nearly proportional to the atomic weight, and is about 100 for gold. The effect due to scattering of the individual corpuscles expressed by the second term of the equation is consequently small for heavy atoms compared with that due to the distributed electric field.

Scattering of α and β Particles by Matter.   679

Neglecting the second term, the average deflexion per atom is $\dfrac{3\pi b}{8R}$. We are now in a position to consider the relative effects on the distribution of particles due to single and to compound scattering. Following J. J. Thomson's argument, the average deflexion θ_t after passing through a thickness t of matter is proportional to the square root of the number of encounters and is given by

$$\theta_t = \frac{3\pi b}{8R}\,\sqrt{\pi R^2 . n . t} = \frac{3\pi b}{8}\,\sqrt{\pi n t},$$

where n as before is equal to the number of atoms per unit volume.

The probability p_1 for compound scattering that the deflexion of the particle is greater than ϕ is equal to $e^{-\phi^2/\theta_t^2}$.

Consequently $\phi^2 = -\dfrac{9\pi^3}{64}\,b^2\,nt\log p_1.$

Next suppose that single scattering alone is operative. We have seen (§ 3) that the probability p_2 of a deflexion greater than ϕ is given by

$$p_2 = \frac{\pi}{1}\,b^2 . n . t\cot^2\phi/2.$$

By comparing these two equations

$$p_2\log p_1 = -\cdot 181\phi^2\cot^2\phi/2,$$

ϕ is sufficiently small that

$$\tan\phi/2 = \phi/2,$$

$$p_2\log p_1 = -\cdot 72.$$

If we suppose $p_2 = \cdot 5$, then $p_1 = \cdot 24.$

If $p_2 = \cdot 1$, $p_1 = \cdot 0004.$

It is evident from this comparison, that the probability for any given deflexion is always greater for single than for compound scattering. The difference is especially marked when only a small fraction of the particles are scattered through any given angle. It follows from this result that the distribution of particles due to encounters with the atoms is for small thicknesses mainly governed by single scattering. No doubt compound scattering produces some effect in equalizing the distribution of the scattered particles ; but its effect becomes relatively smaller, the smaller the fraction of the particles scattered through a given angle.

680   Prof. E. Rutherford *on the*

§ 6. *Comparison of Theory with Experiments.*

On the present theory, the value of the central charge Ne is an important constant, and it is desirable to determine its value for different atoms. This can be most simply done by determining the small fraction of α or β particles of known velocity falling on a thin metal screen, which are scattered between ϕ and $\phi + d\phi$ where ϕ is the angle of deflexion. The influence of compound scattering should be small when this fraction is small.

Experiments in these directions are in progress, but it is desirable at this stage to discuss in the light of the present theory the data already published on scattering of α and β particles.

The following points will be discussed :—

> (*a*) The "diffuse reflexion" of α particles, *i. e.* the scattering of α particles through large angles (Geiger and Marsden).
>
> (*b*) The variation of diffuse reflexion with atomic weight of the radiator (Geiger and Marsden).
>
> (*c*) The average scattering of a pencil of α rays transmitted through a thin metal plate (Geiger).
>
> (*d*) The experiments of Crowther on the scattering of β rays of different velocities by various metals.

(*a*) In the paper of Geiger and Marsden (*loc. cit.*) on the diffuse reflexion of α particles falling on various substances it was shown that about 1/8000 of the α particles from radium C falling on a thick plate of platinum are scattered back in the direction of the incidence. This fraction is deduced on the assumption that the α particles are uniformly scattered in all directions, the observations being made for a deflexion of about 90°. The form of experiment is not very suited for accurate calculation, but from the data available it can be shown that the scattering observed is about that to be expected on the theory if the atom of platinum has a central charge of about 100 e.

(*b*) In their experiments on this subject, Geiger and Marsden gave the relative number of α particles diffusely reflected from thick layers of different metals, under similar conditions. The numbers obtained by them are given in the table below, where z represents the relative number of scattered particles, measured by the number of scintillations per minute on a zinc sulphide screen.

Metal.	Atomic weight.	z.	$z/A^{3/2}$.
Lead	207	62	208
Gold	197	67	242
Platinum	195	63	232
Tin	119	34	226
Silver	108	27	241
Copper	64	14·5	225
Iron	56	10·2	250
Aluminium ...	27	3·4	243
			Average 233

On the theory of single scattering, the fraction of the total number of α particles scattered through any given angle in passing through a thickness t is proportional to $n.A^2t$, assuming that the central charge is proportional to the atomic weight A. In the present case, the thickness of matter from which the scattered α particles are able to emerge and affect the zinc sulphide screen depends on the metal. Since Bragg has shown that the stopping power of an atom for an α particle is proportional to the square root of its atomic weight, the value of nt for different elements is proportional to $1/\sqrt{A}$. In this case t represents the greatest depth from which the scattered α particles emerge. The number z of α particles scattered back from a thick layer is consequently proportional to $A^{3/2}$ or $z/A^{3/2}$ should be a constant.

To compare this deduction with experiment, the relative values of the latter quotient are given in the last column. Considering the difficulty of the experiments, the agreement between theory and experiment is reasonably good [*].

The single large scattering of α particles will obviously affect to some extent the shape of the Bragg ionization curve for a pencil of α rays. This effect of large scattering should be marked when the α rays have traversed screens of metals of high atomic weight, but should be small for atoms of light atomic weight.

(c) Geiger made a careful determination of the scattering of α particles passing through thin metal foils, by the scintillation method, and deduced the most probable angle

[*] The effect of change of velocity in an atomic encounter is neglected in this calculation.

through which the α particles are deflected in passing through known thicknesses of different kinds of matter.

A narrow pencil of homogeneous α rays was used as a source. After passing through the scattering foil, the total number of α particles deflected through different angles was directly measured. The angle for which the number of scattered particles was a maximum was taken as the most probable angle. The variation of the most probable angle with thickness of matter was determined, but calculation from these data is somewhat complicated by the variation of velocity of the α particles in their passage through the scattering material. A consideration of the curve of distribution of the α particles given in the paper (*loc. cit.* p. 496) shows that the angle through which half the particles are scattered is about 20 per cent greater than the most probable angle.

We have already seen that compound scattering may become important when about half the particles are scattered through a given angle, and it is difficult to disentangle in such cases the relative effects due to the two kinds of scattering. An approximate estimate can be made in the following way :— From (§ 5) the relation between the probabilities p_1 and p_2 for compound and single scattering respectively is given by

$$p_2 \log p_1 = -\cdot721.$$

The probability q of the combined effects may as a first approximation be taken as

$$q = (p_1{}^2 + p_2{}^2)^{1/2}.$$

If $q = \cdot5$, it follows that

$$p_1 = \cdot2 \quad \text{and} \quad p_2 = \cdot46.$$

We have seen that the probability p_2 of a single deflexion greater than ϕ is given by

$$p_2 = \frac{\pi}{4} n \cdot t \cdot b^2 \cot^2 \phi/2.$$

Since in the experiments considered ϕ is comparatively small

$$\phi \frac{\sqrt{p_2}}{\sqrt{\pi n t}} = b = \frac{2 \mathrm{N} e \mathrm{E}}{m u^2}.$$

Geiger found that the most probable angle of scattering of the α rays in passing through a thickness of gold equivalent in stopping power to about ·76 cm. of air was 1° 40′. The angle ϕ through which half the α particles are turned thus corresponds to 2° nearly.

$$t = \cdot00017 \text{ cm.} \; ; \; n = 6\cdot07 \times 10^{22} \; ;$$

$$u \text{ (average value)} = 1\cdot8 \times 10^9.$$

$$\mathrm{E}/m = 1\cdot5 \times 10^{14} \text{, E.s. units} \; ; \; e = 4\cdot65 \times 10^{-10}.$$

Taking the probability of single scattering $=·46$ and substituting the above values in the formula, the value of N for gold comes out to be 97.

For a thickness of gold equivalent in stopping power to 2·12 cms. of air, Geiger found the most probable angle to be $3° 40'$. In this case $t=·00047$, $\phi=4°·4$, and average $u= 1·7 \times 10^9$, and N comes out to be 114.

Geiger showed that the most probable angle of deflexion for an atom was nearly proportional to its atomic weight. It consequently follows that the value of N for different atoms should be nearly proportional to their atomic weights, at any rate for atomic weights between gold and aluminium.

Since the atomic weight of platinum is nearly equal to that of gold, it follows from these considerations that the magnitude of the diffuse reflexion of α particles through more than 90° from gold and the magnitude of the average small angle scattering of a pencil of rays in passing through gold-foil are both explained on the hypothesis of single scattering by supposing the atom of gold has a central charge of about $100\,e$.

(d) Experiments of Crowther on scattering of β rays.— We shall now consider how far the experimental results of Crowther on scattering of β particles of different velocities by various materials can be explained on the general theory of single scattering. On this theory, the fraction of β particles p turned through an angle greater than ϕ is given by

$$p=\frac{\pi}{4}n.t.b^2 \cot^2 \phi/2.$$

In most of Crowther's experiments ϕ is sufficiently small that $\tan \phi/2$ may be put equal to $\phi/2$ without much error. Consequently

$$\phi^2=2\pi n.t.b^2 \quad \text{if } p=1/2.$$

On the theory of compound scattering, we have already seen that the chance p_1 that the deflexion of the particles is greater than ϕ is given by

$$\phi^2/\log p_1 = -\frac{9\pi^3}{64}n.t.b^2.$$

Since in the experiments of Crowther the thickness t of matter was determined for which $p_1=1/2$,

$$\phi^2=·96\pi n t b^2.$$

For a probability of 1/2, the theories of single and compound

Prof. E. Rutherford *on the*

scattering are thus identical in general form, but differ by a numerical constant. It is thus clear that the main relations on the theory of compound scattering of Sir J. J. Thomson, which were verified experimentally by Crowther, hold equally well on the theory of single scattering.

For example, if t_m be the thickness for which half the particles are scattered through an angle ϕ, Crowther showed that $\phi/\sqrt{t_m}$ and also $\dfrac{mu^2}{E} \cdot \sqrt{t_m}$ were constants for a given material when ϕ was fixed. These relations hold also on the theory of single scattering. Notwithstanding this apparent similarity in form, the two theories are fundamentally different. In one case, the effects observed are due to cumulative effects of small deflexions, while in the other the large deflexions are supposed to result from a single encounter. The distribution of scattered particles is entirely different on the two theories when the probability of deflexion greater than ϕ is small.

We have already seen that the distribution of scattered α particles at various angles has been found by Geiger to be in substantial agreement with the theory of single scattering, but cannot be explained on the theory of compound scattering alone. Since there is every reason to believe that the laws of scattering of α and β particles are very similar, the law of distribution of scattered β particles should be the same as for α particles for small thicknesses of matter. Since the value of mu^2/E for the β particles is in most cases much smaller than the corresponding value for the α particles, the chance of large single deflexions for β particles in passing through a given thickness of matter is much greater than for α particles. Since on the theory of single scattering the fraction of the number of particles which are deflected through a given angle is proportional to kt, where t is the thickness supposed small and k a constant, the number of particles which are undeflected through this angle is proportional to $1-kt$. From considerations based on the theory of compound scattering, Sir J. J. Thomson deduced that the probability of deflexion less than ϕ is proportional to $1-e^{-\mu/t}$ where μ is a constant for any given value of ϕ.

The correctness of this latter formula was tested by Crowther by measuring electrically the fraction I/I_0 of the scattered β particles which passed through a circular opening subtending an angle of 36° with the scattering material. If

$$I/I_0 = 1 - e^{-\mu/t},$$

the value of I should decrease very slowly at first with

Scattering of α and β Particles by Matter.   685

increase of t. Crowther, using aluminium as scattering
material, states that the variation of I/I_0 was in good accord
with this theory for small values of t. On the other hand,
if single scattering be present, as it undoubtedly is for α rays,
the curve showing the relation between I/I_0 and t should be
nearly linear in the initial stages. The experiments of
Madsen [*] on scattering of β rays, although not made with
quite so small a thickness of aluminium as that used by
Crowther, certainly support such a conclusion. Considering
the importance of the point at issue, further experiments on
this question are desirable.

From the table given by Crowther of the value $\phi/\sqrt{t_m}$ for
different elements for β rays of velocity $2\cdot68\times10^{10}$ cms.
per second, the values of the central charge Ne can be
calculated on the theory of single scattering. It is supposed,
as in the case of the α rays, that for the given value of
$\phi/\sqrt{t_m}$ the fraction of the β particles deflected by single
scattering through an angle greater than ϕ is $\cdot46$ instead
of $\cdot5$.

The values of N calculated from Crowther's data are
given below.

Element.	Atomic weight.	$\phi/\sqrt{t_m}$	N.
Aluminium	27	4·25	22
Copper	63·2	10·0	42
Silver	103	15·4	78
Platinum	194	29·0	138

It will be remembered that the values of N for gold
deduced from scattering of the α rays were in two calculations 97 and 114. These numbers are somewhat smaller
than the values given above for platinum (viz. 138), whose
atomic weight is not very different from gold. Taking into
account the uncertainties involved in the calculation from
the experimental data, the agreement is sufficiently close to
indicate that the same general laws of scattering hold for the
α and β particles, notwithstanding the wide differences in
the relative velocity and mass of these particles.

As in the case of the α rays, the value of N should be
most simply determined for any given element by measuring

* Phil. Mag. xviii. p. 909 (1909).

686   Prof. E. Rutherford *on the*

the small fraction of the incident β particles scattered through a large angle. In this way, possible errors due to small scattering will be avoided.

The scattering data for the β rays, as well as for the α rays, indicate that the central charge in an atom is approximately proportional to its atomic weight. This falls in with the experimental deductions of Schmidt *. In his theory of absorption of β rays, he supposed that in traversing a thin sheet of matter, a small fraction α of the particles are stopped, and a small fraction β are reflected or scattered back in the direction of incidence. From comparison of the absorption curves of different elements, he deduced that the value of the constant β for different elements is proportional to $n\mathrm{A}^2$ where n is the number of atoms per unit volume and A the atomic weight of the element. This is exactly the relation to be expected on the theory of single scattering if the central charge on an atom is proportional to its atomic weight.

§ 7. *General Considerations.*

In comparing the theory outlined in this paper with the experimental results, it has been supposed that the atom consists of a central charge supposed concentrated at a point, and that the large single deflexions of the α and β particles are mainly due to their passage through the strong central field. The effect of the equal and opposite compensating charge supposed distributed uniformly throughout a sphere has been neglected. Some of the evidence in support of these assumptions will now be briefly considered. For concreteness, consider the passage of a high speed α particle through an atom having a positive central charge $N\!e$, and surrounded by a compensating charge of N electrons. Remembering that the mass, momentum, and kinetic energy of the α particle are very large compared with the corresponding values for an electron in rapid motion, it does not seem possible from dynamic considerations that an α particle can be deflected through a large angle by a close approach to an electron, even if the latter be in rapid motion and constrained by strong electrical forces. It seems reasonable to suppose that the chance of single deflexions through a large angle due to this cause, if not zero, must be exceedingly small compared with that due to the central charge.

It is of interest to examine how far the experimental evidence throws light on the question of the extent of the

* *Annal. d. Phys.* iv. 23. p. 671 (1907).

distribution of the central charge. Suppose, for example, the central charge to be composed of N unit charges distributed over such a volume that the large single deflexions are mainly due to the constituent charges and not to the external field produced by the distribution. It has been shown (§ 3) that the fraction of the α particles scattered through a large angle is proportional to $(NeE)^2$, where Ne is the central charge concentrated at a point and E the charge on the deflected particle. If, however, this charge is distributed in single units, the fraction of the α particles scattered through a given angle is proportional to Ne^2 instead of N^2e^2. In this calculation, the influence of mass of the constituent particle has been neglected, and account has only been taken of its electric field. Since it has been shown that the value of the central point charge for gold must be about 100, the value of the distributed charge required to produce the same proportion of single deflexions through a large angle should be at least 10,000. Under these conditions the mass of the constituent particle would be small compared with that of the α particle, and the difficulty arises of the production of large single deflexions at all. In addition, with such a large distributed charge, the effect of compound scattering is relatively more important than that of single scattering. For example, the probable small angle of deflexion of a pencil of α particles passing through a thin gold foil would be much greater than that experimentally observed by Geiger (§ *b–c*). The large and small angle scattering could not then be explained by the assumption of a central charge of the same value. Considering the evidence as a whole, it seems simplest to suppose that the atom contains a central charge distributed through a very small volume, and that the large single deflexions are due to the central charge as a whole, and not to its constituents. At the same time, the experimental evidence is not precise enough to negative the possibility that a small fraction of the positive charge may be carried by satellites extending some distance from the centre. Evidence on this point could be obtained by examining whether the same central charge is required to explain the large single deflexions of α and β particles; for the α particle must approach much closer to the centre of the atom than the β particle of average speed to suffer the same large deflexion.

The general data available indicate that the value of this central charge for different atoms is approximately proportional to their atomic weights, at any rate for atoms heavier than aluminium. It will be of great interest to examine

688 *Scattering of α and β Particles by Matter.*

experimentally whether such a simple relation holds also for the lighter atoms. In cases where the mass of the deflecting atom (for example, hydrogen, helium, lithium) is not very different from that of the α particle, the general theory of single scattering will require modification, for it is necessary to take into account the movements of the atom itself (see § 4).

It is of interest to note that Nagaoka * has mathematically considered the properties of a "Saturnian" atom which he supposed to consist of a central attracting mass surrounded by rings of rotating electrons. He showed that such a system was stable if the attractive force was large. From the point of view considered in this paper, the chance of large deflexion would practically be unaltered, whether the atom is considered to be a disk or a sphere. It may be remarked that the approximate value found for the central charge of the atom of gold $(100\,e)$ is about that to be expected if the atom of gold consisted of 49 atoms of helium, each carrying a charge $2\,e$. This may be only a coincidence, but it is certainly suggestive in view of the expulsion of helium atoms carrying two unit charges from radioactive matter.

The deductions from the theory so far considered are independent of the sign of the central charge, and it has not so far been found possible to obtain definite evidence to determine whether it be positive or negative. It may be possible to settle the question of sign by consideration of the difference of the laws of absorption of the β particle to be expected on the two hypotheses, for the effect of radiation in reducing the velocity of the β particle should be far more marked with a positive than with a negative centre. If the central charge be positive, it is easily seen that a positively charged mass if released from the centre of a heavy atom, would acquire a great velocity in moving through the electric field. It may be possible in this way to account for the high velocity of expulsion of α particles without supposing that they are initially in rapid motion within the atom.

Further consideration of the application of this theory to these and other questions will be reserved for a later paper, when the main deductions of the theory have been tested experimentally. Experiments in this direction are already in progress by Geiger and Marsden.

University of Manchester,
    April 1911.

* Nagaoka, Phil. Mag. vii. p. 445 (1904).

604   Dr. H. Geiger *and* Mr. E. Marsden *on the Laws of*

But whatever value is attached to the numerical evidence by those competent to judge, the broad conclusion from the *direction* of the changes is unaffected.

I should like also to reply to one criticism which has been made. It is not necessary to the hypothesis that the atomic free frequency should be absolutely invariable throughout all chemical changes. If it were the case that a slight change (say 1 per cent.) in the atomic free frequency would account for all the observed changes of refractivity and dispersion, the criticism would have force. But any one who examines the figures in the table for (*e. g.*) hydrogen, nitrogen, and ammonia, will see that no slight change in a frequency can possibly account for the observed changes. For hydrogen $n_0^2 \times 10^{-27} = 12409$, for nitrogen 17095. For ammonia the number which expresses its average value is 8135, an enormous drop. At the same time the refractivity has gone up only $3\frac{1}{2}$ per cent. But if the number of vibrators had remained constant while the average free frequency decreased, the increase of the refractivity must have been much greater than this. Hence, to account for the observed changes, one must also assume that the number of vibrators (? electrons) has fallen off in about the same proportion as the frequency.

These two hypotheses seem much more improbable than that which is here put forward.

LXI. *The Laws of Deflexion of* α *Particles through Large Angles**. *By* Dr. H. GEIGER *and* E. MARSDEN †.

IN a former paper ‡ one of us has shown that in the passage of α particles through matter the deflexions are, on the average, small and of the order of a few degrees only. In the experiments a narrow pencil of α particles fell on a zinc-sulphide screen in vacuum, and the distribution of the scintillations on the screen was observed when different metal foils were placed in the path of the α particles. From the distribution obtained, the most probable angle of scattering could be deduced, and it was shown that the results could be explained on the assumption that the deflexion of a single α particle is the resultant of a large number of very small deflexions caused by the passage of the α particle through the successive individual atoms of the scattering substance.

* Communicated to *k. d.-k. Akad. d. Wiss. Wien.*
† Communicated by Prof. E. Rutherford, F.R.S.
‡ H. Geiger, Roy. Soc. Proc. vol. lxxxiii. p. 492 (1910); vol. lxxxvi. p. 235 (1912).

Deflexion of α Particles through Large Angles. 605

In an earlier paper *, however, we pointed out that α particles are sometimes turned through very large angles. This was made evident by the fact that when α particles fall on a metal plate, a small fraction of them, about 1/8000 in the case of platinum, appears to be diffusely reflected. This amount of reflexion, although small, is, however, too large to be explained on the above simple theory of scattering. It is easy to calculate from the experimental data that the probability of a deflexion through an angle of 90° is vanishingly small, and of a different order to the value found experimentally.

Professor Rutherford † has recently developed a theory to account for the scattering of α particles through these large angles, the assumption being that the deflexions are the result of an intimate encounter of an α particle with a single atom of the matter traversed. In this theory an atom is supposed to consist of a strong positive or negative central charge concentrated within a sphere of less than about 3×10^{-12} cm. radius, and surrounded by electricity of the opposite sign distributed throughout the remainder of the atom of about 10^{-8} cm. radius. In considering the deflexion of an α particle directed against such an atom, the main deflexion-effect can be supposed to be due to the central concentrated charge which will cause the α particle to describe an hyperbola with the centre of the atom as one focus.

The angle between the directions of the α particle before and after deflexion will depend on the perpendicular distance of the initial trajectory from the centre of the atom. The fraction of the α particles whose paths are sufficiently near to the centre of the atom will, however, be small, so that the probability of an α particle suffering a large deflexion of this nature will be correspondingly small. Thus, assuming a narrow pencil of α particles directed against a thin sheet of matter containing atoms distributed at random throughout its volume, if the scattered particles are counted by the scintillations they produce on a zinc-sulphide screen distance r from the point of incidence of the pencil in a direction making an angle ϕ with it, the number of α particles falling on unit area of the screen per second is deduced to be equal to

$$\frac{Q n t b^2 \operatorname{cosec}^4 \phi/2}{16 r^2},$$

where Q is the number of α particles per second in the

* H. Geiger and E. Marsden, Roy. Soc. Proc. vol. lxxxii. p. 495 (1909).

† E. Rutherford, Phil. Mag. vol. xxi. p. 669 (1911).

original pencil, n the number of atoms in unit volume of the material, and t the thickness of the foil. The quantity

$$b = \frac{2NeE}{mu^2},$$

where Ne is the central charge of the atom, and m, E, and u are the respective mass, charge, and velocity of the α particle.

The number of deflected α particles is thus proportional to (1) $\operatorname{cosec}^4 \phi/2$, (2) thickness of scattering material t if the thickness is small, (3) the square of the central charge Ne of the atoms of the particular matter employed to scatter the particles, (4) the inverse fourth power of the velocity u of the incident α particles.

At the suggestion of Prof. Rutherford, we have carried out experiments to test the main conclusions of the above theory. The following points were investigated :—

(1) Variation with angle.
(2) Variation with thickness of scattering material.
(3) Variation with atomic weight of scattering material.
(4) Variation with velocity of incident α particles.
(5) The fraction of particles scattered through a definite angle.

The main difficulty of the experiments has arisen from the necessity of using a very intense and narrow source of α particles owing to the smallness of the scattering effect. All the measurements have been carried out by observing the scintillations due to the scattered α particles on a zinc-sulphide screen, and during the course of the experiments over 100,000 scintillations have been counted. It may be mentioned in anticipation that all the results of our investigation are in good agreement with the theoretical deductions of Prof. Rutherford, and afford strong evidence of the correctness of the underlying assumption that an atom contains a strong charge at the centre of dimensions, small compared with the diameter of the atom.

(1) *Variation of Scattering with Angle.*

We have already pointed out that to obtain measurable effects an intense pencil of α particles is required. It is further necessary that the path of the α particles should be in an evacuated chamber to avoid complications due to the absorption and scattering of the air. The apparatus used is shown in fig. 1, and mainly consisted of a strong cylindrical metal box B, which contained the source of α particles R.

Deflexion of α Particles through Large Angles.   607

the scattering foil F, and a microscope M to which the zinc-sulphide screen S was rigidly attached. The box was fastened down to a graduated circular platform A, which could be rotated by means of a conical airtight joint C. By rotating the platform the box and microscope moved with it, whilst the scattering foil and radiating source remained in position, being attached to the tube T, which was fastened to the standard L. The box B was closed by the ground-glass plate P, and could be exhausted through the tube T.

Fig. 1.

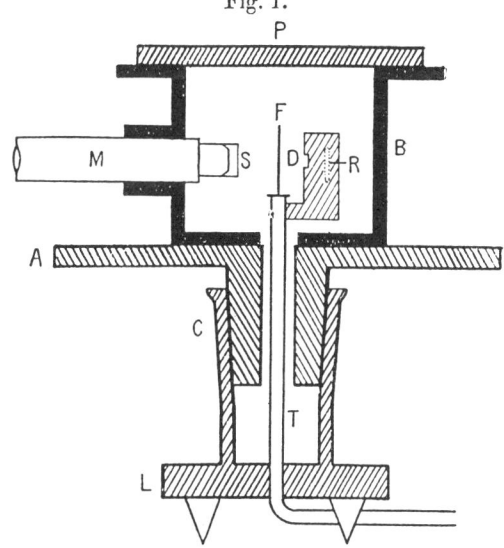

The source of α particles employed was similar to that used originally by Rutherford and Royds * in their experiments on the nature of the α particle. It consisted of a small thin-walled glass tube about 1 mm. in diameter, containing a large quantity of well purified radium emanation. The α particles emitted by the emanation and its active deposit could pass through the glass walls without much reduction of range. For these experiments the unhomogeneity of the source, due to the different α particles from the emanation, Ra A and Ra C, does not interfere with the application of the law of scattering with angle as deduced from the theory, as each group of α particles is scattered according to the same law.

By means of a diaphragm placed at D, a pencil of α particles was directed normally on to the scattering foil F. By

* E. Rutherford and T. Royds, Phil. Mag. vol. xvii. p. 281 (1909).

608   Dr. H. Geiger *and* Mr. E. Marsden *on the Laws of*

rotating the microscope the α particles scattered in different directions could be observed on the screen S.   Although over 100 millicuries of radium emanation were available for the experiments, the smallness of the effect for the larger angles of deflexion necessitated short distances of screen and source from the scattering foil.   In some experiments the distance between the source and scattering foil was 2·5 cm., and the screen moved in a circle of 1·6 cm. radius, while in other experiments these distances were increased.   Observations were taken in various experiments for angles of deflexion from 5° to 150°.   When measuring the scattering through large angles the zinc-sulphide screen had to be turned very near to the source, and the β and γ rays produced a considerable luminescence on it, thus making countings of the scintillations difficult.   The effect of the β rays was reduced as far as possible by enclosing the source in a lead box shown shaded in the diagram.   The amount of lead was, however, limited by considerations of the space taken up by it, and consequently observations could not be made for angles of deflexion between 150° and 180°.

In the investigation of the scattering through relatively small angles the distances of source and screen from the scattering foil were increased considerably in order to obtain beams of smaller solid angle.

The number of particles scattered through different angles was found to decrease extremely rapidly with increase of angle, and as it is not possible to count with certainty more than 90 scintillations per minute or less than about 5 per minute, measurements could only be made over a relatively small range of angles at the same time.   The number of α particles scattered through large angles was first measured, and as the emanation decayed it was possible to take measurements for smaller and smaller angles, and from the known decay of the emanation measurements taken at different times could be corrected for the decrease of activity.

Even when no scattering foil was used a few scintillations were always observed on the screen.   They were obviously due to scattered radiation from the walls of the vessel and from the edge of the diaphragm limiting the beam.   The effect was reduced as far as possible by lining the box with paper and by using a substance of low atomic weight, viz. aluminium, for the diaphragm.   The number of stray α particles was determined for different positions of the microscope by removing the scattering foil so that the necessary corrections could be applied with certainty.

In order to make the best use of the emanation available,

Deflexion of α Particles through Large Angles.   609

measurements were made simultaneously with different foils. These foils were attached to frames which fitted into a slot in the tube T in such a way that they could be exchanged and accurately replaced in position.   Table I. gives an example of a particular set of countings, when a silver foil was used to scatter the α particles.

TABLE I.—Variation of Scattering with Angle. (Example of a set of measurements.) Silver Foil. Time elapsed since filling of emanation tube, 51 hours. Correction for decay, 0·683.

Angle φ.	Scintillations per minute.				$\dfrac{1}{\sin^4 \phi/2}$.	N × sin⁴ φ/2.
	Without foil.	With foil.	Corrected for effect without foil.	Corrected for decay, N.		
150...	0·2	4·95	4·75	6·95	1·15	6·0
135 ..	2·6	8·3	5·7	8·35	1·38	6·1
120...	3·8	10·3	6·5	9·5	1·79	5·3
105...	0·6	10·6	10·0	14·6	2·53	5·8
75...	0·0	28·6	28·6	41·9	7·25	5·8
60...	0·3	69·2	68·9	101	16·0	6·3

In this set about 2500 scintillations were counted. After a few days had elapsed the measurements for the smaller angles were repeated and the range of angles extended. Proceeding in this way the whole range of angles was investigated in the course of a few weeks. When measuring relatively large angles of deflexion a wide beam of about 15° radius had to be used in order to obtain a suitable number of scintillations, but for the smaller angles the aperture of the diaphragm confining the beam was reduced considerably, so that the angle at which the scintillations were counted was always large compared with the angular radius of the beam. When changing over from one diaphragm to another comparative measurements for different angles were made so as to obtain an accurate value of the reduction constant.

Table II. gives the collected results for two series of experiments with foils of silver and gold. The thicknesses of the foils were in the first series equivalent to 0·45 and 0·3 cm. air, and in the second series 0·45 and 0·1 cm. air for silver and gold respectively. Col. I. gives the values of the

Phil. Mag. S. 6. Vol. 25. No. 148. *April* 1913.   2 T

610   Dr. H. Geiger *and* Mr. E. Marsden *on the Laws of*

TABLE II.

Variation of Scattering with Angle.   (Collected results.)

I. Angle of deflexion, ϕ.	II. $\dfrac{1}{\sin^4 \phi/2}$	III. Number of scintillations, N.	IV. $\dfrac{N}{\sin^4 \phi/2}$	V. Number of scintillations, N.	VI. $\dfrac{N}{\sin^4 \phi/2}$
		SILVER.		GOLD.	
150	1·15	22·2	19·3	33·1	28·8
135	1·38	27·4	19·8	43·0	31·2
120	1·79	33·0	18·4	51·9	29·0
105	2·53	47·3	18·7	69·5	27·5
75	7·25	136	18·8	211	29·1
60	16·0	320	20·0	477	29·8
45	46·6	989	21·2	1435	30·8
37·5	93·7	1760	18·8	3300	35·3
30	223	5260	23·6	7800	35·0
22·5	690	20300	29·4	27300	39·6
15	3445	105400	30·6	132000	38·4
30	223	5·3	0·024	3·1	0·014
22·5	690	16·6	0·024	8·4	0·012
15	3445	93·0	0·027	48·2	0·014
10	17330	508	0·029	200	0·0115
7·5	54650	1710	0·031	607	0·011
5	276300	...	...	3320	0·012

angles ϕ between the direction of the beam and the direction in which the scattered α particles were counted.   Col. II. gives the values of $\dfrac{1}{\sin^4 \phi/2}$.   In Cols. III. and V. the numbers of scintillations are entered which were observed for the silver and gold respectively.   Corrections are made for the decay of the emanation, for the natural effect, and for change of diaphragm.   For the smaller angles corrections have been applied (in no case exceeding 20 per cent.) owing to the fact that the beam of α particles was of finite dimensions and not negligible compared with the angle of deflexion. These corrections were calculated from geometrical considerations.   In Cols. IV. and VI. the ratios of the numbers of scintillations to $\dfrac{1}{\sin^4 \phi/2}$ are entered.   It will be seen that in both sets the values are approximately constant.   The deviations are somewhat systematic, the ratio increasing with decreasing angle.   However, any slight asymmetry in the apparatus and other causes would affect the results in a systematic way so that, fitting on the two sets of observations and considering the enormous variation in the numbers of scattered particles, from 1 to 250,000, the deviations from constancy of the ratio are probably well within the experimental

Deflexion of α Particles through Large Angles.   611

error. The experiments, therefore, prove that the number of α particles scattered in a definite direction varies as cosec⁴ φ/2.

Variation with Thickness of Material.

In investigating the variation of scattering with thickness of material, it seemed necessary to use a homogeneous source of α particles, for according to the theory the effect of the change of velocity with increasing thickness will be very appreciable for α particles of low velocity. In the experiments on " compound scattering " by one of us, a source was used consisting of Ra C deposited from radium emanation *in situ* in a small conical tube fitted with a mica window, the emanation being withdrawn when measurements were taken by expanding into a large volume connected to it. In our first experiments we used such a source, but the observations eventually showed it to be unsuitable. After expansion some emanation remains clinging to the walls of the glass tube. This emanation and the Ra A associated with it gives α particles of considerably lower velocity than the α particles of Ra C, and although the number of α particles so contributed was of the order of only a few per cent. of the number from the Ra C, yet owing to the fact that the amount of scattering increases very rapidly with decreasing velocity, the disturbances caused by the slower α particles were so large as to render the source unsuitable for the present work.

Fig. 2.

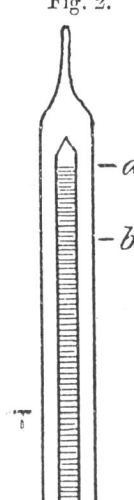

The source finally adopted was prepared as shown in fig. 2. About 80 millicuries of radium emanation were very highly purified and pressed into the conical end of the glass tube T of about 1 mm. internal diameter. After the emanation had remained in position for a sufficient time to attain equilibrium with Ra C, it was expanded into a bulb below, and a small part of the capillary tube was drawn off at *b*. About 1 mm. of the end of the tube which was coated with the Ra C was then cut off (at *a*) and freed from occluded emanation by washing with alcohol and by heating. The resulting source of Ra C was used in the experiments, and with due care its decay was found to be in agreement with theory, at least for the first 80 minutes.

The arrangement used for the comparison of the scattering

2 T 2

612   Dr. H. Geiger *and* Mr. E. Marsden *on the Laws of*

of different thicknesses of metal foils is shown diagrammatically in fig. 3. It consists essentially of a source of
α radiation R, a diaphragm D, a scattering foil F, and a zinc-
sulphide screen Z on which the scattered α particles were
observed. The main part of the apparatus was enclosed in
a cylindrical brass ring A, the ends of which were planed so
that they could be closed airtight by the two glass plates
B and C. The depth of the ring was 3·5 cm., and its
internal and external diameters 5·5 and 7·5 cm. respectively.
Two holes were drilled through the glass plate B, one in the
centre and the other 1·65 cm. excentric. The source of
radiation R was placed directly against a sheet of mica
which was waxed over and closed the opening E. By
placing the source outside the apparatus, any small amount
of emanation associated with it was prevented from entering
the chamber and disturbing the measurements.

Fig. 3.

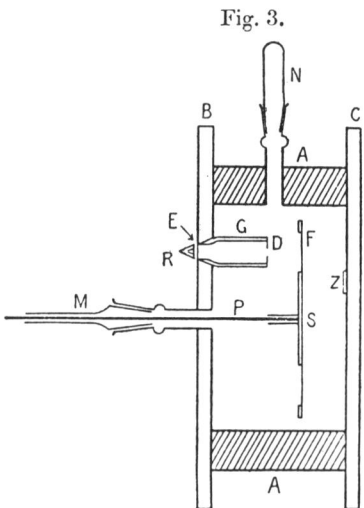

By means of the diaphragm D a narrow pencil of α particles could be directed on to the scattering foil. The
different foils were attached to the disk S and covered five of
six holes drilled through it at equal distances from its centre.
The uncovered opening was used to determine the natural
effect. The disk could be fitted on to the rod P, which was
fastened to the ground-glass joint M so that it could be
rotated and the different foils brought in front of the
diaphragm. The scattered α particles were observed by
means of a microscope on the zinc-sulphide screen Z fixed
inside the glass plate.

Deflexion of α Particles through Large Angles.   613

In making the observations the disk carrying the foils was placed in position about 1·2 cm. from the glass plate C. The apparatus was then completely exhausted through a tube not shown in the diagram, charcoal cooled by liquid air being used for the final exhaustion. After the source of radiation had been placed in position, the microscope was adjusted at that part of the zinc-sulphide screen where the scintillations appeared at a rate convenient for counting. With a source of 30 millicuries of Ra C this was usually the case for an angle of deflexion of from 20° to 30°. The area of the screen visible through the microscope was about 1 sq. mm., whilst the main beam of α particles covered an area of about 3 sq. mm.

As soon as the Ra A in the source had decayed completely (*i. e.* after 20 minutes) countings were commenced. Measurements were first taken with the layers of foils of smaller thickness, and as the source decayed they were extended to the thicker foils. From the known decay of the active deposit of radium the measurements could all be corrected for the variation in activity of the source, the results being verified by making observations on the same foils at different times. An experiment generally extended for about 80 minutes. After that time the decay corrections for the source were not always reliable owing to small quantities of radium emanation associated with it, as has been mentioned above. Owing to the relatively short time available in each experiment for the completion of the measurements, only about 100 to 200 scintillations could be counted with each foil.

As in the experiments on the variation of scattering with angle, some scintillations appeared on the zinc-sulphide screen even when no scattering foil was interposed. It was found that these scintillations were due to α particles which had been scattered from the edges of the diaphragm limiting the beam. Experiments were made with paper diaphragms and with aluminium diaphragms of only $\frac{1}{10}$ mm. thickness, whilst a diaphragm D′ (fig. 4) was also introduced to prevent scattering from the inside of the glass tube G carrying the main diaphragm D. Even with these precautions the effect was still so large that accurate experiments with foils of low atomic weight would have been impossible. The difficulty was, however, successfully overcome by intercepting the stray α particles by a screen K, which could be turned by means of a ground-glass joint (N in fig. 3) about a vertical axis passing through A so as to be just outside the main pencil. The adjustment was made by observation of the scintillations produced by the main beam on the zinc-sulphide

614   Dr. H. Geiger *and* Mr. E. Marsden *on the Laws of*

screen Z, which was temporarily placed at Z'. The magnitude of the effect may be judged from the following figures obtained in a particular experiment with an aluminium diaphragm :—The number of scintillations without both the screen K and the scattering foil F was 60 per minute, whilst by bringing the screen K into position the number was reduced to 0·5 per minute. With the screen K in position and an aluminium foil equivalent to 0·5 cm. air as scattering foil, the number of scintillations was 14 per minute, or about one quarter the effect without screen or scattering foil.

Fig. 4.

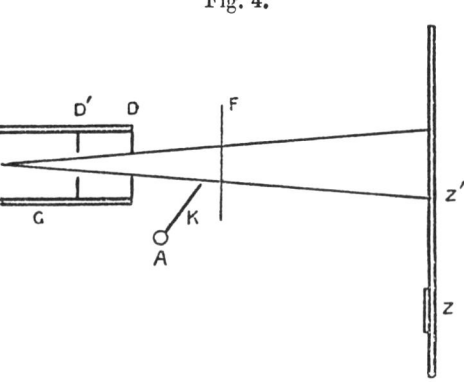

In the following table the results of an experiment with gold foils are tabulated. Column I. gives the number of foils and column II. the thicknesses expressed as the stopping power of α particles in centimetres of air as determined by the scintillation method. The figures given in column III. represent the number of scintillations observed on the zinc-sulphide screen. These figures are corrected for the variation of activity with time of the source. A slight correction has been made due to the increase of scattering on account of the decrease of velocity of the α particles in passing through the foils. The magnitude of this correction could be calculated from the results given in the last section of the present paper, and amounted to 9 per cent. in this experiment for the thickest foil used. The last column of the table gives the ratio of the corrected number of scintillations to the thickness. The values are constant within the limits of the experimental error. The variations exhibited by the figures are well within the probability errors, owing to the relatively small number of scintillations which could be counted in the time available.

Deflexion of α Particles through Large Angles.   615

TABLE III.

Gold.—Variation of Scattering with Thickness.

I.	II.	III.	IV.
Number of Foils.	Air equivalent. T in cm.	Number N of scintillations per minute.	Ratio $\frac{N}{T}$.
1	0·11	21·9	200
2	0·22	38·4	175
5	0·51	84·3	165
8	0·81	121·5	150
9	0·90	145	160

Similar experiments were carrried out with foils of tin, silver, copper, and aluminium. In each set about 1000 scintillations were counted. The results are plotted in fig. 5, where the abscissæ represent the thickness of the scattering foil expressed in centimetres of air equivalent and the ordinates the number of scattered particles. Similar corrections to the above have been introduced in each case.

Fig. 5.

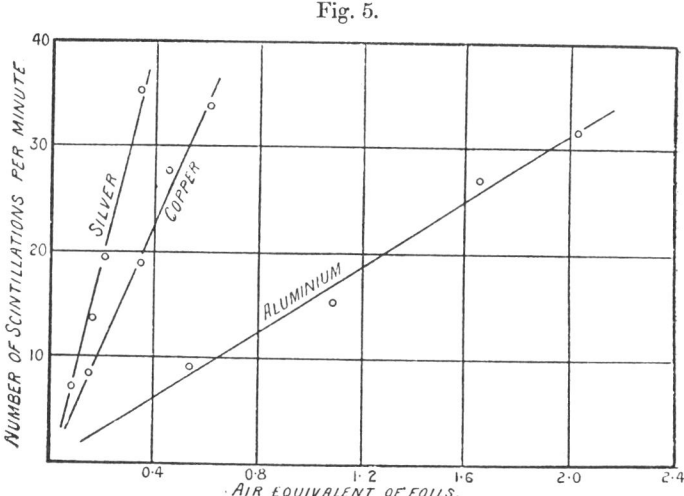

For all the metals examined the points lie on straight lines which pass through the origin. The experiments therefore prove that for small thicknesses of matter the scattering is proportional to the thickness. If there is any appreciable

616   Dr. H. Geiger *and* Mr. E. Marsden *on the Laws of*

diminution in velocity of 'the α particles in passing through the foils, the number of scattered particles increases somewhat more rapidly than the thickness.

Variation with Atomic Weight.

Assuming that the magnitude of the central charge of the atom is proportional to the atomic weight A, Professor Rutherford has shown that the number of α particles scattered by different foils containing the same number of atoms should be proportional to A^2. With the thin foils which had to be used experimentally, it was found impracticable to calculate the number of atoms per unit area by weighing the foils. It proved much more reliable to deduce the required number of atoms from the air equivalent as found by the reduction of the range of α particles by the scintillation method. This method had the advantage that the thickness was determined at the exact part of the foil which served to scatter the α particles, thus eliminating any errors due to variations in the thickness of the foils. Bragg and others have given numbers connecting the thicknesses of foils of various materials and their stopping power, and it has been shown that for different foils of the same air equivalent the numbers of atoms per unit area are inversely proportional to the square roots of the atomic weights. Consequently if the scattering per atom of atomic weight A is proportional to A^2, the scattering per centimetre air equivalent will be proportional to $A^2 \times A^{-\frac{1}{2}}$, *i. e.* to $A^{3/2}$.

In the experimental investigation the same apparatus was used as in the previous experiments on the variation of scattering with thickness of material. The openings in the disk S were covered with thin foils of different materials, and their thicknesses chosen in such a way that they gave approximately the same effect of scattering. A number of different sets of experiments were made, the foils being varied in each experiment. The results in a particular experiment are given in Table IV. Columns I. and II. give the foils used and their respective atomic weights. In column III. the air equivalents of the foils are entered. Column IV. gives the number of scintillations observed after correction for the variation in activity of the source and the loss of velocity of the α particles in the foil. Column V. gives the number of scintillations per unit air equivalent of material. In column VI. the values of $A^{3/2}$ are given, and in column VII. the ratios of the numbers of scintillations to $A^{3/2}$ are calculated. The figures are constant within the experimental error.

Deflexion of α Particles through Large Angles.   617

TABLE IV.

Variation of Scattering with Atomic Weight.   (Example of a set of measurements.)

I.	II.	III.	IV.	V.	VI	VII.
Substance.	Atomic weight. A.	Air equivalent in cm.	Number of scintillations per minute corrected for decay.	Number N of scintillations per cm. air equivalent.	$A^{3/2}$.	$N \times A^{2/3}$.
Gold	197	·229	133	581	2770	0·21
Tin	119	·441	119	270	1300	0·21
Silver	107·9	·262	51·7	198	1120	0·18
Copper	63·6	·616	71	115	507	0·23
Aluminium..	27·1	2·05	71	34·6	141	0·24

The combined results of four experiments are given in Table V. In the last column are given the ratios of the numbers of scintillations per centimetre equivalent to $A^{3/2}$. This ratio should be constant according to theory. The experimental values show a slight increase with decreasing atomic weight.

TABLE V.

Variation of Scattering with Atomic Weight.   (Collected results using Ra C.)

Substance.	Total number of scintillations counted for each material.	$A^{3/2}$.	Ratio of scintillations per cm. air equivalent to $A^{3/2}$ *.
Gold	850	2770	95
Platinum	200	2730	99
Tin	700	1300	96
Silver	800	1120	98
Copper	600	507	104
Aluminium	700	144	110

* *Note* 1.—Since these experiments were carried out, Richardson and one of us (Phil. Mag. vol. xxv. p. 184 (1913)) have determined the masses per unit area per cm. air equivalent for different metals, using the scintillation method. Introducing the results, and calculating the values of the ratio of the scattering per atom divided by A^2, the following are obtained :—Au 3·4, Pt 3·2, Sn 3·3, Ag 3·6, Cu 3·7, Al 3·6. These numbers show better agreement than those in the last column above, which are calculated on the assumption of Bragg's law.

618   Dr. H. Geiger *and* Mr. E. Marsden *on the Laws oj*

On account of the importance of these experiments further measurements were made under somewhat different conditions. The main difficulty in the previous experiments arose from the fact, that owing to the rapid decay of the source it was impossible to count in each case a sufficient number of scintillations to obtain a true average value. In the following set of measurements radium emanation in equilibrium with its active deposit was used as source of radiation. The source consisted of a conical glass tube (fig. 6) of about $1\frac{1}{2}$ mm. internal diameter at its widest part, the height of the cone being about 2·5 mm.

Fig. 6.

The end of the tube was closed airtight by a sheet of mica of 0·62 cm. air equivalent. This tube was filled with about 30 millicuries of highly purified emanation and placed at R (fig. 3, p. 612) directly against the mica window E, the air equivalent of which was also 0·62 cm.

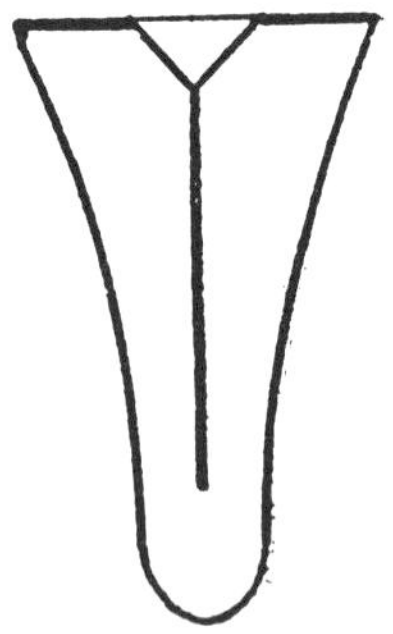

The difficulty introduced by the employment of α particles of different velocities (emanation, Ra A, and Ra C) was eliminated by using foils of approximately the same air equivalent. The α particles therefore suffered the same reduction in velocity in each foil, and the numbers of scattered particles were therefore directly comparable. It was of course impossible to obtain foils of exactly the same air equivalent, but this difficulty was easily overcome by determining the scattering for two foils of the same material, one slightly smaller and the other slightly larger than a standard thickness of 0·6 cm. air equivalent.

Owing to the large variation with atomic weight of the amount of scattering, the foils could not be all directly compared with each other at the same angle. They were therefore compared in sets, the angle being chosen smaller for the sets of lower atomic weight. Column VI. in the following table gives the mean results of the ratio of the number of scattered particles to $A^{3/2}$.

The scattering of carbon was obtained by using thin sheets of paraffin wax which contained about 85·2 per cent. carbon and 14·8 per cent. hydrogen. The air equivalent of the carbon was calculated from Bragg's law to be about 78 per cent. of the whole stopping power, and on account of the low atomic weight of hydrogen all the scattering effect was assumed due to the carbon. The measurements of the scattering were made by comparison with that due to aluminium foils of the same air equivalent.

Deflexion of α Particles through Large Angles.   619

TABLE VI.

Variation of Scattering with Atomic Weight.   (Collected results using Radium emanation.)

I. Substance.	II. Air equiva- lents of foils used.	III. Total number of scintilla- tions counted for each substance.	IV. Number N of scintillations at same angle and for same air equivalent.	V. $A^{3/2}$.	VI. $N \times A^{2/3}$.
Gold	·52, ·68	1200	2400	2770	·85
Platinum	·54, ·625	1000	2900	2730	1·05
Tin............ ...	·51, 1·15	1400	1290	1300	·99
Silver	·38, ·435	600	1060	1120	·95
Copper	·495, ·61	1300	570	507	1·12
Aluminium ...	·45, ·52, 1·06	1600	151	144	1·05
Carbon	·55, ·57	400	57	41·6	1·37

Note 2.—Introducing the new data for the mass per unit area of foils of the same air equivalent, as in note 1, the following are the values for the ratio of the scattering per atom divided by A^2:—Au 3·1, Pt 3·4, Sn 3·4, Ag 3·4, Cu 3·95, Al 3·4.

It will be seen from the table that, although the experimental conditions were very different from those in the previous experiments, the results are similar, and indicate the essential correctness of the assumption that the scattering per atom is proportional to the square of the atomic weight. The deviations from constancy of the ratio (see notes 1 and 2) are nearly within the experimental error.

The measurements have not so far been extended to substances of lower atomic weight than carbon. When the atomic weight is small and comparable with the mass of the α particle, the laws of scattering will require some modification to take into account the relative motion of the atom itself when a collision occurs.

Variation of Scattering with Velocity.

In order to determine the variation of scattering with velocity the apparatus was somewhat modified. A conical glass tube coated with active deposit was again used as source of radiation. This source was placed about 1 mm. from the mica window (E, fig. 3), so that it was possible to insert additional sheets of mica between the source and the window to reduce the velocity of the α particles. Mica sheets were used for this purpose on account of their uniformity of

620   Dr. H. Geiger *and* Mr. E. Marsden *on the Laws of*

thickness in comparison with metal foils. The micas were
attached to a cardboard disk which could be rotated to bring
the different sheets successively in position.   The α particles
were scattered by a foil of gold or silver, of stopping power
about 3 mm. of air, which was attached to a rod passing
through the ground glass N.   This made it possible to turn
the foil away from the main beam during an experiment in
order to test the natural effect.   The disk S, in this case,
rotated in a plane very close to the glass plate C and carried
sheets of mica of different thicknesses.   By rotating the
ground-glass joint the micas could be placed directly in
front of the zinc-sulphide screen, making it possible to test
the homogeneity of the α particles after they had been
scattered.
    The results are given in Table VII. Column I. gives the
number of mica sheets which were interposed in addition to
the mica window, and column II. the ranges of the α particles
incident on the scattering foil. The values of the velocities v
were calculated from these ranges R by use of the formula
$v^3 = aR$ previously found by one of us *.   The relative values

TABLE VII.

Variation of Scattering with Velocity.

I.	II.	III.	IV.	V.
Number of sheets of mica.	Range R of α particles after leaving mica.	Relative values of $1/v^4$.	Number N of scintillations per minute.	Nv^4.
0	5·5	1·0	24·7	25
1	4·76	1·21	29·0	24
2	4·05	1·50	33·4	22
3	3·32	1·91	44	23
4	2·51	2·84	81	28
5	1·84	4·32	101	23
6	1·04	9·22	255	28

of $1/v^4$ are given in column III. The number of scintillations
per minute N are entered in column IV., and in column V.
relative values of $N \times v^4$ are given. Over the range examined
the number of scintillations varies in the ratio $1 : 10$, while
it will be seen that the product Nv^4 remains sensibly
constant. Several experiments were made, and in every case

    * H. Geiger, Roy. Soc. Proc. A. vol. lxxxiii. p. 506 (1910).

Deflexion of α Particles through Large Angles.   621

the scattering was found to vary at a rate more nearly proportional to the inverse fourth power of the velocity than to any other integral power.   Owing to the comparative uncertainty of the values of the velocity for small ranges, however, the error of experiment may be somewhat greater than appears from column V. of the table.

In these experiments it proved essential to use a source possessing a high degree of homogeneity.   In earlier experiments, where we were not able to fulfil this condition, the scattering apparently increased much more rapidly than the inverse fourth power of the velocity of the Ra C α particles. Even with a source of Ra C with which only a small quantity of emanation was associated, the amount of scattering first rapidly increased on interposing the sheets of mica, then showed a slight decrease, and finally increased again.   This irregularity was due to the α particles of the emanation and Ra A, which are of shorter range than those of Ra C, and therefore more easily scattered.

The measurements could not easily be extended to α particles of lower velocity than corresponds to a range of about 1 centimetre, owing to the difficulty of observing the faint scintillations at lower ranges.   However, in one particular experiment, by adding sheets of mica to cut down the velocity the number of scattered α particles appearing on the screen was increased 25 times, showing how easily the α particles of low velocity are scattered.

The results of the examination of the homogeneity of the scattered α particles showed that at least in the case of gold they remained practically homogeneous after the scattering. Experiments of this nature in the case of scattering foils of low atomic weight would be very interesting, but are somewhat difficult.

Determination of Absolute Number of Scattered α Particles.

In the previous sections we have completely verified the theory given by Prof. Rutherford.   Since, according to this theory, the large deflexion of an α particle is the result of a close encounter with a single atom of matter, it is possible to calculate the magnitude of the central charge of the atom when the fraction of α particles scattered under definite conditions is determined.   We have made several attempts under different conditions to obtain a quantitative estimate of the scattered particles, but the results so far have only given us an approximate value.   The main difficulty arises from the fact that the scattered particles consist of such a small fraction of the original beam that different methods of measurement have to be employed in the two cases. The number of scattered α particles was determined from the number of scintillations observed on the zinc-sulphide screen, a correction being necessary owing to the fact that with the particular screens used only about 85 per cent. of the incident α particles produce scintillations. The number of α particles in the main beam was in one case in which an emanation tube was used (as shown in fig. 1, p. 607) determined directly by the scintillation method, several weeks being allowed to elapse, so that the emanation had decayed to a small value. In other experiments Ra C deposited on the inside of a conical glass tube (as in fig. 2, p. 611) was used, and the number of α particles was calculated from its γ-ray activity and the distance and area of the diaphragm determining the beam.

The results showed that, using a gold foil of air equivalent 1 mm. (actual thickness $2 \cdot 1 \times 10^{-5}$ cm.), the fraction of incident Ra C α particles ($v = 2 \cdot 06 \times 10^9$ cm./sec.) scattered through an angle of 45° and observed on an area of 1 sq. mm. placed normally at a distance of 1 cm. from the point of incidence of the beam, was $3 \cdot 7 \times 10^{-7}$. Substituting this value in the equation given at the commencement of this paper, it can be calculated that the value of the number of elementary electric charges composing the central charge of the gold atom is about half the atomic weight. This result is probably correct to 20 per cent., and agrees with the deduction of Prof. Rutherford from the less definite data given in our previous paper.

From the results of this and the previous sections it is possible to calculate the probability of an α particle being scattered through any angle under any specified conditions. For materials of atomic weight greater than that of aluminium, it is sufficiently accurate to put N equal to half the atomic weight in the equation given at the commencement of the paper.

It will be seen that the laws of " single scattering " found in this paper are quite distinct from the laws of " compound scattering " previously deduced by Geiger. It must be remembered, however, that the experiments are not directly comparable. In the present paper we are dealing with very thin sheets of matter, and are measuring the very small fraction of α particles which are deflected by single collisions through relatively large angles. The experiments of Geiger, however, deal with larger thicknesses of scattering foils and angles of deflexion of a few degrees only. Under these conditions the scattering is due to the combination of a large number of deflexions not only by the central charges of the·

Deflexion of α Particles through Large Angles.   623

atoms, but probably also by the electronic charges distributed throughout the remainder of their volumes.

Summary.

The experiments described in the foregoing paper were carried out to test a theory of the atom proposed by Prof. Rutherford, the main feature of which is that there exists at the centre of the atom an intense highly concentrated electrical charge. The verification is based on the laws of scattering which were deduced from this theory. The following relations have been verified experimentally :—

(1) The number of α particles emerging from a scattering foil at an angle ϕ with the original beam varies as $1/\sin^4 \phi/2$, when the α particles are counted on a definite area at a constant distance from the foil. This relation has been tested for angles varying from 5° to 150°, and over this range the number of α particles varied from 1 to 250,000 in good agreement with the theory.

(2) The number of α particles scattered in a definite direction is directly proportional to the thickness of the scattering foil for small thicknesses. For larger thicknesses the decrease of velocity of the α particles in the foil causes a somewhat more rapid increase in the amount of scattering.

(3) The scattering per atom of foils of different materials varies approximately as the square of the atomic weight. This relation was tested for foils of atomic weight from that of carbon to that of gold.

(4) The amount of scattering by a given foil is approximately proportional to the inverse fourth power of the velocity of the incident α particles. This relation was tested over a range of velocities such that the number of scattered particles varied as $1:10$.

(5) Quantitative experiments show that the fraction of α particles of Ra C, which is scattered through an angle of 45° by a gold foil of 1 mm. air equivalent ($2 \cdot 1 \times 10^{-5}$ cm.), is $3 \cdot 7 \times 10^{-7}$ when the scattered particles are counted on a screen of 1 sq. mm. area placed at a distance of 1 cm. from the scattering foil. From this figure and the foregoing results, it can be calculated that the number of elementary charges composing the centre of the atom is equal to half the atomic weight.

We are indebted to Prof. Rutherford for his kind interest in these experiments, and for placing at our disposal the large quantities of radium emanation necessary. We are also indebted to the Government Grant Committee of the Royal Society for a grant to one of us, out of which part of the expenses has been paid.

THE

LONDON, EDINBURGH, AND DUBLIN

PHILOSOPHICAL MAGAZINE

AND

JOURNAL OF SCIENCE.

[SIXTH SERIES.]

JULY 1913.

I. *On the Constitution of Atoms and Molecules.*
   By N. BOHR, *Dr. phil. Copenhagen*[*].

Introduction.

IN order to explain the results of experiments on scattering of α rays by matter Prof. Rutherford[†] has given a theory of the structure of atoms. According to this theory, the atoms consist of a positively charged nucleus surrounded by a system of electrons kept together by attractive forces from the nucleus; the total negative charge of the electrons is equal to the positive charge of the nucleus. Further, the nucleus is assumed to be the seat of the essential part of the mass of the atom, and to have linear dimensions exceedingly small compared with the linear dimensions of the whole atom. The number of electrons in an atom is deduced to be approximately equal to half the atomic weight. Great interest is to be attributed to this atom-model; for, as Rutherford has shown, the assumption of the existence of nuclei, as those in question, seems to be necessary in order to account for the results of the experiments on large angle scattering of the α rays[‡].

In an attempt to explain some of the properties of matter on the basis of this atom-model we meet, however, with difficulties of a serious nature arising from the apparent

[*] Communicated by Prof. E. Rutherford, F.R.S.
[†] E. Rutherford, Phil. Mag. xxi. p. 669 (1911).
[‡] See also Geiger and Marsden, Phil. Mag. April 1913.

Phil. Mag. S. 6. Vol. 26. No. 151. *July* 1913.   B

2   Dr. N. Bohr *on the Constitution*

instability of the system of electrons : difficulties purposely avoided in atom-models previously considered, for instance, in the one proposed by Sir J. J. Thomson *. According to the theory of the latter the atom consists of a sphere of uniform positive electrification, inside which the electrons move in circular orbits.

The principal difference between the atom-models proposed by Thomson and Rutherford consists in the circumstance that the forces acting on the electrons in the atom-model of Thomson allow of certain configurations and motions of the electrons for which the system is in a stable equilibrium ; such configurations, however, apparently do not exist for the second atom-model. The nature of the difference in question will perhaps be most clearly seen by noticing that among the quantities characterizing the first atom a quantity appears—the radius of the positive sphere—of dimensions of a length and of the same order of magnitude as the linear extension of the atom, while such a length does not appear among the quantities characterizing the second atom, viz. the charges and masses of the electrons and the positive nucleus ; nor can it be determined solely by help of the latter quantities.

The way of considering a problem of this kind has, how-ever, undergone essential alterations in recent years owing to the development of the theory of the energy radiation, and the direct affirmation of the new assumptions introduced in this theory, found by experiments on very different phe-nomena such as specific heats, photoelectric effect, Röntgen-rays, &c. The result of the discussion of these questions seems to be a general acknowledgment of the inadequacy of the classical electrodynamics in describing the behaviour of systems of atomic size †. Whatever the alteration in the laws of motion of the electrons may be, it seems necessary to introduce in the laws in question a quantity foreign to the classical electrodynamics, *i. e.* Planck's constant, or as it often is called the elementary quantum of action. By the introduction of this quantity the question of the stable con-figuration of the electrons in the atoms is essentially changed, as this constant is of such dimensions and magnitude that it, together with the mass and charge of the particles, can determine a length of the order of magnitude required.

This paper is an attempt to show that the application of the above ideas to Rutherford's atom-model affords a basis

* J. J. Thomson, Phil. Mag. vii. p. 237 (1904).
† See f. inst., ' Théorie du rayonnement et les quanta.' Rapports de la réunion à Bruxelles, Nov. 1911. Paris, 1912.

for a theory of the constitution of atoms. It will further be shown that from this theory we are led to a theory of the constitution of molecules.

In the present first part of the paper the mechanism of the binding of electrons by a positive nucleus is discussed in relation to Planck's theory. It will be shown that it is possible from the point of view taken to account in a simple way for the law of the line spectrum of hydrogen. Further, reasons are given for a principal hypothesis on which the considerations contained in the following parts are based.

I wish here to express my thanks to Prof. Rutherford for his kind and encouraging interest in this work.

Part I.—Binding of Electrons by Positive Nuclei.

§ 1. *General Considerations.*

The inadequacy of the classical electrodynamics in accounting for the properties of atoms from an atom-model as Rutherford's, will appear very clearly if we consider a simple system consisting of a positively charged nucleus of very small dimensions and an electron describing closed orbits around it. For simplicity, let us assume that the mass of the electron is negligibly small in comparison with that of the nucleus, and further, that the velocity of the electron is small compared with that of light.

Let us at first assume that there is no energy radiation. In this case the electron will describe stationary elliptical orbits. The frequency of revolution ω and the major-axis of the orbit $2a$ will depend on the amount of energy W which must be transferred to the system in order to remove the electron to an infinitely great distance apart from the nucleus. Denoting the charge of the electron and of the nucleus by $-e$ and E respectively and the mass of the electron by m, we thus get

$$\omega = \frac{\sqrt{2}}{\pi}\frac{W^{\frac{3}{2}}}{eE\sqrt{m}}, \quad 2a = \frac{eE}{W}. \quad \cdot \quad \cdot \quad \cdot \quad (1)$$

Further, it can easily be shown that the mean value of the kinetic energy of the electron taken for a whole revolution is equal to W. We see that if the value of W is not given, there will be no values of ω and a characteristic for the system in question.

Let us now, however, take the effect of the energy radiation into account, calculated in the ordinary way from the acceleration of the electron. In this case the electron will

4   Dr. N. Bohr *on the Constitution*

no longer describe stationary orbits. W will continuously increase, and the electron will approach the nucleus describing orbits of smaller and smaller dimensions, and with greater and greater frequency ; the electron on the average gaining in kinetic energy at the same time as the whole system loses energy. This process will go on until the dimensions of the orbit are of the same order of magnitude as the dimensions of the electron or those of the nucleus. A simple calculation shows that the energy radiated out during the process considered will be enormously great compared with that radiated out by ordinary molecular processes.

It is obvious that the behaviour of such a system will be very different from that of an atomic system occurring in nature. In the first place, the actual atoms in their permanent state seem to have absolutely fixed dimensions and frequencies. Further, if we consider any molecular process, the result seems always to be that after a certain amount of energy characteristic for the systems in question is radiated out, the systems will again settle down in a stable state of equilibrium, in which the distances apart of the particles are of the same order of magnitude as before the process.

Now the essential point in Planck's theory of radiation is that the energy radiation from an atomic system does not take place in the continuous way assumed in the ordinary electrodynamics, but that it, on the contrary, takes place in distinctly separated emissions, the amount of energy radiated out from an atomic vibrator of frequency ν in a single emission being equal to $\tau h \nu$, where τ is an entire number, and h is a universal constant*.

Returning to the simple case of an electron and a positive nucleus considered above, let us assume that the electron at the beginning of the interaction with the nucleus was at a great distance apart from the nucleus, and had no sensible velocity relative to the latter. Let us further assume that the electron after the interaction has taken place has settled down in a stationary orbit around the nucleus. We shall, for reasons referred to later, assume that the orbit in question is circular : this assumption will, however, make no alteration in the calculations for systems containing only a single electron.

Let us now assume that, during the binding of the electron, a homogeneous radiation is emitted of a frequency ν, equal to half the frequency of revolution of the electron in its final

* See f. inst., M. Planck, *Ann. d. Phys.* xxxi. p. 758 (1910) ; xxxvii. p. 642 (1912) ; *Verh. deutsch. Phys. Ges.* 1911, p. 138.

orbit ; then, from Planck's theory, we might expect that the amount of energy emitted by the process considered is equal to $\tau h\nu$, where h is Planck's constant and τ an entire number. If we assume that the radiation emitted is homogeneous, the second assumption concerning the frequency of the radiation suggests itself, since the frequency of revolution of the electron at the beginning of the emission is 0. The question, however, of the rigorous validity of both assumptions, and also of the application made of Planck's theory, will be more closely discussed in § 3.

Putting
$$W = \tau h \frac{\omega}{2}, \quad \cdots \cdots \quad (2)$$

we get by help of the formula (1)

$$W = \frac{2\pi^2 m e^2 E^2}{\tau^2 h^2}, \quad \omega = \frac{4\pi^2 m e^2 E^2}{\tau^3 h^3}, \quad 2a = \frac{\tau^2 h^2}{2\pi^2 m e E} \cdots \quad (3)$$

If in these expressions we give τ different values, we get a series of values for W, ω, and a corresponding to a series of configurations of the system. According to the above considerations, we are led to assume that these configurations will correspond to states of the system in which there is no radiation of energy; states which consequently will be stationary as long as the system is not disturbed from outside. We see that the value of W is greatest if τ has its smallest value 1. This case will therefore correspond to the most stable state of the system, *i. e.* will correspond to the binding of the electron for the breaking up of which the greatest amount of energy is required.

Putting in the above expressions $\tau = 1$ and $E = e$, and introducing the experimental values

$$e = 4\cdot7 \cdot 10^{-10}, \quad \frac{e}{m} = 5\cdot31 \cdot 10^{17}, \quad h = 6\cdot5 \cdot 10^{-27},$$

we get

$$2a = 1\cdot1 \cdot 10^{-8} \text{ cm.,} \quad \omega = 6\cdot2 \cdot 10^{15} \frac{1}{\text{sec.}}, \quad \frac{W}{e} = 13 \text{ volt.}$$

We see that these values are of the same order of magnitude as the linear dimensions of the atoms, the optical frequencies, and the ionization-potentials.

The general importance of Planck's theory for the discussion of the behaviour of atomic systems was originally pointed out by Einstein*. The considerations of Einstein

* A. Einstein, *Ann. d. Phys.* xvii. p. 132 (1905); xx. p. 199 (1906); xxii. p. 180 (1907).

6   Dr. N. Bohr *on the Constitution*

have been developed and applied on a number of different phenomena, especially by Stark, Nernst, and Sommerfield. The agreement as to the order of magnitude between values observed for the frequencies and dimensions of the atoms, and values for these quantities calculated by considerations similar to those given above, has been the subject of much discussion. It was first pointed out by Haas *, in an attempt to explain the meaning and the value of Planck's constant on the basis of J. J. Thomson's atom-model, by help of the linear dimensions and frequency of an hydrogen atom.

Systems of the kind considered in this paper, in which the forces between the particles vary inversely as the square of the distance, are discussed in relation to Planck's theory by J. W. Nicholson †. In a series of papers this author has shown that it seems to be possible to account for lines of hitherto unknown origin in the spectra of the stellar nebulæ and that of the solar corona, by assuming the presence in these bodies of certain hypothetical elements of exactly indicated constitution. The atoms of these elements are supposed to consist simply of a ring of a few electrons surrounding a positive nucleus of negligibly small dimensions. The ratios between the frequencies corresponding to the lines in question are compared with the ratios between the frequencies corresponding to different modes of vibration of the ring of electrons. Nicholson has obtained a relation to Planck's theory showing that the ratios between the wave-length of different sets of lines of the coronal spectrum can be accounted for with great accuracy by assuming that the ratio between the energy of the system and the frequency of rotation of the ring is equal to an entire multiple of Planck's constant. The quantity Nicholson refers to as the energy is equal to twice the quantity which we have denoted above by W. In the latest paper cited Nicholson has found it necessary to give the theory a more complicated form, still, however, representing the ratio of energy to frequency by a simple function of whole numbers.

The excellent agreement between the calculated and observed values of the ratios between the wave-lengths in question seems a strong argument in favour of the validity of the foundation of Nicholson's calculations. Serious

* A. E. Haas, *Jahrb. d. Rad. u. El.* vii. p. 261 (1910). See further, A. Schidlof, *Ann. d. Phys.* xxxv. p. 90 (1911); E. Wertheimer, *Phys. Zeitschr.* xii. p. 409 (1911), *Verh. deutsch. Phys. Ges.* 1912, p. 431; F. A. Lindemann, *Verh. deutsch. Phys. Ges.* 1911, pp. 482, 1107; F. Haber, *Verh. deutsch. Phys. Ges.* 1911, p. 1117.

† J. W. Nicholson, Month. Not. Roy. Astr. Soc. lxxii. pp. 49, 139, 677, 693, 729 (1912).

of Atoms and Molecules.   7

objections, however, may be raised against the theory. These objections are intimately connected with the problem of the homogeneity of the radiation emitted. In Nicholson's calculations the frequency of lines in a line-spectrum is identified with the frequency of vibration of a mechanical system in a distinctly indicated state of equilibrium. As a relation from Planck's theory is used, we might expect that the radiation is sent out in quanta; but systems like those considered, in which the frequency is a function of the energy, cannot emit a finite amount of a homogeneous radiation; for, as soon as the emission of radiation is started, the energy and also the frequency of the system are altered. Further, according to the calculation of Nicholson, the systems are unstable for some modes of vibration. Apart from such objections—which may be only formal (see p. 23)—it must be remarked, that the theory in the form given does not seem to be able to account for the well-known laws of Balmer and Rydberg connecting the frequencies of the lines in the line-spectra of the ordinary elements.

It will now be attempted to show that the difficulties in question disappear if we consider the problems from the point of view taken in this paper. Before proceeding it may be useful to restate briefly the ideas characterizing the calculations on p. 5. The principal assumptions used are :

(1) That the dynamical equilibrium of the systems in the stationary states can be discussed by help of the ordinary mechanics, while the passing of the systems between different stationary states cannot be treated on that basis.

(2) That the latter process is followed by the emission of a *homogeneous* radiation, for which the relation between the frequency and the amount of energy emitted is the one given by Planck's theory.

The first assumption seems to present itself; for it is known that the ordinary mechanics cannot have an absolute validity, but will only hold in calculations of certain mean values of the motion of the electrons. On the other hand, in the calculations of the dynamical equilibrium in a stationary state in which there is no relative displacement of the particles, we need not distinguish between the actual motions and their mean values. The second assumption is in obvious contrast to the ordinary ideas of electrodynamics, but appears to be necessary in order to account for experimental facts.

In the calculations on page 5 we have further made use

8   Dr. N. Bohr *on the Constitution*

of the more special assumptions, viz. that the different
stationary states correspond to the emission of a different
number of Planck's energy-quanta, and that the frequency of
the radiation emitted during the passing of the system from
a state in which no energy is yet radiated out to one of the
stationary states, is equal to half the frequency of revolution
of the electron in the latter state. We can, however (see
§ 3), also arrive at the expressions (3) for the stationary
states by using assumptions of somewhat different form.
We shall, therefore, postpone the discussion of the special
assumptions, and first show how by the help of the above
principal assumptions, and of the expressions (3) for the
stationary states, we can account for the line-spectrum of
hydrogen.

§ 2. *Emission of Line-spectra.*

Spectrum of Hydrogen.—General evidence indicates that
an atom of hydrogen consists simply of a single electron
rotating round a positive nucleus of charge e *. The re-
formation of a hydrogen atom, when the electron has been
removed to great distances away from the nucleus—*e. g.* by
the effect of electrical discharge in a vacuum tube—will
accordingly correspond to the binding of an electron by a
positive nucleus considered on p. 5. If in (3) we put $E = e$,
we get for the total amount of energy radiated out by the
formation of one of the stationary states,

$$W_\tau = \frac{2\pi^2 m e^4}{h^2 \tau^2}.$$

The amount of energy emitted by the passing of the
system from a state corresponding to $\tau = \tau_1$ to one corresponding to $\tau = \tau_2$, is consequently

$$W_{\tau_2} - W_{\tau_1} = \frac{2\pi^2 m e^4}{h^2} \left(\frac{1}{\tau_2^2} - \frac{1}{\tau_1^2} \right).$$

If now we suppose that the radiation in question is homogeneous, and that the amount of energy emitted is equal to
$h\nu$, where ν is the frequency of the radiation, we get

$$W_{\tau_2} - W_{\tau_1} = h\nu,$$

* See f. inst. N. Bohr, Phil. Mag. xxv. p. 24 (1913). The conclusion
drawn in the paper cited is strongly supported by the fact that hydrogen,
in the experiments on positive rays of Sir J. J. Thomson, is the only
element which never occurs with a positive charge corresponding to the
loss of more than one electron (comp. Phil. Mag. xxiv. p. 672 (1912)).

and from this

$$\nu = \frac{2\pi^2 m e^4}{h^3}\left(\frac{1}{\tau_2^2} - \frac{1}{\tau_1^2}\right). \quad . \quad . \quad . \quad . \quad . \quad (4)$$

We see that this expression accounts for the law connecting the lines in the spectrum of hydrogen. If we put $\tau_2 = 2$ and let τ_1 vary, we get the ordinary Balmer series. If we put $\tau_2 = 3$, we get the series in the ultra-red observed by Paschen[*] and previously suspected by Ritz. If we put $\tau_2 = 1$ and $\tau_2 = 4, 5, ..$, we get series respectively in the extreme ultra-violet and the extreme ultra-red, which are not observed, but the existence of which may be expected.

The agreement in question is quantitative as well as qualitative. Putting

$$e = 4\cdot7 \cdot 10^{-10}, \quad \frac{e}{m} = 5\cdot31 \cdot 10^{17}, \quad \text{and} \quad h = 6\cdot5 \cdot 10^{-27},$$

we get

$$\frac{2\pi^2 m e^4}{h^3} = 3\cdot1 \cdot 10^{15}.$$

The observed value for the factor outside the bracket in the formula (4) is

$$3\cdot290 \cdot 10^{15}.$$

The agreement between the theoretical and observed values is inside the uncertainty due to experimental errors in the constants entering in the expression for the theoretical value. We shall in § 3 return to consider the possible importance of the agreement in question.

It may be remarked that the fact, that it has not been possible to observe more than 12 lines of the Balmer series in experiments with vacuum tubes, while 33 lines are observed in the spectra of some celestial bodies, is just what we should expect from the above theory. According to the equation (3) the diameter of the orbit of the electron in the different stationary states is proportional to τ^2. For $\tau = 12$ the diameter is equal to $1\cdot6 \cdot 10^{-6}$ cm., or equal to the mean distance between the molecules in a gas at a pressure of about 7 mm. mercury; for $\tau = 33$ the diameter is equal to $1\cdot2 \cdot 10^{-5}$ cm., corresponding to the mean distance of the molecules at a pressure of about 0·02 mm. mercury. According to the theory the necessary condition for the appearance of a great number of lines is therefore a very small density of the gas; for simultaneously to obtain an

* F. Paschen, *Ann. d. Phys.* xxvii. p. 565 (1908).

10　　　　　Dr. N. Bohr *on the Constitution*

intensity sufficient for observation the space filled with the gas must be very great. If the theory is right, we may therefore never expect to be able in experiments with vacuum tubes to observe the lines corresponding to high numbers of the Balmer series of the emission spectrum of hydrogen; it might, however, be possible to observe the lines by investigation of the absorption spectrum of this gas (see § 4).

It will be observed that we in the above way do not obtain other series of lines, generally ascribed to hydrogen; for instance, the series first observed by Pickering* in the spectrum of the star ζ Puppis, and the set of series recently found by Fowler† by experiments with vacuum tubes containing a mixture of hydrogen and helium. We shall, however, see that, by help of the above theory, we can account naturally for these series of lines if we ascribe them to helium.

A neutral atom of the latter element consists, according to Rutherford's theory, of a positive nucleus of charge $2e$ and two electrons. Now considering the binding of a single electron by a helium nucleus, we get, putting $E = 2e$ in the expressions (3) on page 5, and proceeding in exactly the same way as above,

$$\nu = \frac{8\pi^2 m e^4}{h^3}\left(\frac{1}{\tau_2^2} - \frac{1}{\tau_1^2}\right) = \frac{2\pi^2 m e^4}{h^3}\left(\frac{1}{\left(\frac{\tau_2}{2}\right)^2} - \frac{1}{\left(\frac{\tau_1}{2}\right)^2}\right).$$

If we in this formula put $\tau_2 = 1$ or $\tau_2 = 2$, we get series of lines in the extreme ultra-violet. If we put $\tau_2 = 3$, and let τ_1 vary, we get a series which includes 2 of the series observed by Fowler, and denoted by him as the first and second principal series of the hydrogen spectrum. If we put $\tau_2 = 4$, we get the series observed by Pickering in the spectrum of ζ Puppis. Every second of the lines in this series is identical with a line in the Balmer series of the hydrogen spectrum; the presence of hydrogen in the star in question may therefore account for the fact that these lines are of a greater intensity than the rest of the lines in the series. The series is also observed in the experiments of Fowler, and denoted in his paper as the Sharp series of the hydrogen spectrum. If we finally in the above formula put $\tau_2 = 5, 6, ..$, we get series, the strong lines of which are to be expected in the ultra-red.

The reason why the spectrum considered is not observed in

* E. C. Pickering, Astrophys. J. iv. p. 369 (1896); v. p. 92 (1897).
† A. Fowler, Month. Not. Roy. Astr. Soc. lxxiii. Dec. 1912.

ordinary helium tubes may be that in such tubes the ionization of helium is not so complete as in the star considered or in the experiments of Fowler, where a strong discharge was sent through a mixture of hydrogen and helium. The condition for the appearance of the spectrum is, according to the above theory, that helium atoms are present in a state in which they have lost both their electrons. Now we must assume that the amount of energy to be used in removing the second electron from a helium atom is much greater than that to be used in removing the first. Further, it is known from experiments on positive rays, that hydrogen atoms can acquire a negative charge ; therefore the presence of hydrogen in the experiments of Fowler may effect that more electrons are removed from some of the helium atoms than would be the case if only helium were present.

Spectra of other substances.—In case of systems containing more electrons we must—in conformity with the result of experiments—expect more complicated laws for the line-spectra than those considered. I shall try to show that the point of view taken above allows, at any rate, a certain understanding of the laws observed.

According to Rydberg's theory—with the generalization given by Ritz [*]—the frequency corresponding to the lines of the spectrum of an element can be expressed by

$$\nu = F_r(\tau_1) - F_s(\tau_2),$$

where τ_1 and τ_2 are entire numbers, and F_1, F_2, F_3, ... are functions of τ which approximately are equal to $\dfrac{K}{(\tau+a_1)^2}$, $\dfrac{K}{(\tau+a_2)^2}$, ... K is a universal constant, equal to the factor outside the bracket in the formula (4) for the spectrum of hydrogen. The different series appear if we put τ_1 or τ_2 equal to a fixed number and let the other vary.

The circumstance that the frequency can be written as a difference between two functions of entire numbers suggests an origin of the lines in the spectra in question similar to the one we have assumed for hydrogen; *i. e.* that the lines correspond to a radiation emitted during the passing of the system between two different stationary states. For systems containing more than one electron the detailed discussion may be very complicated, as there will be many different configurations of the electrons which can be taken into consideration as stationary states. This may account for the different sets of series in the line spectra emitted from the

[*] W. Ritz, *Phys. Zeitschr.* ix. p. 521 (1908).

12   Dr. N. Bohr *on the Constitution*

substances in question. Here I shall only try to show how, by help of the theory, it can be simply explained that the constant K entering in Rydberg's formula is the same for all substances.

Let us assume that the spectrum in question corresponds to the radiation emitted during the binding of an electron ; and let us further assume that the system including the electron considered is neutral. The force on the electron, when at a great distance apart from the nucleus and the electrons previously bound, will be very nearly the same as in the above case of the binding of an electron by a hydrogen nucleus. The energy corresponding to one of the stationary states will therefore for τ great be very nearly equal to that given by the expression (3) on p. 5, if we put $E = e$. For τ great we consequently get

$$\lim (\tau^2 . F_1(\tau)) = \lim (\tau^2 . F_2(\tau)) = \ldots = \frac{2\pi^2 m e^4}{h^3},$$

in conformity with Rydberg's theory.

§ 3. *General Considerations continued.*

We shall now return to the discussion (see p. 7) of the special assumptions used in deducing the expressions (3) on p. 5 for the stationary states of a system consisting of an electron rotating round a nucleus.

For one, we have assumed that the different stationary states correspond to an emission of a different number of energy-quanta. Considering systems in which the frequency is a function of the energy, this assumption, however, may be regarded as improbable; for as soon as one quantum is sent out the frequency is altered. We shall now see that we can leave the assumption used and still retain the equation (2) on p. 5, and thereby the formal analogy with Planck's theory.

Firstly, it will be observed that it has not been necessary, in order to account for the law of the spectra by help of the expressions (3) for the stationary states, to assume that in any case a radiation is sent out corresponding to more than a single energy-quantum, $h\nu$. Further information on the frequency of the radiation may be obtained by comparing calculations of the energy radiation in the region of slow vibrations based on the above assumptions with calculations based on the ordinary mechanics. As is known, calculations on the latter basis are in agreement with experiments on the energy radiation in the named region.

Let us assume that the ratio between the total amount of

of *Atoms and Molecules.* 13

energy emitted and the frequency of revolution of the electron for the different stationary states is given by the equation $W = f(\tau) \cdot h\omega$, instead of by the equation (2). Proceeding in the same way as above, we get in this case instead of (3)

$$W = \frac{\pi^2 m e^2 E^2{}_!}{2h^2 f^2(\tau)}, \quad \omega = \frac{\pi^2 m e^2 E^2}{2h^3 f^3(\tau)}.$$

Assuming as above that the amount of energy emitted during the passing of the system from a state corresponding to $\tau = \tau_1$ to one for which $\tau = \tau_2$ is equal to $h\nu$, we get instead of (4)

$$\nu = \frac{\pi^2 m e^2 E^2}{2h^3} \left(\frac{1}{f^2(\tau_2)} - \frac{1}{f^2(\tau_1)} \right).$$

We see that in order to get an expression of the same form as the Balmer series we must put $f(\tau) = c\tau$.

In order to determine c let us now consider the passing of the system between two successive stationary states corresponding to $\tau = N$ and $\tau = N-1$; introducing $f(\tau) = c\tau$, we get for the frequency of the radiation emitted

$$\nu = \frac{\pi^2 m e^2 E^2}{2c^2 h^3} \cdot \frac{2N-1}{N^2(N-1)^2}.$$

For the frequency of revolution of the electron before and after the emission we have

$$\omega_N = \frac{\pi^2 m e^2 E^2}{2c^3 h^3 N^3} \quad \text{and} \quad \omega_{N-1} = \frac{\pi^2 m e^2 E^2}{2c^3 h^3 (N-1)^3}.$$

If N is great the ratio between the frequency before and after the emission will be very near equal to 1; and according to the ordinary electrodynamics we should therefore expect that the ratio between the frequency of radiation and the frequency of revolution also is very nearly equal to 1. This condition will only be satisfied if $c = \frac{1}{2}$. Putting $f(\tau) = \frac{\tau}{2}$, we, however, again arrive at the equation (2) and consequently at the expression (3) for the stationary states.

If we consider the passing of the system between two states corresponding to $\tau = N$ and $\tau = N-n$, where n is small compared with N, we get with the same approximation as above, putting $f(\tau) = \frac{\tau}{2}$,

$$\nu = n\omega.$$

14   Dr. N. Bohr *on the Constitution*

The possibility of an emission of a radiation of such a frequency may also be interpreted from analogy with the ordinary electrodynamics, as an electron rotating round a nucleus in an elliptical orbit will emit a radiation which according to Fourier's theorem can be resolved into homogeneous components, the frequencies of which are $n\omega$, if ω is the frequency of revolution of the electron.

We are thus led to assume that the interpretation of the equation (2) is not that the different stationary states correspond to an emission of different numbers of energy-quanta, but that the frequency of the energy emitted during the passing of the system from a state in which no energy is yet radiated out to one of the different stationary states, is equal to different multiples of $\frac{\omega}{2}$, where ω is the frequency of revolution of the electron in the state considered. From this assumption we get exactly the same expressions as before for the stationary states, and from these by help of the principal assumptions on p. 7 the same expression for the law of the hydrogen spectrum. Consequently we may regard our preliminary considerations on p. 5 only as a simple form of representing the results of the theory.

Before we leave the discussion of this question, we shall for a moment return to the question of the significance of the agreement between the observed and calculated values of the constant entering in the expressions (4) for the Balmer series of the hydrogen spectrum. From the above consideration it will follow that, taking the starting-point in the form of the law of the hydrogen spectrum and assuming that the different lines correspond to a homogeneous radiation emitted during the passing between different stationary states, we shall arrive at exactly the same expression for the constant in question as that given by (4), if we only assume (1) that the radiation is sent out in quanta $h\nu$, and (2) that the frequency of the radiation emitted during the passing of the system between successive stationary states will coincide with the frequency of revolution of the electron in the region of slow vibrations.

As all the assumptions used in this latter way of representing the theory are of what we may call a qualitative character, we are justified in expecting—if the whole way of considering is a sound one—an absolute agreement between the values calculated and observed for the constant in question, and not only an approximate agreement. The formula (4) may therefore be of value in the discussion of the results of experimental determinations of the constants e, m, and h.

While there obviously can be no question of a mechanical foundation of the calculations given in this paper, it is, however, possible to give a very simple interpretation of the result of the calculation on p. 5 by help of symbols taken from the ordinary mechanics. Denoting the angular momentum of the electron round the nucleus by M, we have immediately for a circular orbit $\pi M = \dfrac{T}{\omega}$, where ω is the frequency of revolution and T the kinetic energy of the electron; for a circular orbit we further have $T = W$ (see p. 3) and from (2), p. 5, we consequently get

$$M = \tau M_0,$$

where
$$M_0 = \frac{h}{2\pi} = 1 \cdot 04 \times 10^{-27}.$$

If we therefore assume that the orbit of the electron in the stationary states is circular, the result of the calculation on p. 5 can be expressed by the simple condition : that the angular momentum of the electron round the nucleus in a stationary state of the system is equal to an entire multiple of a universal value, independent of the charge on the nucleus. The possible importance of the angular momentum in the discussion of atomic systems in relation to Planck's theory is emphasized by Nicholson [*].

The great number of different stationary states we do not observe except by investigation of the emission and absorption of radiation. In most of the other physical phenomena, however, we only observe the atoms of the matter in a single distinct state, *i. e.* the state of the atoms at low temperature. From the preceding considerations we are immediately led to the assumption that the " permanent " state is the one among the stationary states during the formation of which the greatest amount of energy is emitted. According to the equation (3) on p. 5, this state is the one which corresponds to $\tau = 1$.

§ 4. *Absorption of Radiation.*

In order to account for Kirchhoff's law it is necessary to introduce assumptions on the mechanism of absorption of radiation which correspond to those we have used considering the emission. Thus we must assume that a system consisting of a nucleus and an electron rotating round it under certain circumstances can absorb a radiation of a frequency equal to the frequency of the homogeneous radiation emitted during

[*] J. W. Nicholson, *loc. cit.* p. 679.

16   Dr. N. Bohr *on the Constitution*

the passing of the system between different stationary states. Let us consider the radiation emitted during the passing of the system between two stationary states A_1 and A_2 corresponding to values for τ equal to τ_1 and τ_2, $\tau_1 > \tau_2$. As the necessary condition for an emission of the radiation in question was the presence of systems in the state A_1, we must assume that the necessary condition for an absorption of the radiation is the presence of systems in the state A_2.

These considerations seem to be in conformity with experiments on absorption in gases. In hydrogen gas at ordinary conditions for instance there is no absorption of a radiation of a frequency corresponding to the line-spectrum of this gas ; such an absorption is only observed in hydrogen gas in a luminous state. This is what we should expect according to the above. We have on p. 9 assumed that the radiation in question was emitted during the passing of the systems between stationary states corresponding to $\tau \geqq 2$. The state of the atoms in hydrogen gas at ordinary conditions should, however, correspond to $\tau = 1$; furthermore, hydrogen atoms at ordinary conditions combine into molecules, *i. e.* into systems in which the electrons have frequencies different from those in the atoms (see Part III.). From the circumstance that certain substances in a non-luminous state, as, for instance, sodium vapour, absorb radiation corresponding to lines in the line-spectra of the substances, we may, on the other hand, conclude that the lines in question are emitted during the passing of the system between two states, one of which is the permanent state.

How much the above considerations differ from an interpretation based on the ordinary electrodynamics is perhaps most clearly shown by the fact that we have been forced to assume that a system of electrons will absorb a radiation of a frequency different from the frequency of vibration of the electrons calculated in the ordinary way. It may in this connexion be of interest to mention a generalization of the considerations to which we are led by experiments on the photo-electric effect, and which may be able to throw some light on the problem in question. Let us consider a state of the system in which the electron is free, *i. e.* in which the electron possesses kinetic energy sufficient to remove to infinite distances from the nucleus. If we assume that the motion of the electron is governed by the ordinary mechanics and that there is no (sensible) energy radiation, the total energy of the system—as in the above considered stationary states—will be constant. Further, there will be perfect continuity between the two kinds of states, as the difference between

of Atoms and Molecules.   17

frequency and dimensions of the systems in successive
stationary states will diminish without limit if τ increases.
In the following considerations we shall for the sake of
brevity refer to the two kinds of states in question as
" mechanical " states ; by this notation only emphasizing
the assumption that the motion of the electron in both cases
can be accounted for by the ordinary mechanics.

Tracing the analogy between the two kinds of mechanical
states, we might now expect the possibility of an absorption
of radiation, not only corresponding to the passing of the
system between two different stationary states, but also
corresponding to the passing between one of the stationary
states and a state in which the electron is free ; and as above,
we might expect that the frequency of this radiation was determined by the equation $E = h\nu$, where E is the difference
between the total energy of the system in the two states.
As it will be seen, such an absorption of radiation is just
what is observed in experiments on ionization by ultra-violet
light and by Röntgen rays. Obviously, we get in this way
the same expression for the kinetic energy of an electron
ejected from an atom by photo-electric effect as that deduced
by Einstein [*], *i. e.* $T = h\nu - W$, where T is the kinetic energy
of the electron ejected, and W the total amount of energy
emitted during the original binding of the electron.

The above considerations may further account for the
result of some experiments of R. W. Wood [†] on absorption
of light by sodium vapour. In these experiments, an
absorption corresponding to a very great number of lines in
the principal series of the sodium spectrum is observed, and
in addition a continuous absorption which begins at the head
of the series and extends to the extreme ultra-violet. This
is exactly what we should expect according to the analogy in
question, and, as we shall see, a closer consideration of the
above experiments allows us to trace the analogy still
further. As mentioned on p. 9 the radii of the orbits of
the electrons will for stationary states corresponding to high
values for τ be very great compared with ordinary atomic
dimensions. This circumstance was used as an explanation
of the non-appearance in experiments with vacuum-tubes of
lines corresponding to the higher numbers in the Balmer
series of the hydrogen spectrum. This is also in conformity
with experiments on the emission spectrum of sodium ; in
the principal series of the emission spectrum of this substance

[*] A. Einstein, *Ann. d. Phys.* xvii. p. 146 (1905).
[†] R. W. Wood, Physical Optics, p. 513 (1911).

18   Dr. N. Bohr *on the Constitution*

rather few lines are observed. Now in Wood's experiments the pressure was not very low, and the states corresponding to high values for τ could therefore not appear ; yet in the absorption spectrum about 50 lines were detected. In the experiments in question we consequently observe an absorption of radiation which is not accompanied by a complete transition between two different stationary states. According to the present theory we must assume that this absorption is followed by an emission of energy during which the systems pass back to the original stationary state. If there are no collisions between the different systems this energy will be emitted as a radiation of the same frequency as that absorbed, and there will be no true absorption but only a scattering of the original radiation ; a true absorption will not occur unless the energy in question is transformed by collisions into kinetic energy of free particles. In analogy we may now from the above experiments conclude that a bound electron—also in cases in which there is no ionization—will have an absorbing (scattering) influence on a homogeneous radiation, as soon as the frequency of the radiation is greater than W/h, where W is the total amount of energy emitted during the binding of the electron. This would be highly in favour of a theory of absorption as the one sketched above, as there can in such a case be no question of a coincidence of the frequency of the radiation and a characteristic frequency of vibration of the electron. It will further be seen that the assumption, that there will be an absorption (scattering) of any radiation corresponding to a transition between two different mechanical states, is in perfect analogy with the assumption generally used that a free electron will have an absorbing (scattering) influence on light of any frequency. Corresponding considerations will hold for the emission of radiation.

   In analogy to the assumption used in this paper that the emission of line-spectra is due to the re-formation of atoms after one or more of the lightly bound electrons are removed, we may assume that the homogeneous Röntgen radiation is emitted during the settling down of the systems after one of the firmly bound electrons escapes, *e. g.* by impact of cathode particles [*]. In the next part of this paper, dealing with the constitution of atoms, we shall consider the question more closely and try to show that a calculation based on this assumption is in quantitative agreement with the results of experiments : here we shall only mention briefly a problem with which we meet in such a calculation.

   [*] Compare J. J. Thomson, Phil. Mag. xxiii. p. 456 (1912).

Experiments on the phenomena of X-rays suggest that not only the emission and absorption of radiation cannot be treated by the help of the ordinary electrodynamics, but not even the result of a collision between two electrons of which the one is bound in an atom. This is perhaps most clearly shown by some very instructive calculations on the energy of β-particles emitted from radioactive substances recently published by Rutherford *. These calculations strongly suggest that an electron of great velocity in passing through an atom and colliding with the electrons bound will loose energy in distinct finite quanta. As is immediately seen, this is very different from what we might expect if the result of the collisions was governed by the usual mechanical laws. The failure of the classical mechanics in such a problem might also be expected beforehand from the absence of anything like equipartition of kinetic energy between free electrons and electrons bound in atoms. From the point of view of the "mechanical" states we see, however, that the following assumption—which is in accord with the above analogy—might be able to account for the result of Rutherford's calculation and for the absence of equipartition of kinetic energy : two colliding electrons, bound or free, will, after the collision as well as before, be in mechanical states. Obviously, the introduction of such an assumption would not make any alteration necessary in the classical treatment of a collision between two free particles. But, considering a collision between a free and a bound electron, it would follow that the bound electron by the collision could not acquire a less amount of energy than the difference in energy corresponding to successive stationary states, and consequently that the free electron which collides with it could not lose a less amount.

The preliminary and hypothetical character of the above considerations needs not to be emphasized. The intention, however, has been to show that the sketched generalization of the theory of the stationary states possibly may afford a simple basis of representing a number of experimental facts which cannot be explained by help of the ordinary electrodynamics, and that the assumptions used do not seem to be inconsistent with experiments on phenomena for which a satisfactory explanation has been given by the classical dynamics and the wave theory of light.

* E. Rutherford, Phil. Mag. xxiv. pp. 453 & 893 (1912).

C 2

20   Dr. N. Bohr *on the Constitution*

§ 5. *The permanent State of an Atomic System.*

We shall now return to the main object of this paper—the discussion of the " permanent " state of a system consisting of nuclei and bound electrons. For a system consisting of a nucleus and an electron rotating round it, this state is, according to the above, determined by the condition that the angular momentum of the electron round the nucleus is equal to $\dfrac{h}{2\pi}$.

On the theory of this paper the only neutral atom which contains a single electron is the hydrogen atom. The permanent state of this atom should correspond to the values of a and ω calculated on p. 5. Unfortunately, however, we know very little of the behaviour of hydrogen atoms on account of the small dissociation of hydrogen molecules at ordinary temperatures. In order to get a closer comparison with experiments, it is necessary to consider more complicated systems.

Considering systems in which more electrons are bound by a positive nucleus, a configuration of the electrons which presents itself as a permanent state is one in which the electrons are arranged in a ring round the nucleus. In the discussion of this problem on the basis of the ordinary electrodynamics, we meet—apart from the question of the energy radiation—with new difficulties due to the question of the stability of the ring. Disregarding for a moment this latter difficulty, we shall first consider the dimensions and frequency of the systems in relation to Planck's theory of radiation.

Let us consider a ring consisting of n electrons rotating round a nucleus of charge E, the electrons being arranged at equal angular intervals around the circumference of a circle of radius a.

The total potential energy of the system consisting of the electrons and the nucleus is

$$P = -\frac{ne}{a}\,(E - es_n),$$

where

$$s_n = \frac{1}{4}\sum_{s=1}^{s=n-1} \operatorname{cosec} \frac{s\pi}{n}.$$

For the radial force exerted on an electron by the nucleus and the other electrons we get

$$F = -\frac{1}{n}\frac{dP}{da} = -\frac{e}{a^2}\,(E - es_n).$$

Denoting the kinetic energy of an electron by T and neglecting the electromagnetic forces due to the motion of the electrons (see Part II.), we get, putting the centrifugal force on an electron equal to the radial force,

$$\frac{2T}{a} = \frac{e}{a^2}(E - es_n),$$

or

$$T = \frac{e}{2a}(E - es_n).$$

From this we get for the frequency of revolution

$$\omega = \frac{1}{2\pi}\sqrt{\frac{e(E - es_n)}{ma^3}}.$$

The total amount of energy W necessary transferred to the system in order to remove the electrons to infinite distances apart from the nucleus and from each other is

$$W = -P - nT = \frac{ne}{2a}(E - es_n) = nT,$$

equal to the total kinetic energy of the electrons.

We see that the only difference in the above formula and those holding for the motion of a single electron in a circular orbit round a nucleus is the exchange of E for $E - es_n$. It is also immediately seen that corresponding to the motion of an electron in an elliptical orbit round a nucleus, there will be a motion of the n electrons in which each rotates in an elliptical orbit with the nucleus in the focus, and the n electrons at any moment are situated at equal angular intervals on a circle with the nucleus as the centre. The major axis and frequency of the orbit of the single electrons will for this motion be given by the expressions (1) on p. 3 if we replace E by $E - es_n$ and W by $\frac{W}{n}$. Let us now suppose that the system of n electrons rotating in a ring round a nucleus is formed in a way analogous to the one assumed for a single electron rotating round a nucleus. It will thus be assumed that the electrons, before the binding by the nucleus, were at a great distance apart from the latter and possessed no sensible velocities, and also that during the binding a homogeneous radiation is emitted. As in the case of a single electron, we have here that the total amount of energy emitted during the formation of the system is equal to the final kinetic energy of the electrons. If we now suppose that during the formation of the system the electrons at any moment are situated at equal angular intervals on the circumference of a circle with the nucleus in the centre, from analogy with the considerations on p. 5 we are here led to assume the existence of a series of stationary configurations in which the kinetic energy per electron is equal to $\tau h \dfrac{\omega}{2}$, where τ is an entire number, h Planck's constant, and ω the frequency of revolution. The configuration in which the greatest amount of energy is emitted is, as before, the one in which $\tau = 1$. This configuration we shall assume to be the permanent state of the system if the electrons in this state are arranged in a single ring. As for the case of a single electron we get that the angular momentum of each of the electrons is equal to $\dfrac{h}{2\pi}$. It may be remarked that instead of considering the single electrons we might have considered the ring as an entity. This would, however, lead to the same result, for in this case the frequency of revolution ω will be replaced by the frequency $n\omega$ of the radiation from the whole ring calculated from the ordinary electrodynamics, and T by the total kinetic energy nT.

There may be many other stationary states corresponding to other ways of forming the system. The assumption of the existence of such states seems necessary in order to account for the line-spectra of systems containing more than one electron (p. 11); it is also suggested by the theory of Nicholson mentioned on p. 6, to which we shall return in a moment. The consideration of the spectra, however, gives, as far as I can see, no indication of the existence of stationary states in which all the electrons are arranged in a ring and which correspond to greater values for the total energy emitted than the one we above have assumed to be the permanent state.

Further, there may be stationary configurations of a system of n electrons and a nucleus of charge E in which all the electrons are not arranged in a single ring. The question, however, of the existence of such stationary configurations is not essential for our determination of the permanent state, as long as we assume that the electrons in this state of the system are arranged in a single ring. Systems corresponding to more complicated configurations will be discussed on p. 24.

Using the relation $T = h \dfrac{\omega}{2}$ we get, by help of the above expressions for T and ω, values for a and ω corresponding to the permanent state of the system which only differ from those given by the equations (3) on p. 5, by exchange of E for $E - es_n$.

The question of stability of a ring of electrons rotating round a positive charge is discussed in great detail by Sir J. J. Thomson *. An adaption of Thomson's analysis for the case here considered of a ring rotating round a nucleus of negligibly small linear dimensions is given by Nicholson †. The investigation of the problem in question naturally divides in two parts : one concerning the stability for displacements of the electrons in the plane of the ring ; one concerning displacements perpendicular to this plane. As Nicholson's calculations show, the answer to the question of stability differs very much in the two cases in question. While the ring for the latter displacements in general is stable if the number of electrons is not great ; the ring is in no case considered by Nicholson stable for displacements of the first kind.

According, however, to the point of view taken in this paper, the question of stability for displacements of the electrons in the plane of the ring is most intimately connected with the question of the mechanism of the binding of the electrons, and like the latter cannot be treated on the basis of the ordinary dynamics. The hypothesis of which we shall make use in the following is that the stability of a ring of electrons rotating round a nucleus is secured through the above condition of the universal constancy of the angular momentum, together with the further condition that the configuration of the particles is the one by the formation of which the greatest amount of energy is emitted. As will be shown, this hypothesis is, concerning the question of stability for a displacement of the electrons perpendicular to the plane of the ring, equivalent to that used in ordinary mechanical calculations.

Returning to the theory of Nicholson on the origin of lines observed in the spectrum of the solar corona, we shall now see that the difficulties mentioned on p. 7 may be only formal. In the first place, from the point of view considered above the objection as to the instability of the systems for displacements of the electrons in the plane of the ring may not be valid. Further, the objection as to the emission of the radiation in quanta will not have reference to the calculations in question, if we assume that in the coronal spectrum we are not dealing with a true emission but only with a scattering of radiation. This assumption seems probable if we consider

* *Loc. cit.*   † *Loc. cit.*

24   *On the Constitution of Atoms and Molecules.*

the conditions in the celestial body in question; for on account of the enormous rarefaction of the matter there may be comparatively few collisions to disturb the stationary states and to cause a true emission of light corresponding to the transition between different stationary states; on the other hand there will in the solar corona be intense illumination of light of all frequencies which may excite the natural vibrations of the systems in the different stationary states. If the above assumption is correct, we immediately understand the entirely different form for the laws connecting the lines discussed by Nicholson and those connecting the ordinary line-spectra considered in this paper.

Proceeding to consider systems of a more complicated constitution, we shall make use of the following theorem, which can be very simply proved :—

" In every system consisting of electrons and positive nuclei, in which the nuclei are at rest and the electrons move in circular orbits with a velocity small compared with the velocity of light, the kinetic energy will be numerically equal to half the potential energy."

By help of this theorem we get—as in the previous cases of a single electron or of a ring rotating round a nucleus— that the total amount of energy emitted, by the formation of the systems from a configuration in which the distances apart of the particles are infinitely great and in which the particles have no velocities relative to each other, is equal to the kinetic energy of the electrons in the final configuration.

In analogy with the case of a single ring we are here led to assume that corresponding to any configuration of equilibrium a series of geometrically similar, stationary configurations of the system will exist in which the kinetic energy of every electron is equal to the frequency of revolution multiplied by $\frac{\tau}{2}h$ where τ is an entire number and h Planck's constant. In any such series of stationary configurations the one corresponding to the greatest amount of energy emitted will be the one in which τ for every electron is equal to 1. Considering that the ratio of kinetic energy to frequency for a particle rotating in a circular orbit is equal to π times the angular momentum round the centre of the orbit, we are therefore led to the following simple generalization of the hypotheses mentioned on pp. 15 and 22.

" *In any molecular system consisting of positive nuclei and electrons in which the nuclei are at rest relative to each other and the electrons move in circular orbits, the angular momentum* *of every electron round the centre of its orbit will in the permanent state of the system be equal to $\dfrac{h}{2\pi}$, where h is Planck's constant"* *.

In analogy with the considerations on p. 23, we shall assume that a configuration satisfying this condition is stable if the total energy of the system is less than in any neighbouring configuration satisfying the same condition of the angular momentum of the electrons.

As mentioned in the introduction, the above hypothesis will be used in a following communication as a basis for a theory of the constitution of atoms and molecules. It will be shown that it leads to results which seem to be in conformity with experiments on a number of different phenomena.

The foundation of the hypothesis has been sought entirely in its relation with Planck's theory of radiation ; by help of considerations given later it will be attempted to throw some further light on the foundation of it from another point of view.

April 5, 1913.

1024   Mr. H. G. J. Moseley *on the*

effect in the case of many metals and alloys are subject to variations as great as an octave and more.   This difficulty is all the more real in that as yet we are not in a position to determine what influences on and in the extremely thin bounding surface of the metal in which the light absorption takes place are the determining factors in this displacement.

Berlin, Physikalisches Institut
    der Universität, July 1913.

XCIII. *The High-Frequency Spectra of the Elements.* By H. G. J. MOSELEY, *M.A.**

[Plate XXIII.]

IN the absence of any available method of spectrum analysis, the characteristic types of X radiation, which an atom emits when suitably excited, have hitherto been described in terms of their absorption in aluminium †.   The interference phenomena exhibited by X rays when scattered by a crystal have now, however, made possible the accurate determination of the frequencies of the various types of radiation.   This was shown by W. H. and W. L. Bragg‡, who by this method analysed the line spectrum emitted by the platinum target of an X-ray tube.   C. G. Darwin and the author§ extended this analysis and also examined the continuous spectrum, which in this case constitutes the greater part of the radiation.   Recently Prof. Bragg‖ has also determined the wave-lengths of the strongest lines in the spectra of nickel, tungsten, and rhodium.   The electrical methods which have hitherto been employed are, however, only successful where a constant source of radiation is available.   The present paper contains a description of a method of photographing these spectra, which makes the analysis of the X rays as simple as any other branch of spectroscopy. The author intends first to make a general survey of the principal types of high-frequency radiation, and then to examine the spectra of a few elements in greater detail and with greater accuracy.   The results already obtained show that such data have an important bearing on the question of

* Communicated by Prof. E. Rutherford, F.R.S.
† *Cf.* Barkla, Phil. Mag. xxii. p. 396 (1911).
‡ Proc. Roy. Soc. A. lxxxviii. p. 428 (1913).
§ Phil. Mag. xxvi. p. 210 (1913).
‖ Proc. Roy. Soc. A. lxxxix. p. 246 (1913).

High-Frequency Spectra of the Elements.   1025

the internal structure of the atom, and strongly support the views of Rutherford[*] and of Bohr[†].

Kaye[‡] has shown that an element excited by a stream of sufficiently fast cathode rays emits its characteristic X radiation. He used as targets a number of substances mounted on a truck inside an exhausted tube. A magnetic device enabled each target to be brought in turn into the line of fire. This apparatus was modified to suit the present work. The cathode stream was concentrated on to a small area of the target, and a platinum plate furnished with a fine vertical slit placed immediately in front of the part bombarded. The tube was exhausted by a Gaede mercury pump, charcoal in liquid air being also sometimes used to remove water vapour. The X rays, after passing through the slit marked S in fig. 1,

Fig. 1.

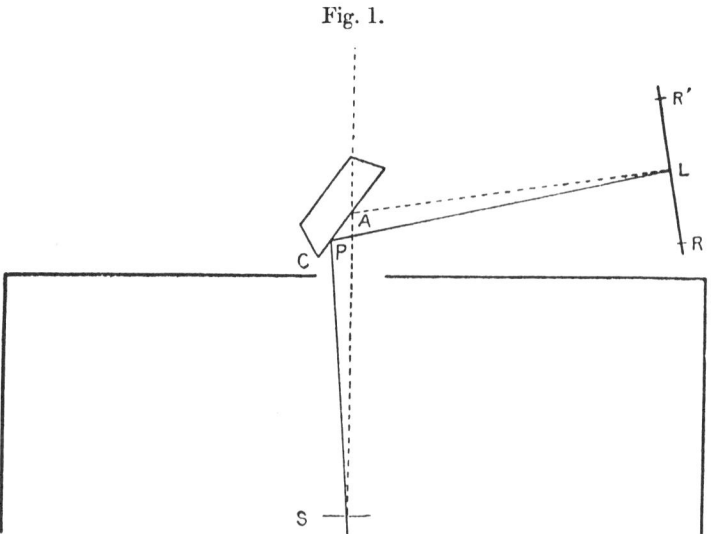

emerged through an aluminium window ·02 mm. thick. The rest of the radiation was shut off by a lead box which surrounded the tube. The rays fell on the cleavage face, C, of a crystal of potassium ferrocyanide which was mounted on the prism-table of a spectrometer. The surface of the crystal was vertical and contained the geometrical axis of the spectrometer.

[*] Phil. Mag. xxi. p. 669 (1911).
[†] Phil. Mag. xxvi. pp. 1, 476, & 857 (1913).
[‡] Phil. Trans. Roy. Soc. A. ccix. p. 123 (1909).

1026   Mr. H. G. J. Moseley *on the*

Now it is known* that X rays consist in general of two types, the heterogeneous radiation and characteristic radiations of definite frequency. The former of these is reflected from such a surface at all angles of incidence, but at the large angles used in the present work the reflexion is of very little intensity. The radiations of definite frequency, on the other hand, are reflected only when they strike the surface at definite angles, the glancing angle of incidence θ, the wavelength λ, and the "grating constant" d of the crystal being connected by the relation

$$n\lambda = 2d \sin \theta, \quad \ldots \quad \ldots \quad (1)$$

where n, an integer, may be called the "order" in which the reflexion occurs. The particular crystal used, which was a fine specimen with face 6 cm. square, was known to give strong reflexions in the first three orders, the third order being the most prominent.

If then a radiation of definite wave-length happens to strike any part P of the crystal at a suitable angle, a small part of it is reflected. Assuming for the moment that the source of the radiation is a point, the locus of P is obviously the arc of a circle, and the reflected rays will travel along the generating lines of a cone with apex at the image of the source. The effect on a photographic plate L will take the form of the arc of an hyperbola, curving away from the direction of the direct beam. With a fine slit at S, the arc becomes a fine line which is slightly curved in the direction indicated.

The photographic plate was mounted on the spectrometer arm, and both the plate and the slit were 17 cm. from the axis. The importance of this arrangement lies in a geometrical property, for when these two distances are equal the point L at which a beam reflected at a definite angle strikes the plate is independent of the position of P on the crystal surface. The angle at which the crystal is set is then immaterial so long as a ray can strike some part of the surface at the required angle. The angle θ can be obtained from the relation $2\theta = 180° - SPL = 180° - SAL$.

The following method was used for measuring the angle SAL. Before taking a photograph a reference line R was made at both ends of the plate by replacing the crystal by a lead screen furnished with a fine slit which coincided with the axis of the spectrometer. A few seconds' exposure to the X rays then gave a line R on the plate, and so defined on it

* Moseley and Darwin, *loc. cit.*

the line joining S and A. A second line R' was made in the same way after turning the spectrometer arm through a definite angle. The arm was then turned to the position required to catch the reflected beam and the angles LAP for any lines which were subsequently found on the plate deduced from the known value of RAP and the position of the lines on the plate. The angle LAR was measured with an error of not more than $0°\cdot1$, by superposing on the negative a plate on which reference lines had been marked in the same way at intervals of 1°. In finding from this the glancing angle of reflexion two small corrections were necessary in practice, since neither the face of the crystal nor the lead slit coincided accurately with the axis of the spectrometer. Wave-lengths varying over a range of about 30 per cent. could be reflected for a given position of the crystal.

In almost all cases the time of exposure was five minutes. Ilford X-ray plates were used and were developed with rodinal. The plates were mounted in a plate-holder, the front of which was covered with black paper. In order to determine the wave-length from the reflexion angle θ it is necessary to know both the order n in which the reflexion occurs and the grating constant d. n was determined by photographing every spectrum both in the second order and the third. This also gave a useful check on the accuracy of the measurements ; d cannot be calculated directly for the complicated crystal potassium ferrocyanide. The grating constant of this particular crystal had, however, previously [*] been accurately compared with d', the constant of a specimen of rocksalt. It was found that

$$d = 3d' \frac{\cdot1988}{\cdot1985}.$$

Now W. L. Bragg [†] has shown that the atoms in a rocksalt crystal are in simple cubic array. Hence the number of atoms per c.c.

$$2 \frac{N\sigma}{M} = \frac{1}{(d')^3} :$$

N, the number of molecules in a gram-mol., $= 6\cdot05 \times 10^{23}$, assuming the charge on an electron to be $4\cdot89 \times 10^{-10}$: σ, the density of this crystal of rocksalt, was $2\cdot167$, and M the molecular weight $= 58\cdot46$.

[*] Moseley & Darwin, *loc. cit.*
[†] Proc. Roy. Soc. A. lxxxix. p. 248 (1913).

Table I.

Element.	Line.	θ_α.	λ.	θ_β.	λ.	$\lambda_\alpha/\lambda_\beta$.	$Q=(\nu/\tfrac{3}{4}\nu_0)^{\frac{1}{2}}$	N atomic number.	Atomic weight.
Calcium.	α	23.4°	3.357×10⁻⁸	36.7°	3.368×10⁻⁸	1.089	19.00	20	40.09
	β	21.4	3.035	33.3	3.094				
Scandium.	α	...	...	...	...	...	...	21	44.1
	β								
Titanium.	α	19.1	2.766	29.3	2.758	1.093	20.99	22	48.1
	β	17.4	2.528	26.6	2.524				
Vanadium.	α	17.35	2.521	26.55	2.519	1.097	21.96	23	51.06
	β	15.8	2.302	24.05	2.297				
Chromium.	α	15.75	2.295	24.1	2.301	1.100	22.98	24	52.0
	β	14.3	2.088	21.8	2.093				
Manganese.	α	14.5	2.117	22.0	2.111	1.101	23.99	25	54.93
	β	13.15	1.923	19.9	1.918				
Iron.	α	13.3	1.945	20.2	1.946	1.103	24.99	26	55.85
	β	12.05	1.765	18.25	1.765				
Cobalt.	α	12.25	1.794	18.6	1.798	1.104	26.00	27	58.97
	β	11.15	1.635	16.8	1.629				
Nickel.	α	11.35	1.664	17.15	1.662	1.104	27.04	28	58.68
	β	10.25	1.504	15.5	1.506				
Copper.	α	10.55	1.548	15.95	1.549	1.105	28.01	29	63.57
	β	9.55	1.403	14.4	1.402				
Zinc.	α	9.85	1.446	14.85	1.445	1.106	29.01	30	65.37
	β	not	found	13.4	1.306				

This gives $d' = 2\cdot814 \times 10^{-8}$ and $d = 8\cdot454 \times 10^{-8}$ cm. It is seen that the determination of wave-length depends on $e^{\frac{1}{3}}$, so that the effect of uncertainty in the value of this quantity will not be serious. Lack of homogeneity in the crystal is a more likely source of error, as minute inclusions of water would make the true density greater than that found experimentally.

Twelve elements have so far been examined. The ten given in Table I. were chosen as forming a continuous series with only one gap. It was hoped in this way to bring out clearly any systematic results. The inclusion of nickel was of special interest owing to its anomalous position in the periodic system. Radiations from these substances are readily excited, and the large angles of reflexion make it easy to measure the wave-lengths with accuracy. Calcium alone gave any trouble. In this case, owing to the high absorption coefficient of the principal radiation—about 1200 cm.$^{-1}$ in aluminium—the X-ray tube was provided with a window of goldbeaters' skin and the air between the crystal and the photographic plate displaced by hydrogen. The layer of lime which covered the surface of the metal gave off such a quantity of gas that the X rays could only be excited for a second or two at a time. Brass was substituted for zinc to avoid volatilization by the intense heat generated at the point struck by the cathode rays. Ferro-vanadium (35 per cent. V) and ferro-titanium (23 per cent. Ti), for which I am indebted to the International Vanadium Co., proved convenient substitutes for the pure elements, which are not easily obtained in the solid form.

Plate XXIII. shows the spectra in the third order placed approximately in register. Those parts of the photographs which represent the same angle of reflexion are in the same vertical line. The actual angles can be taken from Table I. It is to be seen that the spectrum of each element consists of two lines. Of these the stronger has been called α in the table, and the weaker β. The lines found on any of the plates besides α and β were almost certainly all due to impurities. Thus in both the third and second order the cobalt spectrum shows Ni α very strongly and Fe α faintly. In the third order the nickel spectrum shows Mn α_2 faintly. The brass spectra naturally show α and β both of Cu and of Zn, but Zn β_2 has not yet been found. In the second order the ferro-vanadium and ferro-titanium spectra show very intense third-order Fe lines, and the former also shows Cu α_3 faintly. The Co contained Ni and 0·8 per cent. Fe, the Ni 2·2 per cent. Mn, and the V only a trace of Cu. No other lines have been found; but a search over a wide range of wave-lengths has been made only for one or two elements, and perhaps prolonged exposures, which have not yet been attempted, will show more complex spectra. The prevalence of lines due to impurities suggests that this may prove a powerful method of chemical analysis. Its advantage over ordinary spectroscopic methods lies in the simplicity of the spectra and the impossibility of one substance masking the radiation from another. It may even lead to the discovery of missing elements, as it will be possible to predict the position of their characteristic lines.

It will be seen from Table I. that the wave-lengths calculated from the two orders are in good agreement. The third order gives the stronger reflexion, and as the angles dealt with are the larger these results are the more accurate. The similarity of the different spectra is shown by the fact that the two lines α and β remain approximately constant, not only in relative intensity but also in relative wavelength. The frequency of β increases, however, slightly faster than that of α. The same two lines α strong and β weak constitute the rhodium spectrum examined by Bragg *, and they are obviously in some way closely related. One or two photographs taken with the radiation from platinum gave results in good agreement with those obtained by the electrical method, and no trace of the elaborate system of bands described by de Broglie † in the reflexion from rocksalt was encountered. The three lines found by Herweg ‡ in the reflexion from selenite doubtless represent part of the Pt spectrum in the second order. The actual breadth of the lines and certain minute details in their structure will not be considered here, as discussion would take too much space and more experiments are needed. The only other element examined was tantalum. In this case the radiation belongs to the L series, and the spectrum consists of a strong line of wave-length $1·525 \times 10^{-8}$ cm., two others of less intensity at $1·330$ and $1·287 \times 10^{-8}$ cm., and probably some very faint lines also.

A discussion will now be given of the meaning of the wave-lengths found for the principal spectrum-line α. In Table I. the values are given of the quantity

$$Q = \sqrt{\frac{\nu}{\frac{3}{4}\nu_0}},$$

* Proc. Roy. Soc. A. lxxxix. p. 277 (1913).
† *Le Radium*, x. pp. 186 & 245 (1913).
‡ *Deutsch. Phys. Ges. Verh.* xv. 13, p. 555 (1913).

ν being the frequency of the radiation α, and ν_0 the fundamental frequency of ordinary line spectra. The latter is obtained from Rydberg's wave-number, $N_0 = \dfrac{\nu}{c} = 109{,}720$.

The reason for introducing this particular constant will be given later. It is at once evident that Q increases by a constant amount as we pass from one element to the next, using the chemical order of the elements in the periodic system. Except in the case of nickel and cobalt *, this is also the order of the atomic weights. While, however, Q increases uniformly the atomic weights vary in an apparently arbitrary manner, so that an exception in their order does not come as a surprise. We have here a proof that there is in the atom a fundamental quantity, which increases by regular steps as we pass from one element to the next. This quantity can only be the charge on the central positive nucleus, of the existence of which we already have definite proof. Rutherford has shown, from the magnitude of the scattering of α particles by matter, that this nucleus carries a $+$ charge approximately equal to that of $\dfrac{A}{2}$ electrons, where A is the atomic weight. Barkla, from the scattering of X rays by matter, has shown that the number of electrons in an atom is roughly $\dfrac{A}{2}$, which for an electrically neutral atom comes to the same thing. Now atomic weights increase on the average by about 2 units at a time, and this strongly suggests the view that N increases from atom to atom always by a single electronic unit. We are therefore led by experiment to the view that N is the same as the number of the place occupied by the element in the periodic system. This atomic number is then for H 1 for He 2 for Li 3 ... for Ca 20 ... for Zn 30, &c. This theory was originated by Broek † and since used by Bohr ‡. We can confidently predict that in the few cases in which the order of the atomic weights A clashes with the chemical order of the periodic system, the chemical properties are governed by N ; while A is itself probably a complicated function of N. The very close similarity between the X-ray spectra of the different elements shows that these radiations originate inside the atom, and have no direct connexion with the complicated light-spectra and chemical properties which are governed by the structure of its surface.

* *Cf.* Barkla, Phil. Mag. xiv. p. 408 (1907).
† *Phys. Zeit.* xiv. p. 32 (1913).   ‡ *Loc. cit.*

1032   Mr. H. G. J. Moseley *on the*

We will now examine the relation

$$Q = \sqrt{\frac{\nu}{\frac{3}{4}\nu_0}}$$

more closely. So far the argument has relied on the fact that Q is a quantity which increases from atom to atom by equal steps. Now Q has been obtained by multiplying $\nu^{\frac{1}{2}}$ by a constant factor so chosen as to make the steps equal to unity. We have, therefore,

$$Q = N - k,$$

where k is a constant. Hence the frequency ν varies as $(N-k)^2$. If N for calcium is really 20 then $k=1$.

There is good reason to believe that the X-ray spectra with which we are now dealing come from the innermost ring of electrons [*]. If these electrons are held in equilibrium by mechanical forces, the angular velocity ω with which they are rotating and the radius r of their orbit are connected by

$$m\omega^2 r = \frac{e^2}{r^2}(N - \sigma_n),$$

where σ_n is a small term arising from the influence of the n electrons in the ring on each other, and $\sigma_2 = 0.25$, $\sigma_4 = 0.96$, $\sigma_6 = 1.83$, $\sigma_8 = 2.81$. In obtaining this simple expression the very small effect of other outside rings has been neglected. If then, as we pass from atom to atom, the number of electrons in the central ring remains unaltered,

$$(\omega^2 r^3)_{N+1} - (\omega^2 r^3)_N \text{ remains constant};$$

but these experiments have shown that

$$\nu^{\frac{1}{2}}_{N+1} - \nu^{\frac{1}{2}}_N \text{ is also constant,}$$

and therefore

$$\frac{\omega^2 r^3}{\nu^{\frac{1}{2}}} \text{ is constant.}$$

For the types of radiation considered by Bohr, provided the ring moves from one stationary state to another as a whole, and for the ordinary transverse vibrations of the ring, provided the influence of outer rings can be neglected, ν is proportional to ω.

This gives $\omega^{\frac{3}{2}}r^3$ and therefore $m\omega r^2$, the angular momentum of an electron, the same for all the different atoms. Thus we

---

[*] J. J. Thomson, Phil. Mag. xxiii. p. 456 (1912).

High-Frequency Spectra of the Elements.   1033

have an experimental verification of the principle of the constancy of angular momentum which was first used by Nicholson *, and is the basis of Bohr's theory of the atom.

It is evident that $k = \sigma_n$. If then $k = 1$, it is suggested that the ring contains 4 electrons, for $\sigma_4 = 0.96$.

We are now justified in making a quantitative comparison between the frequency of α and that of the fundamental radiation from such a ring calculated from the theory of Bohr.

We have obtained the experimental result,

$$\nu = \tfrac{3}{4}\nu_0(N - \sigma_n)^2.$$

On his theory, making the assumption that the ring moves as a whole from stationary state 2 to state 1, the frequency of the principal radiation emitted is

$$\nu = \left(\frac{1}{1^2} - \frac{1}{2^2}\right)\frac{2\pi^2 e^4 m}{h^3}(N - \sigma_n)^2,$$

where e is the charge on an electron, m its mass, and h Planck's constant.

The numerical agreement between these two constants ν_0 and $\dfrac{2\pi^2 e^4 m}{h^3}$ is known to be very close, while Bohr's explanation of the Balmer series for hydrogen assumes them to be identical. This numerical agreement between the experimental values and those calculated from a theory designed to explain the ordinary hydrogen spectrum is remarkable, as the wave-lengths dealt with in the two cases differ by a factor of about 2000. The assumption that the whole ring takes part in the radiation introduces, however, a grave difficulty from energy considerations, while no explanation of the faint line β has been forthcoming. Probably further experiments will show that the theory needs some modification.

The results hitherto obtained for the radiations of the L series are too meagre to justify any explanation. As before, the line of longest wave-length is the most prominent, a result similar to that found in ordinary light-spectra. The wave-lengths found for this line in the case of tantalum and platinum suggest that possibly the frequency is here

$$\nu = \left(\frac{1}{2^2} - \frac{1}{3^2}\right)\nu_0(N - \sigma_n)^2.$$

Here N and σ_n are unknown, but it is evident from the periodic system that $N_{Pt} - N_{Ta} = 5$, while probably σ_n remains

* Monthly Notes Roy. Astr. Soc. June 1912.

1034

the same for all elements in the same column. The actual value found for $v_{Pt}^{\frac{1}{3}} - v_{Ta}^{\frac{1}{3}}$ is $1\cdot08 \times 10^8$, and the calculated value is $1\cdot07 \times 10^8$. Whether this relation really holds good can only be decided by further experiment.

In conclusion I wish to express my warm thanks to Prof. Rutherford for the kind interest which he has taken in this work.

Physical Laboratory,
    University of Manchester.

Moseley. Phil. Mag. Ser. 6, Vol. 26, Pl. XXIII.

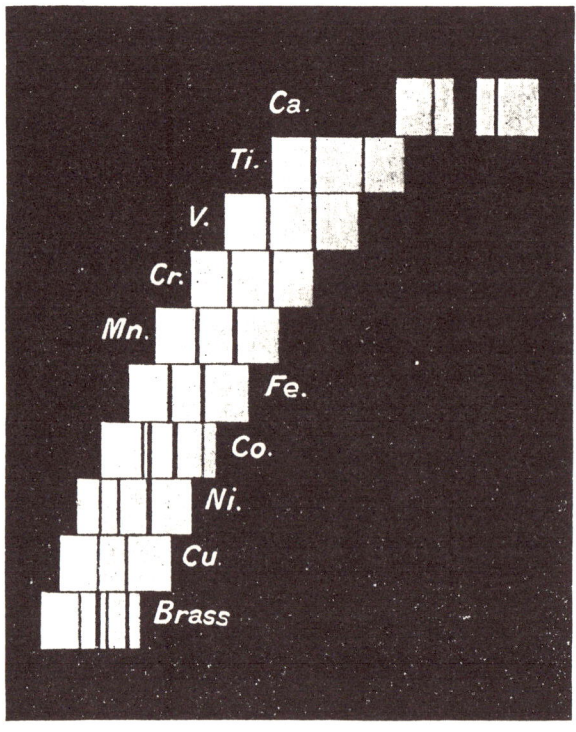

High-Frequency Spectra of the Elements.   705

In view of the fact that the β radiations of thorium D and actinium D are much less penetrating than those of the C_β products, it is somewhat anomalous that the β rays of radium C_2 (equivalent to Th D and Act D) have exactly the same penetrating power as those of radium $C_{1\beta}$. Again, although thorium C and actinium C emit no γ rays, yet radium C undoubtedly appears to ; for it is improbable that the relatively few atoms of radium C_2 can give rise to the whole of the intense γ radiation of radium C. A closer examination of these latter points would be of interest.

In conclusion we wish to express our indebtedness to Prof. Sir E. Rutherford for his kind interest in this research.

LXXX. *The High-Frequency Spectra of the Elements.*
Part II.   *By* H. G. J. MOSELEY, *M.A.**

THE first part † of this paper dealt with a method of photographing X-ray spectra, and included the spectra of a dozen elements. More than thirty other elements have now been investigated, and simple laws have been found which govern the results, and make it possible to predict with confidence the position of the principal lines in the spectrum of any element from aluminium to gold. The present contribution is a general preliminary survey, which claims neither to be complete nor very accurate.

A somewhat different method of photographing these spectra has been developed independently by de Broglie ‡ and by Herveg §. The latter closely confirms the angles given by Moseley and Darwin ‖ for reflexion of Pt rays from selenite. De Broglie finds less satisfactory agreement for the reflexion from rocksalt. De Broglie has also examined the spectra of W and Au, and has obtained for Cu and Fe results similar to those given in Part I.

The general experimental method has remained unaltered, and need not be again described. The same crystal of potassium ferrocyanide has been used as analyser throughout. The sharpness of the lines of short wave-length has

---

* Communicated by the Author.
† Moseley, Phil. Mag. xxvi. p. 1024 (1913).
‡ De Broglie, *C. R.* 17 Nov., 22 Dec., 1913, 19 Jan., 2 Feb., 2 March, 1914.
§ Herveg, *Verh. d. D. Phys. Ges.* xvi. p. 73, Jan. 1914.
‖ Moseley & Darwin, Phil. Mag. xxvi. p. 210 (1913).

3 A 2

704   Mr. H. G. J. Moseley *on the*

been much improved by reducing the breadth of the defining slit to about 0·2 mm. The most convenient type of X-ray tube is drawn to scale in fig. 1. The aluminium trolley which carries the targets can be drawn to and fro by means of silk fishing-line wound on brass bobbins. An iron screen S fastened to the rails is furnished with a fine vertical slit which defines the X-ray beam. The slit should be fixed exactly opposite the focus-spot of the cathode-stream, though a slight error can be remedied by deflecting the cathode rays with a magnet. The X rays escape by a side-tube $2\frac{1}{2}$ cm. diameter closed by an aluminium window 0·022 mm. thick. The X-ray tube, which has a capacity of over 3 litres, was exhausted with a Gaede mercury-pump, for the loan of which I am indebted to Balliol College.

Fig. 1.

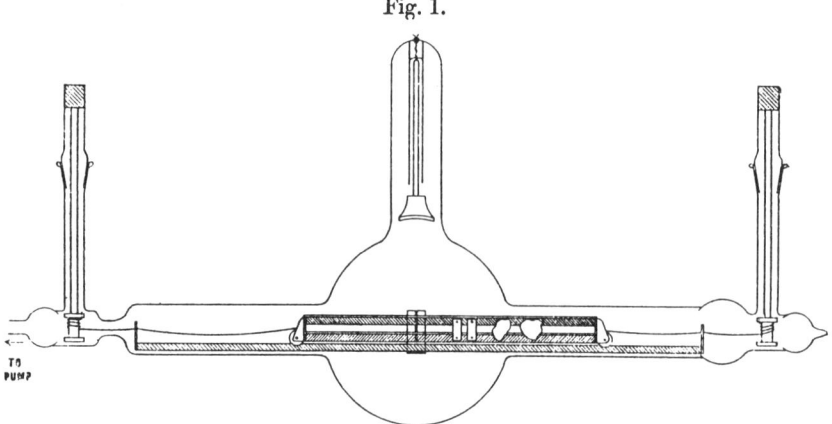

TO
PUMP

The radiations of long wave-length cannot penetrate an aluminium window or more than a centimetre or two of air. The photographs had therefore in this case to be taken inside an exhausted spectrometer. Fig. 2 gives a vertical section to scale of the X-ray tube and spectrometer. The former consists of a bulb containing the cathode, joined by a very large glass T-piece to a long tube of 4 cm. diameter, in which are the rails R and the carriage C. S is the defining-slit and W a window of goldbeaters' skin which separates the tube from the spectrometer. This material, which is usually air-tight, though sometimes it may require varnishing, is extremely transparent to X rays. A circular window of 2 cm. diameter will easily withstand the pressure of the atmosphere if left undisturbed. In these experiments, however, the pressure was relieved

High-Frequency Spectra of the Elements.   705

every time the spectrometer was exhausted, and under such conditions the goldbeaters' skin had frequently to be renewed. The spectrometer, which was specially designed for this work, consists of a strong circular iron box of 30 cm. inside diameter and 8 cm. high, closed by a lid which, when the flange is greased, makes an air-tight joint. Two concentric grooves are cut in the floor of the box. The table A, which carries the plate-holder, rests on three steel balls, of which two run in the outer groove, while the third rests on the floor of the box. The position of the crystal-table B is controlled in like manner by the inner groove. This geometrical construction for a spectrometer is well

Fig. 2.

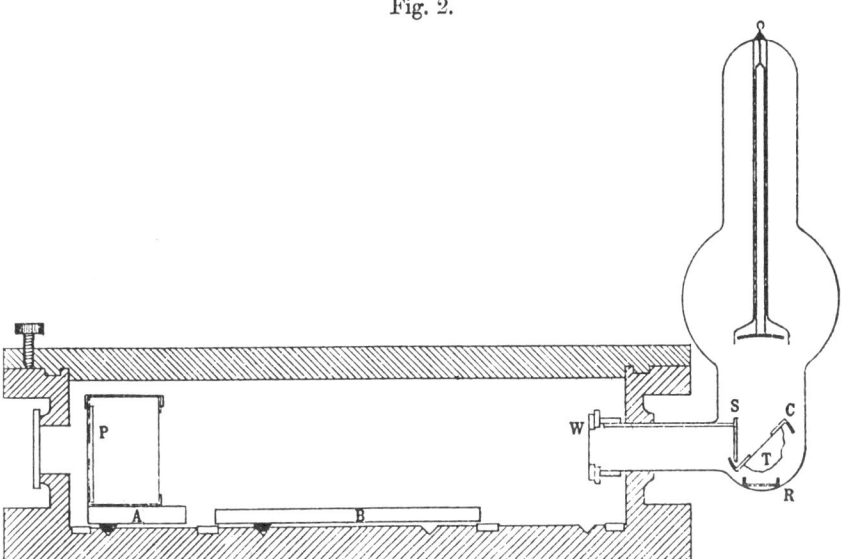

known. The scales are fixed to the box and the verniers to the tables. For these very soft rays the absorption by the black paper front of the plate-holder became serious, and two sheets of black tissue-paper were used instead. Lumps of the pure elements, usually several millimetres thick, were used as targets in the case of Mg, Al, Si, Mo, Ru, Pd, Ag, Sb, Ta. Foils such as Rh, W, Au were either silver-soldered or brazed onto copper. Os was used in the form of a thin chemical deposit on copper. The alloys used were ZrNi (70 per cent.), WFe (50 per cent.), NbTa (50 per cent.), and SnMn (50 per cent.). KCl and the oxides of the rare-earth elements were rubbed onto the surface of nickel plates roughened with coarse emery-powder. The only serious difficulty in the experiments is caused by the heat produced by the cathode ray bombardment, and the consequent liberation of gas and destruction of the surface of the target. This makes it necessary to use the element in a form which is not too volatile and prevents the employment of a very powerful discharge. The total time of an exposure, including rests, varied from three minutes for a substance such as ruthenium, which could safely be heated, to thirty minutes for the rare earth oxides. The importance of using an efficient high-tension valve may again be mentioned.

The oxides of Sa, Eu, Gd, Er were given me by Sir William Crookes, O.M., to whom I wish to express my sincere gratitude. For the loan of the Os and a button of Ru I am indebted to Messrs. Johnson Matthey. The alloys were obtained from the Metallic Compositions Co., and the oxides of La, Ce, Pr, Nd, and Er from Dr. Schuchardt, of Görlitz.

Almost every line was photographed in two different orders, and the double angles of reflexion measured as before to within $0°.1$ and sometimes $0°.05$. In some sets of experiments a small error caused by the crystal surface not being exactly on the spectrometer-axis gave rise to a systematic discrepancy in the results obtained from reflexion in different orders. It was found that this error, which never changed the reflexion-angle by more than $0°.05$, could be measured more accurately from the amount of the discrepancy than from direct observation of the crystal. A more serious correction was necessary when using the long wave-length apparatus. In this case the slit and photograph are not equidistant from the crystal, and the position of the spectrum-lines on the plate is no longer independent of the angle at which the crystal is set. The necessary corrections were calculated geometrically, and verified by photographing the same line for both right-handed and left-handed reflexions and with the crystal set at various angles.

In the work on the very short wave-lengths, the reflexion of the general heterogeneous radiation gave some trouble. This is always an important part of the radiation from an X-ray tube, but with a hard tube it is analysed by reflexion mainly into constituents of very short wave-length, and so usually does not interfere with the line-spectra. It is only with an extremely soft tube, combined with precautions against absorption by the air, that constituents reflected at large angles become prominent. When examining such a spectrum as that of Ag in the K series, the general reflexion

High-Frequency Spectra of the Elements.   707

cannot be avoided. Unfortunately, when photographed it takes the form of irregular fringes, which effectually hide faint spectrum-lines. A change of target, with the position of slit and crystal unaltered, does not affect the appearance of the fringes, a fact which proves that they are due to the general heterogeneous radiation. It is easy to show that the fringes are merely a very foreshortened pattern of patches on the crystal surface which reflect exceptionally well. The way in which they move and spread out laterally as the crystal is turned provides a proof of this, and so does Barkla's * observation that when the crystal is moved sideways the fringes move with it. It is easy to devise methods for getting rid of the fringes. In the first place, narrowing the slit or increasing the distance from the crystal will diminish their intensity compared with that of the line-spectrum. In the second place, turning the crystal will move and blur the fringes, but leave the sharpness of the lines unaffected provided the slit and photograph are equidistant from the reflecting surface †. The quantitative measurements of Moseley and Darwin ‡ on the reflexion of the general radiation must have been little affected by these fringes, as the incident beam was restricted to a very narrow pencil which always impinged on the same part of the crystal.

The results obtained for radiations belonging to Barkla's K series are given in Table I., and for convenience the figures already given in Part I. are included. The wave-length λ has been calculated from the glancing angle of reflexion θ by means of the relation $n\lambda = 2d \sin \theta$, where d has been taken to be $8\cdot454 \times 10^{-8}$ cm. As before, the strongest line is called α and the next line β. The square root of the frequency of each line is plotted in fig. 3, and the wave-lengths can be read off with the help of the scale at the top of the diagram.

The spectrum of Al was photographed in the first order only. The very light elements give several other fainter lines, which have not yet been fully investigated, while the results for Mg and Na are quite complicated, and apparently depart from the simple relations which connect the spectra of the other elements. In the spectra from yttrium onwards only the α line has so far been measured, and further results in these directions will be given in a later paper. The

* Barkla and Martyn, Proc. Phys. Soc. London (1913).
† Moseley, *loc. cit.* p. 1025. See also W. H. and W. L. Bragg, Proc. Roy. Soc. A, lxxxviii. p. 428 (1913).
‡ Moseley and Darwin, *loc. cit.*

708   Mr. H. G. J. Moseley *on the*

TABLE I.

	a line. λ×10⁸ cm.	Qₖ.	N. Atomic Number.	*β* line. λ×10⁸.
Aluminium	8·364	12·05	**13**	7·912
Silicon	7·142	13·04	14	6·729
Chlorine	4·750	16·00	17	......
Potassium	3·759	17·98	19	3·463
Calcium	3·368	19·00	20	3·094
Titanium	2·758	20·99	22	2·524
Vanadium	2·519	21·96	23	2·297
Chromium	2·301	22·98	24	2·093
Manganese	2·111	23·99	25	1·818
Iron	1·946	24·99	26	1·765
Cobalt	1·798	26·00	27	1·629
Nickel	1·662	27·04	28	1·506
Copper	1·549	28·01	29	1·402
Zinc	1·445	29·01	30	1·306
Yttrium	0·838	38·1	39	......
Zirconium	0·794	39·1	40	......
Niobium	0·750	40·2	41	......
Molybdenum........	0·721	41·2	42	......
Ruthenium	0·638	43·6	44	......
Palladium	0·584	45·6	46	......
Silver	0·560	46·6	47	......

spectra both of K and of Cl were obtained by means of
a target of KCl, but it is very improbable that the observed
lines have been attributed to the wrong elements. The
α line for elements from Y onwards appeared to consist of a
very close doublet, an effect previously observed by Bragg *
in the case of rhodium.

The results obtained for the spectra of the L series are
given in Table II. and plotted in fig. 3. These spectra contain five lines, α, β, γ, δ, ε, reckoned in order of decreasing
wave-length and decreasing intensity. There is also always
a faint companion α′ on the long wave-length side of α,
a rather faint line φ between β and γ for the rare earth
elements at least, and a number of very faint lines of wave-
length greater than α. Of these, α, β, φ, and γ have been
systematically measured with the object of finding out how
the spectrum alters from one element to another. The fact
that often values are not given for all these lines merely
indicates the incompleteness of the work. The spectra, so
far as they have been examined, are so entirely similar that
without doubt α, β, and γ at least always exist. Often γ was

* Bragg, 'Nature,' March 12, 1914.

High-Frequency Spectra of the Elements.   709

Fig. 3.

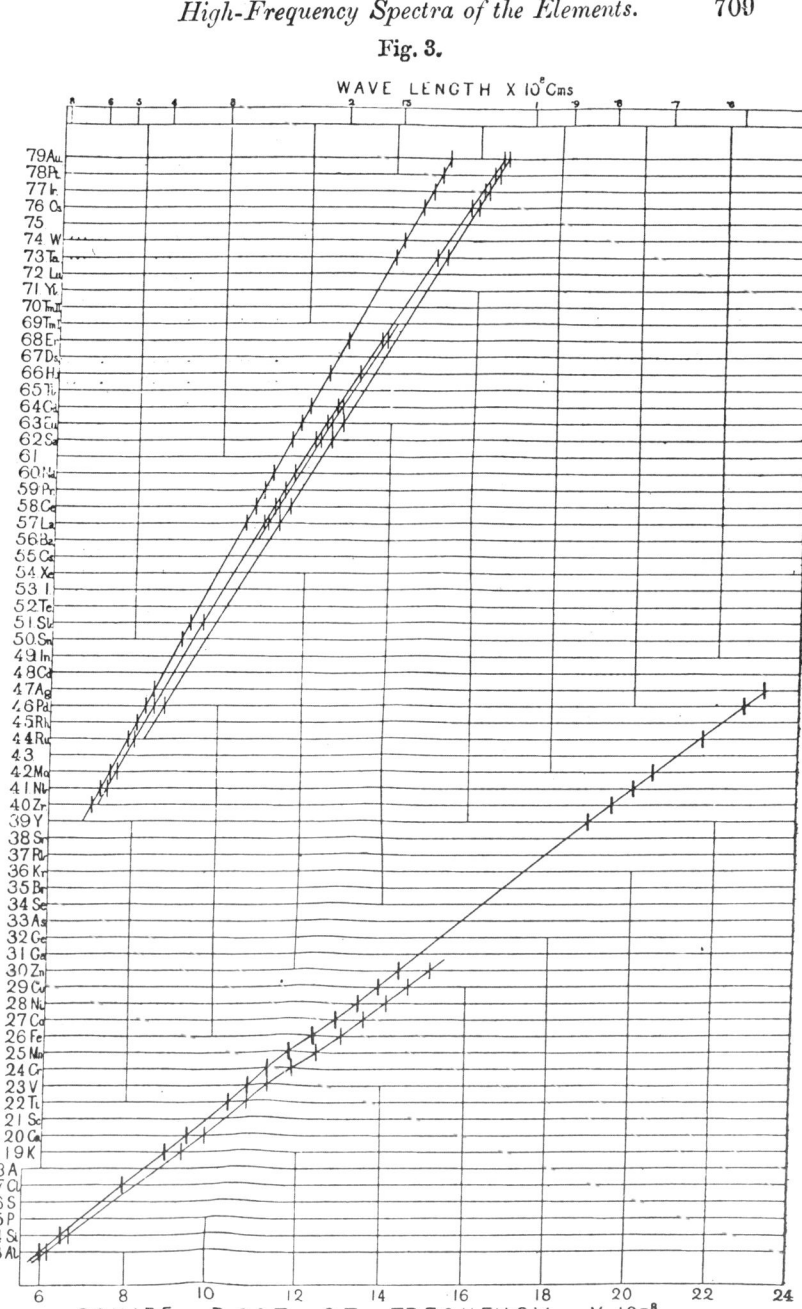

710   Mr. H. G. J. Moseley *on the*

TABLE II.

	a line. $\lambda \times 10^8$ cm.	Q_L.	N. Atomic Number.	β line. $\lambda \times 10^8$.	ϕ line. $\lambda \times 10^8$.	γ line. $\lambda \times 10^8$.
Zirconium	6·091	32·8	40	......	......	......
Niobium	5·749	33·8	41	5·507	......	......
Molybdenum......	5·423	34·8	42	5·187	......	......
Ruthenium	4·861	36·7	44	4·660	......	......
Rhodium	4·622	37·7	45	......	......	......
Palladium	4·385	38·7	46	4·168	......	3·928
Silver	4·170	39·6	47	......	......	......
Tin	3·619	42·6	50	......	......	......
Antimony	3·458	43·6	51	3·245	......	......
Lanthanum	2·676	49·5	57	2·471	2·424	2·313
Cerium	2·567	50·6	58	2·360	2·315	2·209
Praseodymium ...	(2·471)	51·5	59	2·265	......	......
Neodymium	2·382	52·5	60	2·175	......	......
Samarium	2·208	54·5	62	2·008	1·972	1·893
Europium	2·130	55·5	63	1·925	1·888	1·814
Gadolinium	2·057	56·5	64	1·853	1·818	......
Holmium	1·914	58·6	66	1·711	......	......
Erbium	1·790	60·6	68	1·591	1·563	......
Tantalum	1·525	65·6	73	1·330	......	1·287
Tungsten	1·486	66·5	74	......	......	......
Osmium	1·397	68·5	76	1·201	......	1·172
Iridium	1·354	69·6	77	1·155	......	1·138
Platinum	1·316	70·6	78	1·121	......	1·104
Gold	1·287	71·4	79	1·092	......	1·078

not included in the limited range of wave-lengths which can be photographed on one plate. Sometimes lines have not been measured, either on account of faintness or of the confusing proximity of lines due to impurities.

Lines due to impurities were frequently present, but caused little trouble except in the rare earth group. Here two extreme cases occurred. The X-ray spectrum of the praseodymia showed that it consisted roughly of 50 per cent. La, 35 per cent. Ce, and 15 per cent. Pr. Unfortunately the position expected for the a line of Pr coincides with the known position of the β line of La, but the β line of Pr was quite conspicuous, and had precisely the wave-length anticipated. Two specimens of erbia were used. The specimen purchased contained 50 per cent. Er and 50 per cent. of another element, of which the X-ray spectrum coincides with the spectrum calculated for Ho. The erbia given by Sir William Crookes was evidently nearly pure, but showed the a and β lines of Ho quite faintly, and also faint lines agreeing with a and β of Ds and a of Tm I and of Tm II. The Nd was free from La, Ce, and Pr, but contained a fair

High-Frequency Spectra of the Elements. **711**

proportion of Sm. The Sm, Eu, and Gd appeared to be pure. I hope soon to complete the examination of the spectra of this group.

Conclusions.

In fig. 3 the spectra of the elements are arranged on horizontal lines spaced at equal distances. The order chosen for the elements is the order of the atomic weights, except in the cases of A, Co, and Te, where this clashes with the order of the chemical properties. Vacant lines have been left for an element between Mo and Ru, an element between Nd and Sa, and an element between W and Os, none of which are yet known, while Tm, which Welsbach has separated into two constituents, is given two lines. This is equivalent to assigning to successive elements a series of successive characteristic integers. On this principle the integer N for Al, the thirteenth element, has been taken to be 13, and the values of N then assumed by the other elements are given on the left-hand side of fig. 3. This proceeding is justified by the fact that it introduces perfect regularity into the X-ray spectra. Examination of fig. 3 shows that the values of $\nu^{\frac{1}{2}}$ for all the lines examined both in the K and the L series now fall on regular curves which approximate to straight lines. The same thing is shown more clearly by comparing the values of N in Table I. with those of

$$Q_K = \sqrt{\frac{\nu}{\frac{3}{4}\nu_0}},$$

ν being the frequency of the α line and ν_0 the fundamental Rydberg frequency. It is here plain that $Q_K = N - 1$ very approximately, except for the radiations of very short wavelength which gradually diverge from this relation. Again, in Table II. a comparison of N with

$$Q_L = \sqrt{\frac{\nu}{\frac{5}{36}\nu_0}},$$

where ν is the frequency of the L α line, shows that $Q_L = N - 7.4$ approximately, although a systematic deviation clearly shows that the relation is not accurately linear in this case.

Now if either the elements were not characterized by these integers, or any mistake had been made in the order chosen or in the number of places left for unknown elements, these

* Welsbach, *Monatsh.* xxxii. p. 373 (1911).

regularities would at once disappear. We can therefore conclude from the evidence of the X-ray spectra alone, without using any theory of atomic structure, that these integers are really characteristic of the elements. Further, as it is improbable that two different stable elements should have the same integer, three, and only three, more elements are likely to exist between Al and Au. As the X-ray spectra of these elements can be confidently predicted, they should not be difficult to find. The examination of keltium * would be of exceptional interest, as no place has been assigned to this element.

Now Rutherford† has proved that the most important constituent of an atom is its central positively charged nucleus, and van den Broek ‡ has put forward the view that the charge carried by this nucleus is in all cases an integral multiple of the charge on the hydrogen nucleus. There is every reason to suppose that the integer which controls the X-ray spectrum is the same as the number of electrical units in the nucleus, and these experiments therefore give the strongest possible support to the hypothesis of van den Broek. Soddy § has pointed out that the chemical properties of the radio-elements are strong evidence that this hypothesis is true for the elements from thallium to uranium, so that its general validity would now seem to be established.

From the approximate linear relation between $\nu^{\frac{1}{2}}$ and N for each line we obtain the general equation

$$\nu = A (N - b)^2,$$

where A and b are constants characteristic of each line. For the K α line

$$A = \left(\frac{1}{1^2} - \frac{1}{2^2}\right)\nu_0 \quad \text{and} \quad b = 1.$$

For the L α line approximately

$$A = \left(\frac{1}{2^2} - \frac{1}{3^2}\right)\nu_0 \quad \text{and} \quad b = 7 \cdot 4.$$

The fact that the numbers and arrangement of the lines in the K and the L spectra are quite different, strongly suggests that they come from distinct vibrating systems, while the fact that b is much larger for the L lines than for the K lines

---

* Urbain, *C.R.* clii. p. 141 (1911).

† Rutherford, Phil. Mag. xxi. p. 669 (1911), and xxvii. p. 488 (1914).

‡ Van den Broek, *Phys. Zeit.* xiv. p. 32 (1913), and 'Nature,' Nov. 27, Dec. 25, 1913, March 5, 1914.

§ Soddy, *Jahrbuch Rad. und. Elect.* x. p. 193 (1913) ; 'Nature,' Dec. 4, Dec. 18 (1913).

High-Frequency Spectra of the Elements.   **713**

suggests that the L system is situated the further from the nucleus.

It was shown in Part I. * that the linear relation between $\nu^{\frac{1}{2}}$ and $N-b$ was most naturally explained if the vibrating system was a ring of electrons rotating round the central nucleus with an angular momentum which was the same for the different elements. This view has been analysed and put in a more generalised form in a letter to 'Nature 't, written in answer to criticisms made by Lindemann ‡.

Summary.

1. Every element from aluminium to gold is characterized by an integer N which determines its X-ray spectrum. Every detail in the spectrum of an element can therefore be predicted from the spectra of its neighbours.

2. This integer N, the atomic number of the element, is identified with the number of positive units of electricity contained in the atomic nucleus.

3. The atomic numbers for all elements from Al to Au have been tabulated on the assumption that N for Al is 13.

4. The order of the atomic numbers is the same as that of the atomic weights, except where the latter disagrees with the order of the chemical properties.

5. Known elements correspond with all the numbers be-between 13 and 79 except three. There are here three possible elements still undiscovered.

6. The frequency of any line in the X-ray spectrum is approximately proportional to $A(N-b)^2$, where A and b are constants.

I wish to thank Prof. J. S. Townsend, F.R.S., for providing me with every facility for carrying on this work, which has been greatly assisted by a grant from the Institut International de Physique Solvay.

Electrical Laboratory,
   Oxford.

---

* *Loc. cit.* p. 1032.
† Moseley, ' Nature,' Jan. 15 (1914).
‡ F. A. Lindemann, ' Nature,' Jan. 1, Feb. 5, 1914.

488　　　　　Sir E. Rutherford *on the*

include the possibility that the part of ξ thus arising is not negligible we might have defined $\phi(\nu)$ rather differently as the excess at emission of the internal kinetic energy over the mean value ξ. In that case some of the other conclusions would require reconsideration. Also, the results which have been given are not easily harmonized with the values of the specific heats of bodies at low temperatures. For these reasons, the formulation outlined above is to be taken as illustrative rather than final. Another direction in which it is practically certain that the foregoing theory is too much simplified is in the assumption of only one critical frequency ν_0. I hope to be able to return to the discussion of these questions later.

Palmer Physical Laboratory,
　　Princeton, N. J.

LVII. *The Structure of the Atom. By* Sir ERNEST RUTHERFORD, *F.R.S., Professor of Physics, University of Manchester* *.

THE present paper and the accompanying paper by Mr. C. Darwin deal with certain points in connexion with the "nucleus" theory of the atom which were purposely omitted in my first communication on that subject (Phil. Mag. May 1911). A brief account is given of the later investigations which have been made to test the theory and of the deductions which can be drawn from them. At the same time a brief statement is given of recent observations on the passage of α particles through hydrogen, which throw important light on the dimensions of the nucleus.

In my previous paper (*loc. cit.*) I pointed out the importance of the study of the passage of the high speed α and β particles through matter as a means of throwing light on the internal structure of the atom. Attention was drawn to the remarkable fact, first observed by Geiger and Marsden †, that a small fraction of the swift α particles from radioactive substances were able to be deflected through an angle of more than 90° as the results of an encounter with a single atom. It was shown that the type of atom devised by Lord Kelvin and worked out in great detail by Sir J. J. Thomson was unable to produce such large deflexions unless the diameter of the positive sphere was exceedingly small. In order to account for this large angle scattering of α particles, I supposed that the atom consisted of a positively charged nucleus of small dimensions

---

* Communicated by the Author.
† Proc. Roy. Soc. A. lxxxii. p. 495 (1909).

Structure of the Atom. **489**

in which practically all the mass of the atom was concentrated. The nucleus was supposed to be surrounded by a distribution of electrons to make the atom electrically neutral, and extending to distances from the nucleus comparable with the ordinary accepted radius of the atom. Some of the swift α particles passed through the atoms in their path and entered the intense electric field in the neighbourhood of the nucleus and were deflected from their rectilinear path. In order to suffer a deflexion of more than a few degrees, the α particle has to pass very close to the nucleus, and it was assumed that the field of force in this region was not appreciably affected by the external electronic distribution. Supposing that the forces between the nucleus and the α particle are repulsive and follow the law of inverse squares, the α particle describes a hyperbolic orbit round the nucleus and its deflexion can be simply calculated.

It was deduced from this theory that the number of α particles falling normally on unit area of a surface and making an angle ϕ with the direction of the incident rays is proportional to

(1) $\operatorname{cosec}^4 \phi/2$ or $1/\phi^4$ if ϕ be small ;
(2) the number of atoms per unit volume of the scattering material ;
(3) thickness of scattering material t provided this is small;
(4) square of the nucleus charge Ne ;
(5) and is inversely proportional to $(mu^2)^2$, where m is the mass of the α particle and u its velocity.

From the data of scattering on α particles previously given by Geiger [*], it was deduced that the value of the nucleus charge was equal to about half the atomic weight multiplied by the electronic charge. Experiments were begun by Geiger and Marsden [†] to test whether the laws of single scattering of α particles were in agreement with the theory. The general experimental method employed by them consisted in allowing a narrow pencil of α particles to fall normally on a thin film of matter, and observing by the scintillation method the number scattered through different angles. This was a very difficult and laborious piece of work involving the counting of many thousands of particles. They found that their results were in very close accord with the theory. When the thickness of the scattering film was very small, the amount of scattering was directly proportional

[*] Proc. Roy. Soc. A. lxxxiii. p. 492 (1910).
[†] Geiger and Marsden, Phil. Mag. xxv. p. 604 (1913).

490   Sir E. Rutherford *on the*

to the thickness and varied inversely as the fourth power of the velocity of the incident α particles. A special study was made of the number of α particles scattered through angles varying between 5° and 150°. Although over this range the number decreased in the ratio 200,000 to 1, the relation between number and angle agreed with the theory within the limit of experimental error. They found that the scattering of different atoms of matter was approximately proportional to the square of the atomic weight, showing that the charge on the nucleus was nearly proportional to the atomic weight. By determining the number of α particles scattered from thin films of gold, they concluded that the nucleus charge was equal to about half the atomic weight multiplied by the electronic charge. On account of the difficulties of this experiment, the actual number could not be considered correct within more than 20 per cent.

The experimental results of Geiger and Marsden were thus in complete accord with the predictions of the theory, and indicated the essential correctness of this hypothesis of the structure of the atom.

In determining the magnitude of single scattering, I assumed in my previous paper, for simplicity of calculation, that the atom was at rest during an encounter with an α particle. In an accompanying paper, Mr. C. Darwin has worked out the relations to be expected when account is taken of the motion of the recoiling atom. He has shown that no sensible error has been introduced in this way even for atoms of such low atomic weight as carbon. Mr. Darwin has also worked out the scattering to be expected if the law of force is not that of the inverse square, and has shown that it is not in accord with experiment either with regard to the variation of scattering with angle or with the variation of scattering with velocity. The general evidence certainly indicates that the law of force between the α particle and the nucleus is that of the inverse square.

It is of interest to note that C. T. R. Wilson [*], by photographing the trails of the α particle, later showed that the α particle occasionally suffers a sudden deflexion through a large angle. This affords convincing evidence of the correctness of the view that large deflexions do occasionally occur as a result of an encounter with a single atom.

On the theory outlined, the large deflexions of the α particle are supposed to be due to its passage close to the nucleus where the field is very intense and to be not appreciably affected by its passage through the external distribution of

[*] C. T. R. Wilson, Proc. Roy. Soc. A. lxxxvii. p. 277 (1912).

Structure of the Atom.   **491**

electrons. This assumption seems to be legitimate when we remember that the mass and energy of the α particle are very large compared with that of an electron even moving with a velocity comparable with that of light. Simple considerations show that the deflexions which an α particle would experience even in passing through the complex electronic distribution of a heavy atom like gold, must be small compared with the large deflexions actually observed. In fact, the passage of swift α particles through matter affords the most definite and straightforward method of throwing light on the gross structure of the atom, for the α particle is able to penetrate the atom without serious disturbance from the electronic distribution, and thus is only affected by the intense field associated with the nucleus of the atom.

This independence of the large angle scattering on the external distribution of electrons is only true for charged particles whose kinetic energy is very large. It is not to be expected that it will hold for particles moving at very much lower speeds and with much less energy—such, for example, as the ordinary cathode particles or the recoil atoms from active matter. In such cases it is probable that the external electronic distribution plays a far more prominent part in governing the scattering than in the case under consideration.

Scattering of β particles.

It is to be anticipated on the nucleus theory that swift β particles should suffer deflexions through large angles in their passage close to the nucleus. There seems to be no doubt that such large deflexions are actually produced, and I showed in my previous paper that the results of scattering of β particles found by Crowther * could be generally explained on the nucleus theory of atomic structure. It should be borne in mind, however, that there are several important points of distinction between the effects to be expected for an α particle and a β particle. Since the force between the nucleus and β particle is attractive, the β particle increases rapidly in speed in approaching the nucleus. On the ordinary electrodynamics, this entails a loss of energy by radiation, and also an increase of the apparent mass of the electron. Darwin† has worked out mathematically the result of these effects on the orbit of the electron, and has shown that, under certain conditions, the β particle does not escape from the atom but describes a spiral orbit ultimately

* Crowther, Proc. Roy. Soc. A. lxxxiv. p. 226 (1910).
† Darwin, Phil. Mag. xxv. p. 201 (1913).

falling into the nucleus. This result is of great interest, for it may offer an explanation of the disappearance of swift β particles in their passage through matter. In addition, it must be borne in mind that the swiftest β particle expelled from radium C possesses only about one-third of the energy of the corresponding α particle, while the average energy of the β particle is less than one-sixth of that of the α particle. It is thus to be anticipated that the large angle scattering of a β particle by the nucleus will take place in regions where the α particle will only suffer a small deflexion—regions for which the application of the simple theory may not have been accurately tested. For these reasons, it is of great importance to determine the laws of large angle scattering of β particles of different speeds in passing through matter, as it should throw light on a number of important points connected with atomic structure. Experiments are at present in progress in the laboratory to examine the scattering of such swift β particles in detail.

It is obvious that a β particle in passing close to an electron will occasionally suffer a large deflexion. The problem is mathematically similar to that for a close encounter of an α particle with a helium atom of the same mass, which is discussed by Mr. Darwin in the accompanying paper. Such large deflexions due to electronic encounter, however, should be relatively small in number compared with those due to the nucleus of a heavy atom.

Scattering in Hydrogen.

Special interest attaches to the effects to be expected when α particles pass through light gases like hydrogen and helium. In a previous paper by Mr. Nuttall and the author[*], it has been shown that the scattering of α particles in hydrogen and helium is in good agreement with the view that the hydrogen nucleus has one positive charge, while the α particle, or helium, has two. Mr. Darwin has worked out in detail the simple scattering to be anticipated when α particles pass through hydrogen and helium. It is only necessary here to refer to the fact that on the nucleus theory a small number of hydrogen atoms should acquire, as the result of close encounters with α particles, velocities about $1\cdot6$ times that of the velocity of the α particle itself. On account of the fact that the hydrogen atom carries one positive charge while the α particle carries two, it can be calculated that some of the hydrogen atoms should have a range in hydrogen of nearly four times that of the α particle which sets them in motion.

[*] Rutherford and Nuttall, Phil. Mag. xxvi. p. 702 (1913).

Structure of the Atom.   493

Mr. Marsden has kindly made experiments for me to test whether the presence of such hydrogen atoms can be detected. A detailed account of his experiments will appear later, but it suffices to mention here that undoubted evidence has been obtained by him that some of the hydrogen atoms are set in such swift motion that they are able to produce a visible scintillation on a zinc sulphide screen and are able to travel through hydrogen a distance three or four times greater than the colliding α particle. The general method employed was to place a thin α-ray tube containing about 100 millicuries of purified emanation in a tube filled with hydrogen. The scintillations due to the α particle from the tube disappeared in air after traversing a distance of about 5 cm. When the air was displaced by hydrogen, the great majority of the scintillations disappeared at about 20 cm. from the source, which corresponds to the range of the α particle in hydrogen. A small number of scintillations, however, persisted in hydrogen up to a distance of about 90 cm. The scintillations were of less intensity than those due to the ordinary α particle. The number of scintillations observed is of the order of magnitude to be anticipated on the theory of single scattering, supposing that the nucleus in hydrogen and helium has such small dimensions, and that they behave like point charges for distances up to 10^{-13} cm.

There appears to be no doubt that the scintillations observed beyond 20 cm. are due to charged hydrogen atoms which are set in swift motion by a close encounter with an α particle. Experiments are at present in progress by Mr. Marsden to determine the number of hydrogen atoms set in motion, and the variation of the number with the scattering angle.

It does not appear possible to explain the appearance of such swift hydrogen atoms unless it be supposed that the forces of repulsion between the α particle and the hydrogen atom are exceedingly intense. Such intense forces can only arise if the positive nuclei have exceedingly small dimensions, so that a close approach between them is possible.

Dimensions and Constitution of the Nucleus.

In my previous paper I showed that the nucleus must have exceedingly small dimensions, and calculated that in the case of gold its radius was not greater than 3×10^{-12} cm. In order to account for the velocity given to hydrogen atoms by the collision with α particles, it can be simply calculated (see Darwin) that the centres of nuclei of helium and hydrogen must approach within a distance of 1.7×10^{-13} cm. of each other. Supposing for simplicity the nuclei to have dimensions

Phil. Mag. S. 6. Vol. 27. No. 159. *March* 1914.   2 L

and to be spherical in shape, it is clear that the sum of the radii of the hydrogen and helium nuclei is not greater than 1.7×10^{-13} cm. This is an exceedingly small quantity, even *smaller* than the ordinarily accepted value of the diameter of the electron, viz. 2×10^{-13} cm. It is obvious that the method we have considered gives a maximum estimate of the dimensions of the nuclei, and it is not improbable that the hydrogen nucleus itself may have still smaller dimensions. This raises the question whether the hydrogen nucleus is so small that its mass may be accounted for in the same way as the mass of the negative electron.

It is well known from the experiments of Sir J. J. Thomson and others, that no positively charged carrier has been observed of mass less than that of the hydrogen atom. The exceedingly small dimensions found for the hydrogen nucleus add weight to the suggestion that the hydrogen nucleus is the *positive electron*, and that its mass is entirely electromagnetic in origin. According to the electromagnetic theory, the electrical mass of a charged body, supposed spherical, is $\dfrac{2}{3}\dfrac{e^2}{a}$ where e is the charge and a the radius. The hydrogen nucleus consequently must have a radius about 1/1830 of the electron if its mass is to be explained in this way. There is no experimental evidence at present contrary to such an assumption.

The helium nucleus has a mass nearly four times that of hydrogen. If one supposes that the positive electron, *i. e.* the hydrogen atom, is a unit of which all atoms are composed, it is to be anticipated that the helium atom contains four positive electrons and two negative.

It is well known that a helium atom is expelled in many cases in the transformation of radioactive matter, but no evidence has so far been obtained of the expulsion of a hydrogen atom. In conjunction with Mr. Robinson, I have examined whether any other charged atoms are expelled from radioactive matter except helium atoms, and the recoil atoms which accompany the expulsion of α particles. The examination showed that if such particles are expelled, their number is certainly less than 1 in 10,000 of the number of helium atoms. It thus follows that the helium nucleus is a very stable configuration which survives the intense disturbances resulting in its expulsion with high velocity from the radioactive atom, and is one of the units, of which possibly the great majority of the atoms are composed. The radioactive evidence indicates that the atomic weight of successive products decreases by four units consequent on the expulsion of

Structure of the Atom. 495

an α particle, and it has often been pointed out that the atomic weights of many of the permanent atoms differ by about four units.

It will be seen later that the resultant positive charge on the nucleus determines the main physical and chemical properties of the atom. The mass of the atom is, however, dependent on the number and arrangement of the positive and negative electrons constituting the atom. Since the experimental evidence indicates that the nucleus has very small dimensions, the constituent positive and negative electrons must be very closely packed together. As Lorentz has pointed out, the electrical mass of a system of charged particles, if close together, will depend not only on the number of these particles, but on the way their fields interact. For the dimensions of the positive and negative electrons considered, the packing must be very close in order to produce an appreciable alteration in the mass due to this cause. This may, for example, be the explanation of the fact that the helium atom has not quite four times the mass of the hydrogen atom. Until, however, the nucleus theory has been more definitely tested, it would appear premature to discuss the possible structure of the nucleus itself. The general theory would indicate that the nucleus of a heavy atom is an exceedingly complicated system, although its dimensions are very minute.

An important question arises whether the atomic nuclei, which all carry a positive charge, contain negative electrons. This question has been discussed by Bohr *, who concluded from the radioactive evidence that the high speed β particles have their origin in the nucleus. The general radioactive evidence certainly supports such a conclusion. It is well known that the radioactive transformations which are accompanied by the expulsion of high speed β particles are, like the α ray changes, unaffected by wide ranges of temperature or by physical and chemical conditions. On the nucleus theory, there can be no doubt that the α particle has its origin in the nucleus and gains a great part, if not all, of its energy of motion in escaping from the atom. It seems reasonable, therefore, to suppose that a β ray transformation also originates from the expulsion of a negative electron from the nucleus. It is well known that the energy expelled in the form of β and γ rays during the transformation of radium C † is about one-quarter of the energy of the expelled α particle. It does not seem easy to explain this large

* Bohr, Phil. Mag. xxvi. p. 476 (1913).
† See Rutherford and Robinson, Phil. Mag. xxv. p. 301 (1913).

2 L 2

emission of energy by supposing it to have its origin in the electronic distribution. It seems more likely that a very high speed electron is liberated from the nucleus, and in its escape from the atom sets the electronic distribution in violent vibration, giving rise to intense γ rays and also to secondary β particles. The general evidence certainly indicates that many of the high speed electrons from radioactive matter are liberated from the electronic distribution in consequence of the disturbance due to the primary electron escaping from the nucleus.

Charge on the Nucleus.

We have seen that from an examination of the scattering of α particles by matter, it has been found that the positive charge on the nucleus is approximately equal to $\frac{1}{2}Ae$, when A is the atomic weight and e the unit charge. This is equivalent to the statement that the number of electrons in the external distribution is about half the atomic weight in terms of hydrogen. It is of interest to note that this is the value deduced by Barkla * from entirely different evidence, viz. the scattering of X rays in their passage through matter. This is founded on the theory of scattering given by Sir J. J. Thomson, which supposes that each electron in an atom scatters as an independent unit. It seems improbable that the electrons within the nucleus would contribute to this scattering, for they are packed together with positive nuclei and must be held in equilibrium by forces of a different order of magnitude from those which bind the external electrons.

It is obvious from the consideration of the cases of hydrogen and helium, where hydrogen has one electron and helium two, that the number of electrons cannot be exactly half the atomic weight in all cases. This has led to an interesting suggestion by van den Broek† that the number of units of charge on the nucleus, and consequently the number of external electrons, may be equal to the number of the elements when arranged in order of increasing atomic weight. On this view, the nucleus charges of hydrogen, helium, and carbon are 1, 2, 6 respectively, and so on for the other elements, provided there is no gap due to a missing element. This view has been taken by Bohr in his theory of the constitution of simple atoms and molecules.

Recently strong evidence of two distinct kinds has been

* Barkla, Phil. Mag. xxi. p. 648 (1911).
† van den Broek, *Phys. Zeit.* xiv. p. 32 (1913).

Structure of the Atom. 497

brought in support of such a contention. Soddy * has pointed out that the recent generalisation of the relation between the chemical properties of the elements and the radiations can be interpreted by supposing that the atom loses two positive charges by the expulsion of an α particle, and one negative by the expulsion of a high speed electron. From a consideration of the series of products of the three main radioactive branches of uranium, thorium, and actinium, it follows that some of the radioactive elements may be arranged so that the nucleus charge decreases by one unit as we pass from one element to another. It would thus appear that van den Broek's suggestion probably holds for some if not all of the heavy radioactive elements. Recently Moseley † has supplied very valuable evidence that this rule also holds for a number of the lighter elements. By examination of the wave-length of the characteristic X rays emitted by twelve elements varying in atomic weight between calcium (40) and zinc (65·4), he has shown that the variation of wave-length can be simply explained by supposing that the charge on the nucleus increases from element to element by exactly one unit. This holds true for cobalt and nickel, although it has long been known that they occupy an anomalous relative position in the periodic classification of the elements according to atomic weights.

There appears to be no reason why this new and powerful method of analysis, depending on an examination of the frequency of the characteristic X ray spectra of the elements, should not be extended to a large number of elements, so that further definite data on the point may be expected in the near future.

It is clear on the nucleus theory that the physical and chemical properties of the ordinary elements are for the most part dependent entirely on the charge of the nucleus, for the latter determines the number and distribution of the external electrons on which the chemical and physical properties must mainly depend. As Bohr has pointed out, the properties of gravitation and radioactivity, which are entirely uninfluenced by chemical or physical agencies, must be ascribed mainly if not entirely to the nucleus, while the ordinary physical and chemical properties are determined by the number and distribution of the external electrons. On this view, the nucleus charge is a fundamental constant of the atom, while the atomic mass of an atom may be a complicated function of the arrangement of the units which make up the nucleus.

* Soddy, *Jahr. d. Rad.* x. p. 188 (1913).
† Moseley, Phil. Mag. xxvi. p. 1024 (1913).

498   *On the Structure of the Atom.*

It should be borne in mind that there is no inherent impossibility on the nucleus theory that atoms may differ considerably in atomic weight and yet have the same nucleus charge. This is most simply illustrated by radioactive evidence. In the following table the atomic weight and nucleus charge are given for a few of the successive elements arising from the transformation of uranium. The actual nucleus charge of uranium is unknown, but for simplicity it is assumed to be 100.

uccessive Elements ..	$Ur_1 \rightarrow$	$Ur\,X_1 \rightarrow$	$UrX_2 \rightarrow$	$Ur_2 \rightarrow$	$Io \rightarrow$	Ra
Atomic weights	238·5	234·5	234·5	234·5	230·5	226·5
Charge on nucleus	100	98	99	100	98	96

Following the recent theories, it is supposed that the emission of an α particle lowers the nucleus charge by two units, while the emission of a β particle raises it by one unit. It is seen that Ur_1 and Ur_2 have the same nucleus charge although they differ in atomic weight by four units.

If the nucleus is supposed to be composed of a mixture of hydrogen nuclei with one charge and of helium nuclei with two charges, it is *a priori* conceivable that a number of atoms may exist with the same nucleus charge but of different atomic masses. The radioactive evidence certainly supports such a view, but probably only a few of such possible atoms would be stable enough to survive for a measurable time.

Bohr * has drawn attention to the difficulties of constructing atoms on the " nucleus " theory, and has shown that the stable positions of the external electrons cannot be deduced from the classical mechanics. By the introduction of a conception connected with Planck's quantum, he has shown that on certain assumptions it is possible to construct simple atoms and molecules out of positive and negative nuclei, *e. g.* the hydrogen atom and molecule and the helium atom, which behave in many respects like the actual atoms or molecules. While there may be much difference of opinion as to the validity and of the underlying physical meaning of the assumptions made by Bohr, there can be no doubt that the theories of Bohr are of great interest and importance to all physicists as the first definite attempt to construct simple atoms and molecules and to explain their spectra.

University of Manchester,
    February 1914.

* Bohr, Phil. Mag. xxvi. pp. 476, 857 (1913).

The Mass-Spectra of Chemical Elements.　　611

Then we have, from (1) and (2),

$$\frac{\mu}{\sigma}(\alpha^2-\beta^2)\int_a^b U_2 V_2\,dx + \frac{1}{\sigma\mu}\int_a^b (V_2 U_2'' - U_2 V_2'')dx = 0,$$

and

$$(\alpha^2-\beta^2)\int_0^a U_1 V_1\,dx + \int_0^a (V_1 U_1'' - U_1 V_1'')dx = 0.$$

Therefore

$$(\alpha^2-\beta^2)\left[\int_0^a U_1 V_1\,dx + \frac{\mu}{\sigma}\int_a^b U_2 V_2\,dx\right]$$

$$= \int_0^a (U_1 V_1'' - V_1 U_1'')dx + \frac{1}{\sigma\mu}\int_a^b (U_2 V_2'' - V_2 U_2'')dx$$

$$= \left[U_1 V_1' - V_1 U_1'\right]_0^a + \frac{K_2}{K_1}\left[U_2 V_2' - V_2 U_2'\right]_a^b.$$

It follows from (3), (4), and (5), that

$$(\alpha^2-\beta^2)\left[\int_0^a U_1 V_1\,dx + \frac{\mu}{\sigma}\int_a^b U_2 V_2\,dx\right] = 0.$$

Thus $F(\alpha)=0$ cannot have imaginary roots of the form $\xi\pm i\eta$.

LIX. *The Mass-Spectra of Chemical Elements.* *By* F. W. ASTON, *M.A., D.Sc., Clerk Maxwell Student of the University of Cambridge* *.

[Plate XV.]

THE following paper is an account of some results obtained by the analyses of gases by means of the Positive Ray Spectrograph or, as it may be more conveniently termed, Mass-Spectrograph. The principle of the method by which a focussed spectrum is obtained depending solely on the ratio of mass to charge has already been described †, but for the sake of others experimenting in this field it is now proposed to give an account of the actual apparatus in some detail.

* Communicated by the Author.
† F. W. A., Phil. Mag. xxxviii. Dec. 1919, p. 707.

612   Dr. F. W. Aston *on the*

The Discharge Tube.

Fig. 1 is a rough diagram of the present arrangement. The discharge-tube B is an ordinary X-ray bulb 20 cm. in diameter. The ànode A is of aluminium wire 3 mm.

Fig. 1.

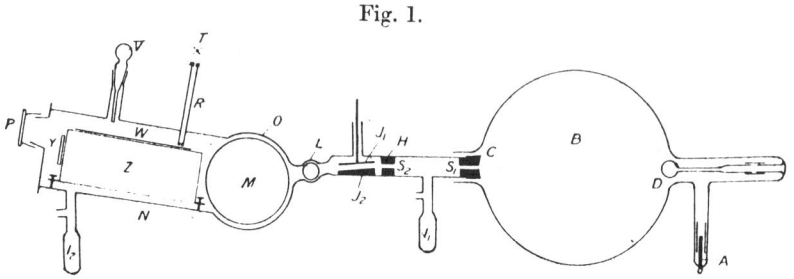

thick surrounded concentrically by an insulated aluminium tube 7 mm. wide to protect the glass walls, as in the Lodge valve.

The aluminium cathode C, 2·5 cm. wide, is concave, about 8 cm. radius of curvature, and is placed just in the neck of the bulb—this shape and position having been adopted after a short preliminary research*. In order to protect the opposite end of the bulb, which would be immediately melted by the very concentrated beam of cathode rays, a silica bulb D about 12 mm. diameter is mounted as indicated. The use of silica as an anticathode was suggested by Prof. Lindemann, and has the great advantage of cutting down the production of undesirable X rays to a minimum.

The discharge is maintained by means of a large induction-coil actuated by a mercury coal-gas break ; about 100 to 150 watts are passed through the primary, and the bulb is arranged to take from 0·5 to 1 milliampere at potentials ranging from 20,000 to 50,000 volts. Owing to the particular shape and position of the electrodes, especially those of the anode, the bulb acts perfectly as its own rectifier.

The method of mounting the cathode will be readily seen from fig. 2, which shows part of the apparatus in greater detail. The neck of the bulb is ground off short and cemented with wax to the flat brass collar E, which forms the mouth of an annular space between a wide outer tube F and the inner tube carrying the cathode.

* F. W. A., Proc. Camb. Phil. Soc. xix. p. 317.

Mass-Spectra of Chemical Elements.   613

The concentric position of the neck is assured by three small ears of brass not shown. The wax joint is kept cool by circulating water through the copper pipe shown in section at G.

Fig. 2.

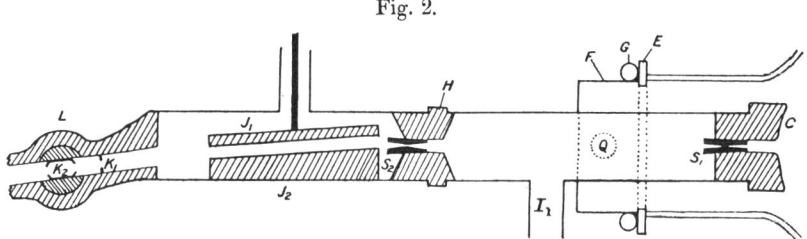

The gas to be analysed is admitted from the customary fine leak into the annular space and so to the discharge by means of the side-tube attached to F shown in dotted section at Q. Exhaustion is performed by a Gaede mercury-pump through a similar tube on the opposite side. The reason for this arrangement is that the space behind the cathode is the only part of the discharge bulb in which the gas is not raised to an extremely high potential. If the inlet or outlet is anywhere in front of the cathode, failing special guards, the discharge is certain to strike to the pump or the gas reservoir. Such special guards have been made in the past by means of dummy cathodes in the bore of the tubes, but, notwithstanding the fact that the gas can only reach the bulb by diffusion, the present arrangement is far more satisfactory and has the additional advantage of enabling the bulb to be dismounted by breaking one joint only.

The Slit System.

The centre of the cathode is pierced with a 3 mm. hole, the back of which is coned out to fit one of the standard slits S_1*. The back of the cathode is turned a gas-light fit in the brass tube 2 cm. diameter carrying it, the other end of which bears the brass plug H which is also coned and fitted with the second slit S_2. The two slits, which are ·05 mm. wide by 2 mm. long, can be accurately adjusted parallel by means of their diffraction patterns. The space between the slits, which are about 10 cm. apart, is kept exhausted to the highest degree by the charcoal tube I_1.

* F. W. A., Phil. Mag. xxxviii. Dec. 1919, p. 714.

614   Dr. F. W. Aston *on the*

By this arrangement it will be seen that not only is loss of rays by collision and neutralization reduced to a minimum, but any serious leak of gas from the bulb to the camera is eliminated altogether.

The Electric Field.

The spreading of the heterogeneous ribbon of rays formed by the slits into an electric spectrum takes place between two parallel flat brass surfaces, J_1, J_2, 5 cm. long, held 2·8 mm. apart by glass distance-pieces, the whole system being wedged immovably in the brass containing-tube in the position shown. The lower surface is cut from a solid cylinder fitting the tube and connected to it and earth. The upper surface is a thick brass plate, which can be raised to the desired potential by means of a set of small storage-cells. In order to have the plates as near together as possible, they are sloped at 1 in 20—*i. e.*, half the angle of slope of the mean ray of the part of the spectrum which is to be selected by the diaphragms. Of these there are two : one, K_1, an oblong aperture in a clean brass plate, is fixed just in front of the second movable one, K_2, which is mounted in the bore of a carefully ground stopcock L. The function of the first diaphragm is to prevent any possibility of charged rays striking the greasy surface of the plug of the stopcock when the latter is in any working position. The variable diaphragm is in effect two square apertures sliding past each other as the plug of the stopcock is turned, the fact that they are not in the same plane being irrelevant. When the stopcock is fully open as sketched in fig. 2, the angle of rays passing is a maximum, and may be stopped down to any desired extent by rotation of the plug, becoming zero before any greasy surface is exposed to the rays. Incidentally the stopcock serves another and very convenient use, which is to cut off the camera from the discharge-tube, so that the latter need not be filled with air each time the former is opened to change the plate.

The Magnetic Field.

After leaving the diaphragms the rays pass between the pole-pieces M of a large Du Bois magnet of 2500 turns. The faces of these are circular, 8 cm. diameter, and held 3 mm. apart by brass distance-pieces. The cylindrical pole-pieces themselves are soldered into a brass tube O, which forms part of the camera N. When the latter is built into position, the pole-pieces are drawn by screwed

Mass-Spectra of Chemical Elements.   615

bolts into the arms of the magnet, and so form a structure of great weight and rigidity and provide an admirable foundation for the whole apparatus. Current for the magnet is provided by a special set of large accumulators. The hydrogen lines are brought on to the plate at about 0·2 ampere, and an increase to 5 amperes, which gives practical saturation, only just brings the singly-charged mercury lines into view. The discharge is protected from the strong field of the magnet by the usual soft iron plates, not shown.

The Camera.

The main body of the camera N is made of stout brass tube 6·4 cm. diameter, shaped to fit on to the transverse tube O containing the pole-pieces. The construction of the plate-holder is indicated by the side view in fig. 1 and an end-on view in fig. 3. The rays after being magnetically deflected

Fig. 3.

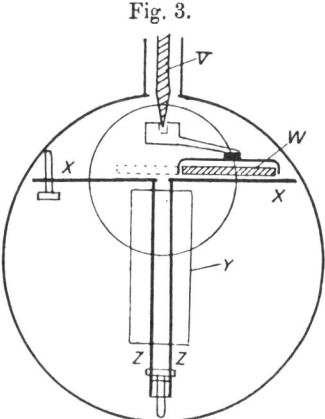

pass between two vertical brass plates Z, Z about 3 mm. apart, and finally reach the photographic plate through a narrow slot 2 mm. wide, 11·8 cm. long, cut in the horizontal metal plate X, X. The three brass plates forming a T-shaped girder are adjusted and locked in position by a set of three levelling-screws at each end ; the right-hand upper one is omitted in fig. 3. The plates Z, Z serve to protect the rays completely from any stray electric field, even that caused by the photographic plate itself becoming charged, until within a few millimetres of their point of impact.

The photographic plate W, which is a 2 cm. strip cut

lengthwise from a 5×4 plate, is supported at its ends on two narrow transverse rails which raise it just clear of the plate X X. Normally it lies to the right of the slot as indicated, and to make an exposure it is moved parallel to itself over the slot by means of a sort of double lazy-tongs carrying wire claws which bracket the ends of the plate as shown. This mechanism, which is not shown in detail, is operated by means of a torque rod V working through a ground glass joint. Y is a small willemite screen.

The adjustment of the plate-holder so that the sensitized surface should be at the best focal plane was done by taking a series of exposures of the bright hydrogen lines with different magnetic fields on a large plate placed in the empty camera at a small inclination to the vertical. On developing this, the actual track of the rays could be seen and the locus of points of maximum concentration determined. The final adjustment was made by trial and error and was exceedingly tedious, as air had to be admitted and a new plate inserted after each tentative small alteration of the levelling-screws.

Experimental procedure.

The plate having been dried in a high vacuum over-night, the whole apparatus is exhausted as completely as possible by the pump with the stopcock L open. I_1 and I_2 are then cut off from the pump by stopcocks and immersed in liquid air for an hour or so. The electric field, which may range from 200 to 500 volts, is then applied and a small current passed through the magnet sufficient to bring the bright hydrogen molecule spot on to the willemite screen Y, where it can be inspected through the plate-glass back of the cap P. In the meantime the leak, pump, and coil have all been started to get the bulb into the desired state.

As soon as this is obtained and has become steady, J_1 is earthed to prevent any rays reaching the camera when the plate is moved over the slot to its first position, which is judged by inspection through P with a non-actinic lamp. The magnet current having been set to the particular value desired and the diaphragm adjusted, the coil is momentarily interrupted while J_1 is raised to the desired potential, after which the exposure starts. During this, preferably both at the beginning and the end, light from a lamp T is admitted for a few seconds down the tube R (fig. 1) the ends of which are pierced with two tiny circular holes. The lower hole is very close to the plate, so that a circular dot or register spot is formed from which the measurements of the lines may be made.

The exposures may range from 20 seconds in the case of hydrogen lines to 30 minutes or more, 15 minutes being usually enough. As soon as it is complete the above procedure is repeated, and the plate moved into the second position. In this way as many as six spectra can be taken on one plate, after which L is shut, I_2 warmed up, and air admitted to the camera. The cap P, which is on a ground joint, can now be removed, and the exposed plate seized and taken out with a special pair of forceps. A fresh plate is now immediately put in, P replaced, and the camera again exhausted, in which state it is left till the next operation.

Form of the Spectrum Lines.

As has been shown (Phil. Mag. Dec. 1919, plate ix.), the shape of the spot formed when undeflected rays from such a slit system strike a photograph surface normally, is somewhat as indicated at *a* (fig. 4). When they strike the plate obliquely the image would be spread out in one direction, as in *b*. This would be the actual form in the apparatus, if the

Fig. 4.

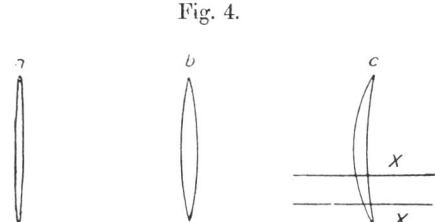

deflexions of the mean and extreme rays (*i. e.*, the rays forming the centre and the tips) were identical. This is true of the magnetic field since each cuts the same number of lines of force ; but it is not so in the case of the electric deflexion. Since the form of the plates, and therefore roughly of the boundaries of the field, is rectangular, the extreme rays passing diagonally will be deflected more than the mean rays and the spot bent into the form shown at *c*. The convex side will be in the direction of the magnetic deflexion, as this is opposed to the deflexion causing the bend. The image on the plate will therefore be the part of this figure falling on the narrow slot in X, X ; and as the apparatus is not exactly symmetrical, its shape in the spectra is the figure lying between the lines X, X in fig. 4, *c.*

Measurement of the Lines.

The plates are measured against a standard Zeiss scale on a comparator designed by the late Dr. Keith Lucas and kindly lent by the Physiological Department. Some of the very faint lines, although easily visible to the unaided eye, were lost even with the lowest power eyepieces obtainable. To measure these, an eyepiece giving a magnification of about $2\frac{1}{2}$ was designed by Dr. Hartridge of King's College.

The general method of deducing mass from position has already been described (Phil. Mag. April 1920, p. 453). Owing to some geometrical cause (probably analogous to a caustic in optics), the more deflected edge of the line is always the brighter and sharper, and it is the distance of this from the register spot which is found to give the most reliable values. For the highest accuracy, owing to halation, one must only compare lines of approximately equal intensity. As this edge is unfortunately not at right angles to the spectrum, measurements can never be regarded as absolute, unless extreme care is taken in the levelling of the spectrum on the comparator. So although theoretically it is sufficient to know the mass of one line to determine (with the correction curve) those of all others, in practice every effort is made to bracket any unknown line by reference lines, and only to trust comparative measurements when the lines are fairly close together. Under these conditions the accuracy claimed for the instrument is about one part in a thousand.

Order of Results and Nomenclature.

The various elements studied will be considered as far as possible in the order in which the experiments were performed. This order is of considerable importance, as in most cases it was impossible to eliminate any element used before the following one was introduced. Evacuation and washing have little effect, as the gases appear to get embedded in the surface of the discharge-bulb and are only released very gradually by subsequent discharge.

The problem of nomenclature became serious when the very complex nature of the heavy elements was apparent,

Mass-Spectra of Chemical Elements. 619

After several possible systems had been discussed it was decided, for the present, to adopt the rather clumsy but definite and elastic one of using the chemical symbol of the mixed element with an index corresponding to its mass : *e.g.*, Ne^{22}, Kr^{84}. This system is made reasonable by the fact that the masses of constituents of mixed elements have all so far proved whole numbers on the scale used.

In cases of particles carrying more than one charge it will be convenient to borrow the nomenclature of optics and refer to the lines given by singly, doubly, and multiply charged particles respectively as lines of the first, second, and higher orders. Thus the molecule of oxygen gives a first order line at 32, and its atom first and second order lines at 16 and 8.

The empirical rule that molecules only give first order lines (J. J. Thomson, 'Rays of Positive Electricity,' p. 54) is very useful in helping to differentiate between elementary atoms and compound molecules of the same mass. Some very recent results give indications that in certain exceptional cases it may break down, so that inferences made from it must not be taken as being absolutely conclusive.

OXYGEN (At. Wt. 16·00) and CARBON (At. Wt. 12·00).

On a mass-spectrum all measurements are relative, and so any known element could be taken as a standard. Oxygen is naturally selected. Its molecule, singly-charged atom, and doubly-charged atom give reference lines at 32, 16, and 8 respectively. The extremely exact integral relation between the atomic weights of oxygen and carbon is itself strong evidence that both are " pure " elements, and so far no evidence appears to have arisen to throw any doubt on this point. Direct comparison of the $C^{.}$ line (12) and the CO line (28) with the above standards shows that the expected whole number relation and additive law hold to the limit of accuracy, *i. e.* one part in a thousand : and this provides standards C^{++} (6), C (12), CO (28), and CO_2 (44). In a similar manner, hydrocarbons give the C_1 and C_2 groups already mentioned (Phil. Mag. April 1920, pp. 452, 453) ; so that a fairly complete scale of reference is immediately available.

NEON (At. Wt. 20·20).

The results obtained with this gas have already been fully dealt with (Phil. Mag. April 1920, p. 449). It has been shown to consist of two isotopes of masses 20 and 22 respectively, with the faint possibility of a third of mass 21.

2 S 2

Spectrum I. on Pl. XV. shows the singly-charged lines of neon, to the left of the C_2 group. It is reproduced here to show the condition of the discharge-tube immediately before compounds of chlorine were introduced.

CHLORINE (At. Wt. 35·46).

Spectra indicating that this element was a mixture of isotopes were first obtained by the use of hydrochloric acid gas, but as this was objectionable on account of its action on mercury, phosgene ($COCl_2$) was substituted. Spectra II., III., and IV. are reproduced from one of the plates taken with this gas. It will be seen that chlorine is characterized by the appearance of four very definite lines in the previously unoccupied space to the right of O_2 (32): measurement shows these lines to correspond exactly to masses 35, 36, 37, and 38. *There is no indication whatever of a line at a point corresponding with the accepted atomic weight* 35·46. On Spectrum II., taken with a small magnetic field, faint lines will be seen at 17·5 and 18·5. These only appeared when chlorine was introduced, and are certainly second order lines corresponding to 35 and 37. These figures seem to leave no possible escape from the conclusion that chlorine is a mixture of isotopes and that two of these have masses 35 and 37. It might be argued that 36 and 38 are also elementary lines and at present there is no evidence to deny this, but it is much more probable that they are the hydrochloric acids HCl^{35} and HCl^{37}. The line 18 is no indication of an element 36, as it is doubtless due to OH_2. Corroborative evidence that Cl^{35} and Cl^{37} are the main if not the only constituents is given by the strong lines 63 and 65 (Spectrum IV.) probably due to $COCl^{35}$ and $COCl^{37}$. If chemical atomic weight is regarded as a statistical average, any lines due to Cl^{35} or its compounds should be considerably stronger than the corresponding ones due to Cl^{37}. This is actually found to be the case. In all spectra taken with chlorine present a faint line is distinguishable corresponding to 39. It is just possible that this is a third isotope.

The unquestionable accuracy of its combining weight on the one hand and the striking whole-number masses given on its mass-spectra by its individual particles on the other, leave little doubt that chlorine is a mixed element, but much critical work will be necessary before its constituents and their relative proportions are decided with certainty.

ARGON (At. Wt. 39·88 Ramsay, 39·91 Leduc).

At the close of the experiments with phosgene the discharge-tube broke down and had to be cleaned and partially rebuilt,

Mass-Spectra of Chemical Elements. 621

so that by the time it had reached suitable working conditions again, all traces of chlorine had disappeared. The tube was run with a mixture of CO_2 and CH_4, and then about 20 per cent. of argon added. The main constituent of the element was at once evident from a very strong line at 40 (Spectrum VI.) reproduced in the second and third orders at 20 and 13·33 (Spectrum V.). The third order line is exceedingly well placed for measurement, and from it the mass of the singly-charged atom is found to be $40 \cdot 00 + \cdot 02$. At first this was thought to be the only constituent, but later a faint companion was seen at 36, which further spectra showed to bear a very definite intensity relation to the 40 line. No evidence drawn from multiple charges is available in this case owing to the probable presence of OH_2 and C; but the above intensity relation and the absence of the line from spectra taken just before argon was introduced, make it extremely likely that it is a true isotope. The presence of about 3 per cent. would account for the fractional atomic weight determined from density.

Nitrogen (At. Wt. 14·01).

This element shows no abnormal characteristics : its atom cannot be distinguished, on the present apparatus, from CH_2 nor its molecule from CO. Its second order line on careful measurement appears to be exactly 7, so it is evidently a pure element, as its chemical combining weight would lead one to expect.

Hydrogen (At. Wt. 1·008) and Helium (At. Wt. 3·99).

The determination of masses so far removed as these from the reference lines offers peculiar difficulties, but, as the lines were expected to approximate to the terms of the geometrical progression 1, 2, 4, 8, etc. the higher terms of which are known, a special method was adopted by which a two to one relation could be tested with some exactness. Two sets of accumulators were selected, each giving very nearly the same potential of about 250 volts. The potentials were then made exactly equal by means of a subsidiary cell and a current-divider, the equality being tested to well within 1 in 1000 by means of a null instrument. If exposures are made with such potentials applied to the electric plates first in parallel and then in series, the magnetic field being kept constant, all masses having an exact two to one relation will be brought into coincidence on the plate (Phil. Mag. April 1920, p. 453). Such

coincidences cannot be detected on the same spectrum photo-
graphically ; but if we first add and then subtract a small
potential from one of the large potentials, two lines will be
obtained which closely bracket the third. To take an actual
instance—with a constant current in the magnet of 0·2
ampere, three exposures were made with a gas containing
hydrogen and helium at potentials of 250, 500+12, and
500−12 volts respectively. The hydrogen molecule line was
found symmetrically bracketed by a pair of atomic lines
(Spectrum VII. *a* and *c*), showing that the mass of the
molecule is exactly double the mass of the atom within
experimental error. When after a suitable increase of the
magnetic field the same procedure was applied to the helium
line and that of the hydrogen molecule, the bracket was no
longer symmetrical (Spectrum VII. *b*), nor was it when the
hydrogen molecule was bracketed by two helium lines (*d*).
Both results show in an unmistakable manner that the mass of
He is less than twice that of H_2. In the same way He was
compared with O^{++}, and H_3, obtained from KOH by Sir J. J.
Thomson's bombardment method, with C^{++}.

The method has some definite advantages and some disadvantages. It is not proposed to discuss these in detail at
present. The values obtained by its use can be checked in
the ordinary way by comparing He with C^{++} and H_3 with
He, these pairs being close enough together for the purpose.
The following table gives the range of values obtained from
the most reliable plates :--

Line	Method.	Mass assumed.	Mass deduced.
He ...... {	bracket	$O^{++}=8$	3·994—3·996
{	direct	$C^{++}=6$	4·005—4·010
H_3 ..... {	bracket	$C^{++}=6$	3·025—3·027
{	direct	He $=4$	3·021—3·030
H_2	bracket	He $=4$	2·012—2·018

From these figures it is safe to conclude that hydrogen is
a " pure" element and that its atomic weight, determined
with such consistency and accuracy by chemical methods
(1·008), is the true mass of its atom.

The above results incidentally appear to settle the nature
of the molecule H_3 beyond doubt.

KRYPTON (At. Wt. 82·92) and XENON (At. Wt. 130·2).

The results with these elements were particularly interesting. The only source available, for which the author is indebted to Sir J. J. Thomson, was the remains of two small samples of gas from evaporated liquid air kindly supplied by Sir James Dewar some years ago for examination by the " parabola " method. Both samples contained nitrogen, oxygen, argon, and krypton, but xenon was only detected in one and its percentage in that must have been quite minute. Krypton is characterized by a remarkable group of five strong lines at 80, 82, 83, 84, 86, and a faint sixth at 78. This group or cluster of isotopes is beautifully reproduced with the same relative values of intensity in the second, and fainter still in the third order. These multiply-charged clusters give most reliable values of mass, as the second order can be compared with A (40) and the third with CO or N_2 (28) with the highest accuracy. It will be noted that one member of each group is obliterated by the reference line, but not the same one. The singly and doubly charged krypton clusters can be seen to the right and left of Spectrum VIII. It will be noticed that krypton is the first element examined which shows unmistakable isotopes differing by one unit only.

On the krypton plates taken with the greatest magnetic field faint, but unmistakable indications of lines in the region of 130 could just be detected. The richest sample was therefore fractionated over liquid air, and the last fraction, a few cubic millimetres, was just sufficient to produce the xenon lines in an unmistakable manner. These can be seen on Spectrum IX., but are somewhat fuzzy owing to the wide diaphragm used to get maximum intensity. They are apparently five in number and appear to follow the integer rule. Until pure xenon is available no final figures can be given, but the values may be taken provisionally as 128, 130, 131, 133, and 135.

MERCURY (At. Wt. 200·6).

Owing to the presence of mercury vapour (which is generally beneficial to the smooth running of the discharge) the multiply-charged particles of this element appear on nearly all the plates taken. They appear as a series of blurred clusters of decreasing intensity around points corresponding to 200, 100, 66·6, 50 ... etc., some of which are indicated in the spectra reproduced. It may be stated provisionally that they indicate a strong component 202, a weak one 204, and a strong band from 197 to 200 containing three or four more unresolvable at present.

624 Dr. F. W. Aston *on the*

Table of Results.

Element.	Atomic number.	Atomic weight.	Minimum number of isotopes.	Mass of isotopes in order of intensity.
H	1	1·008	1	1·008
He	2	3·09	1	4
C	6	12·00	1	12
N	7	14·01	1	14
O	8	16·00	1	16
Ne	10	20·20	2	20, 22, (21)
Cl	17	35·46	2	35, 37, (39)
A	18	39·9	(2)	40, (36)
Kr	36	82·92	6	84, 86, 82, 83, 80, 78
X	54	130·2	5	(128, 131, 130, 133, 135)
Hg	80	200·6	(5)	(197–200, 202, 204)

[Numbers in brackets provisional only.]

The Whole-number Rule.

The most important generalization yielded by these experiments is the remarkable fact that (with the exception of H_1, H_2, and H_3) all masses atomic or molecular, element or compound, so far measured are whole numbers within the accuracy of experiment. It is naturally premature to state that this relation is true for all elements, but the number and variety of those already exhibiting it makes the probability of this extremely high.

On the other hand, it must not be supposed that this would imply that the whole-number rule holds with mathematical exactness, but only that the approximation is of a higher order than that exhibited by the ordinary chemical combining weights and is quite close enough to allow of a theory of atomic structure far simpler than those put forward in the past; for such theories were forced to attempt the explanation of fractions which now appear to be merely fortuitous statistical effects due to the relative quantities of the isotopic constituents.

Thus one may now suppose that an elementary atom of mass m may be changed to one of mass $m + 1$ by the addition of a positive particle and an electron. If both enter the nucleus an isotope results, for the nuclear charge is unaltered. If the positive particle only enters the nucleus, an element of next higher atomic number is formed. In cases where both forms of addition give a stable configuration, the two elements will be isobares.

The electromagnetic theory of mass asserts that mass is not generally additive but only becomes so when the charges are relatively distant from each other. This is certainly the case when the molecules H_2 and H_3 are formed from H_1, so that their masses will be two and three times the mass of H_1 with great exactness. (It must be remembered here that the masses given by these experiments are those of positively charged particles, H_1 being presumably a single particle of positive electricity itself, and that the mass of an electron on the scale used is ·00054 and too small to affect the results.)

In the case of helium, the standard oxygen, and all other elements, this is no longer the case; for the nuclei of these are composed of particles and electrons packed exceedingly close together. The mass of these structures will not be exactly the sum of the masses of their constituents but probably less, so that the unit of mass on the scale chosen will be less than that of a single hydrogen atom.

The Heavier Elements.

The results hold out the probability of great complexity in elements of high atomic number, which has already been proved by entirely different methods in the case of lead. The present apparatus has a resolution factor too low to deal adequately with these; so attention is being given to elements within its scope and to which the analysis can be applied. Results are steadily accumulating, which will be published in due course.

In conclusion the author wishes to express his indebtedness to the Government Grant Committee of the Royal Society for defraying the cost of some of the apparatus employed.

Summary.

A positive ray spectrograph capable of giving a focussed mass-spectrum is fully described in detail and its technique explained.

The results of a provisional analysis of eleven chemical elements—H, He, C, N, O, Ne, Cl, A, Kr, X, Hg—are given, showing that of these the first five only are " pure," the others being apparently composed of various numbers of isotopic constituents, krypton containing no less than six.

With the exception of those due to H_1, H_2, and H_3, all masses measured, allowing for multiple charges, are exactly whole numbers within the error of experiment ($O=16$).

The lines due to hydrogen indicate that the mass of the atom of this element is greater than unity on this scale and in good agreement with the chemical value 1·008. Reasons for this are suggested.

Cavendish Laboratory,
    March 1920.

Aston. Phil. Mag. Ser. 6, Vol. 39. Pl. XV.

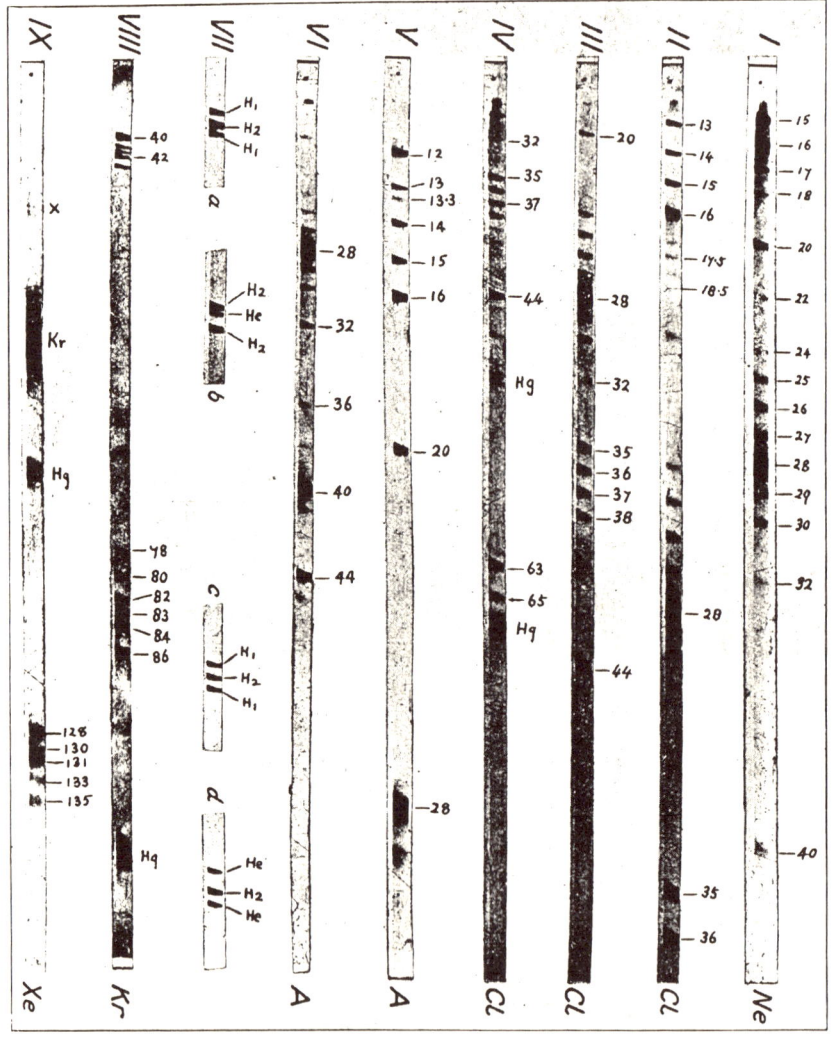

Part Four

Disintegration of the Atom

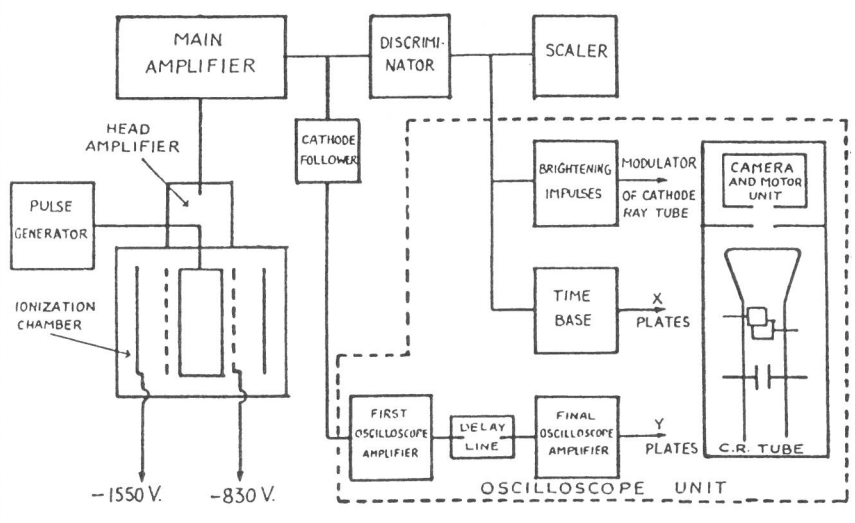

Part Four
Disintegration of the Atom

With the outbreak of war in 1914, many of the young scientists who had thronged to Rutherford's laboratory in Manchester and drawn inspiration from his genius, enlisted for service in the armed forces. Rutherford himself was called to serve on the Admiralty's Board of Invention and Research. It was not until 1917 that he could continue with his own research and then, working almost without assistance, apart from that of William Kay the laboratory steward, he made a discovery as startling and far-reaching in its implications as anything he had achieved previously — the artificial disintegration of an element.

The experiments required ingenuity and the interpretation of the results boldness, two qualities which Rutherford did not lack. From his years of experience he *knew* how atoms behaved. When he observed scintillations on a screen as his trustworthy α particles bombarded nuclei of nitrogen gas, he guessed they were due to hydrogen but nevertheless worked tirelessly to eliminate other possibilities. His paper, published in 1919 — 'An anomalous effect in nitrogen' was the last and shortest of a sequence of four papers published in the *Philosophical Magazine*. In just six pages Rutherford presented what must be regarded as one of the most momentous discoveries in science.

In the same year J. J. Thomson resigned as Cavendish Professor and Rutherford was elected as his successor. On his return to Cambridge, Rutherford was accompanied by James Chadwick, one of his former research students at Manchester. The two were to become close collaborators and together they demonstrated disintegration of the nuclei of elements other than nitrogen, for example boron, fluorine and sodium. Realizing that the α particle was too strongly repelled by the larger number charges in the nuclei of heavier elements to cause disintegration in these, Rutherford had the idea of using high velocity electrons as the projectiles. Even though the electron accelerators of the 1920s could not produce energies to match those of α particles from natural radioactivity, the reasoning was that the electrons, once they had passed through the orbital electrons of the target atoms, would be attracted to the nucleus rather than being repelled. These experiments were not successful. However, following a suggestion by Gamow that positive particles might penetrate the nuclear barrier by quantum mechanical tunnelling, Rutherford proposed the use of protons as the missiles. In 1932, Cockcroft and Walton built a machine to accelerate these particles and succeeded in disintegrating lithium by proton bombardment. This was the first time an atom had been split by controlled means using artificially accelerated particles and the achievement made headline news. The disintegration of heavier elements, however, still resisted attack and required particles accelerated to even higher energies. These were soon to be obtained in the cyclotron, designed and built by Lawrence and Livingstone in California, the forerunner of other giant accelerators which were constructed over subsequent decades.

The year 1932 saw another great event at the Cavendish Laboratory, namely the discovery by Chadwick of the neutron — a particle with mass almost identical to the proton but carrying no charge. The existence of such a particle had been foreseen by Rutherford in 1920 and its discovery solved the mystery of the 'missing mass' of the nuclei of atoms, i.e. the difference between the atomic weight and the number of protons, as well as the mystery of isotopes. Norman Feather showed soon afterwards that the bombardment of atoms by neutrons could produce nuclear disintegration — a process to be used later by Fermi when he produced a sustained nuclear chain reaction by fission and the controlled generation of heat in the first nuclear reactor in 1942.

The development of nuclear physics, and its applications for either peaceful or destructive purposes, is not well chronicled in the pages of the *Philosophical Magazine*. The secrecy surrounding such work, particularly during the Second World War, is one reason for the dearth of papers in this field. The post-war paper by Whitehouse and Galbraith is one of the few to appear and is included here, not because it is a classic, but simply to illustrate the way matters had progressed in the intervening period of 30 years.

1919 37 *Collision of α Particles with Light Atoms. IV. An Anomalous Effect in Nitrogen. By Professor Sir E. Rutherford, F.R.S.*

The discovery of natural radioactivity, and its explanation in terms of the transmutation of one element into another, gave renewed hope to the alchemist's aspiration of producing such transformations by artificial means. All chemical attempts to achieve this goal had failed, individual atoms of elements being, quite simply, impervious to attack by the strongest forces that chemistry could muster.

The first artificially induced transformation (of nitrogen into hydrogen and oxygen) appears to have been discovered almost by accident

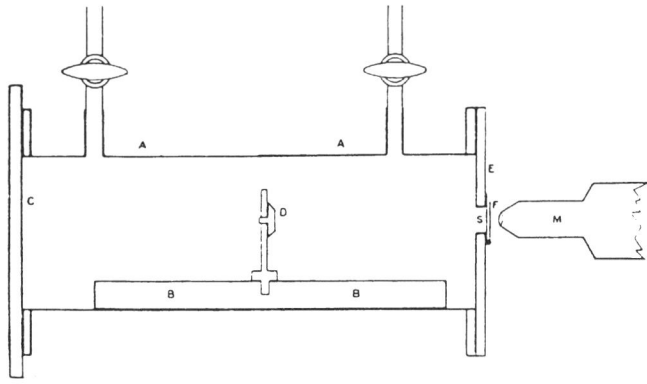

during a series of experiments being undertaken by Rutherford on the collision of α particles with light atoms. It was announced, with measured tone, in this last and shortest of a series of four papers (*Philosophical Magazine* 1919 **37** 537, 562, 571 and 581) describing these investigations. The experimental apparatus used is illustrated in the figure above (taken from the first of the four papers). An intense radioactive source of radium C (now known as ^{214}Po) sits in a rectangular brass box which could be evacuated or filled with a gas at various pressures. A thin foil of silver covers a small opening in one of the end plates and beyond this lies a zinc sulphide screen on which scintillations were observed through a microscope. The experiment consisted of counting the number of scintillations which occurred in a given time due to the arrival of energetic particles. This must have been a tiring business and Rutherford records that the practice was to count for one minute and then to rest for an equal interval, with no more than 1 hour per day being devoted to the task.

In the particular experiments decribed in the present paper, the chamber was first evacuated and the variation in the 'natural' scintillation rate with the amount of an absorbing material (aluminium) placed in front of the screen measured. The 'natural' scintillations were not due to the α particles emitted from the source because even the most energetic of these would have been stopped by the silver foil. Rutherford speculated that they either arose from 'swift H atoms' expelled from the radioactive source itself or else were due to α particles knocking on occluded hydrogen.

Introduction of oxygen or carbon dioxide into the vessel diminished the rate of arrival of scintillations as expected according to the known stopping power of the gases. When dried air or nitrogen was admitted, however, the scintillation rate increased. Furthermore, absorption measurements (see Figure 1 of the paper) demonstrated that the effect was due to collision of α particles with atoms of nitrogen throughout the volume of the gas. 'From the results so far obtained', Rutherford writes, 'it is difficult to avoid the conclusion that the long-range atoms arising from collision of α particles with nitrogen are not nitrogen atoms but probably atoms of hydrogen, or atoms of mass 2. If this be the case, we must conclude that the nitrogen atom is disintegrated under the intense forces developed in a close collision with a swift α particle, and that the hydrogen atom which is liberated formed a constituent part of the nitrogen nucleus.' It was impossible to foresee at the time the far-reaching consequences of this historic discovery.

What Rutherford had observed was the following nuclear transformation (the subscript is the atomic number and the superscript the atomic weight of the elements). The α particle is denoted by ^4_2He.

$$^4_2\text{He} + ^{14}_7\text{N} \rightarrow ^{17}_8\text{O} + ^1_1\text{H}$$

The ejection of a proton (hydrogen) from the nitrogen nucleus leaves behind an oxygen nucleus, although at the time Rutherford made his discovery the oxygen isotope of mass 17 was unknown.

1950   **41**   *Energy Spectrum of Fragments from the Spontaneous Fission of Natural Uranium. By W. J. Whitehouse and W. Galbraith*

While undertaking experiments in Italy in 1934 on the bombardment of uranium with neutrons, Enrico Fermi discovered that the products of the reaction were extremely radioactive. At the time no one thought he had done anything but to create unstable isotopes with atomic numbers greater than uranium, but four years later in France, Irène Juliot-Curie (Marie and Pierre Curie's daughter) and Pavel Savitch claimed, from the results of similar studies, that one of the products might be lanthanum with an atomic number Z of 57 — an element much lighter than uranium ($Z = 92$). Was it conceivable that the uranium atom had been split into two almost equal halves? If so, the implications were far-reaching, as the energy liberated per disintegration (calculated from the mass defect relation $\Delta E = \Delta mc^2$) was much greater than that associated with splitting nuclei of lighter elements.

Nuclear fission of uranium was indeed confirmed in 1939 by Otto Hahn and F. Strassmann, who identified barium ($Z = 56$) — a decay product of lanthanum — as one of the radioisotopes produced when uranium is bombarded with slow neutrons. The terrifying consequences of using the enormous amount of energy released for destructive purposes was realized by Lise Meitner and her nephew Otto Frisch, who jointly published their interpretation of Hahn's experiments in *Nature* (1939 **143** 239), before the start of the Second World War and the imposition of a moratorium on publications relating to the phenomenon.

It was not until 1950 that the first paper relating to nuclear fission appeared in the *Philosophical Magazine*, and then it was a tame one. Here Whitehouse and Galbraith, working at the Atomic Energy Research Establishment at Harwell, UK, report on the kinetic energies of the products resulting from the *spontaneous* fission of uranium (U^{238}) — a process about a million times less likely than the element's radioactive decay by α particle emission. For the mass of sample used, only a few disintegrations per hour were observed, but it was never-theless possible to obtain the energy distribution of the so-called fission fragments. This was found to have peaks at about 60 and 95 MeV, in good agreement with the corresponding curve for neutron-induced fission of U^{235}. The bimodal nature of the distribution is associated with the splitting of the uranium nucleus into many different pairs of lighter elements with unequal masses: the most frequent disintegration yields fission fragments with atomic masses of about 95 and 140.

By the time this paper was published, a controlled nuclear chain reaction had been achieved in America by Fermi in 1942. In the world's first nuclear reactor, neutrons emitted during the disintegration of U^{235} were permitted to cause further disintegration events in a self-sustaining process and heat was generated by dissipation of the kinetic energy of the reaction products. Construction of this prototype of today's power-producing and research nuclear reactors marked the birth of the atomic age, which, unfortunately, also had a sinister side. Under great secrecy, scientists at Los Alamos, a desert site in New Mexico, were developing atomic weapons, eventually to be used with such devastating effects in Japan in 1945.

[581]

LIV. *Collision of α Particles with Light Atoms.* IV. *An Anomalous Effect in Nitrogen.* *By* Professor Sir E. RUTHERFORD, *F.R.S.**

IT has been shown in paper I. that a metal source, coated with a deposit of radium C, always gives rise to a number of scintillations on a zinc sulphide screen far beyond the range of the α particles. The swift atoms causing these scintillations carry a positive charge and are deflected by a magnetic field, and have about the same range and energy as the swift H atoms produced by the passage of α particles through hydrogen. These "natural" scintillations are believed to be due mainly to swift H atoms from the radioactive source, but it is difficult to decide whether they are expelled from the radioactive source itself or are due to the action of α particles on occluded hydrogen.

The apparatus employed to study these "natural" scintillations is the same as that described in paper I. The intense source of radium C was placed inside a metal box about 3 cm. from the end, and an opening in the end of the box was covered with a silver plate of stopping power equal to about 6 cm. of air. The zinc sulphide screen was mounted outside, about 1 mm. distant from the silver plate, to admit of the introduction of absorbing foils between them. The whole apparatus was placed in a strong magnetic field to deflect the β rays. The variation in the number of these "natural" scintillations with absorption in terms of cms. of air is shown in fig. 1, curve A. In this case, the air in the box was exhausted and absorbing foils of aluminium were used. When dried oxygen or carbon dioxide was admitted into the vessel, the number of scintillations diminished to about the amount to be expected from the stopping power of the column of gas.

A surprising effect was noticed, however, when dried air was introduced. Instead of diminishing, the number of scintillations was increased, and for an absorption corresponding to about 19 cm. of air the number was about twice that observed when the air was exhausted. It was clear from this experiment that the α particles in their passage through air gave rise to long-range scintillations which appeared to the eye to be about equal in brightness to H scintillations. A systematic series of observations was undertaken to account for the origin of these scintillations. In the first place we have seen that the passage of α particles through nitrogen and

* Communicated by the Author.

582   Sir E. Rutherford *on Collision of*

oxygen gives rise to numerous bright scintillations which have
a range of about 9 cm. in air.   These scintillations have about
the range to be expected if they are due to swift N or O atoms,
carrying unit charge, produced by collision with α particles.

Fig. 1.

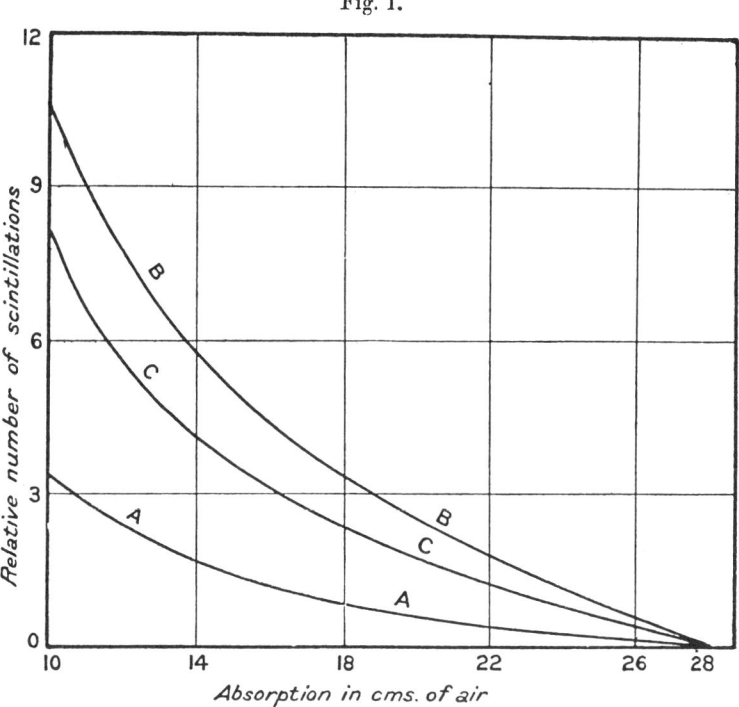

Absorption in cms. of air

All experiments have consequently been made with an absorption greater than 9 cm. of air, so that these atoms are
completely stopped before reaching the zinc sulphide screen.

   It was found that these long-range scintillations could not
be due to the presence of water vapour in the air ; for the
number was only slightly reduced by thoroughly drying
the air.   This is to be expected, since on the average the
number of the additional scintillations due to air was equivalent to the number of H atoms produced by the mixture
of hydrogen at 6 cm. pressure with oxygen.   Since on the
average the vapour pressure of water in air was not more
than 1 cm., the effects of complete drying would not reduce
the number by more than one sixth.   Even when oxygen
and carbon dioxide saturated with water vapour at 20° C.

were introduced in place of dry air, the number of scintillations was much less than with dry air.

It is well known that the amount of hydrogen or gases
containing hydrogen is normally very small in atmospheric
air. No difference was observed whether the air was taken
directly from the room or from outside the laboratory or was
stored for some days over water.

There was the possibility that the effect in air might be due
to liberation of H atoms from the dust nuclei in the air. No
appreciable difference, however, was observed when the dried
air was filtered through long plugs of cotton-wool, or by
storage over water for some days to remove dust nuclei.

Since the anomalous effect was observed in air, but not in
oxygen, or carbon dioxide, it must be due either to nitrogen
or to one of the other gases present in atmospheric air. The
latter possibility was excluded by comparing the effects
produced in air and in chemically prepared nitrogen. The
nitrogen was obtained by the well-known method of adding
ammonium chloride to sodium nitrite, and stored over water.
It was carefully dried before admission to the apparatus.
With pure nitrogen, the number of long-range scintillations
under similar conditions was greater than in air. As a result
of careful experiments, the ratio was found to be 1·25, the
value to be expected if the scintillations are due to nitrogen.

The results so far obtained show that the long-range
scintillations obtained from air must be ascribed to nitrogen,
but it is important, in addition, to show that they are due to
collision of α particles with atoms of nitrogen through the
volume of the gas. In the first place, it was found that
the number of the scintillations varied with the pressure
of the air in the way to be expected if they resulted from
collision of α particles along the column of gas. In addition,
when an absorbing screen of gold or aluminium was placed
close to the source, the range of the scintillations was found
to be reduced by the amount to be expected if the range of
the expelled atom was proportional to the range of the
colliding α particles. These results show that the scintillations arise from the volume of the gas and are not due to
some surface effect in the radioactive source.

In fig. 1 curve A the results of a typical experiment are
given showing the variation in the number of natural scintillations with the amount of absorbing matter in their path
measured in terms of centimetres of air for α particles. In
these experiments carbon dioxide was introduced at a pressure
calculated to give the same absorption of the α rays as ordinary
air. In curve B the corresponding curve is given when air

at N.T.P. is introduced in place of carbon dioxide. The difference curve C shows the corresponding variation of the number of scintillations arising from the nitrogen in the air. It was generally observed that the ratio of the nitrogen effect to the natural effect was somewhat greater for 19 cm. than for 12 cm. absorption.

In order to estimate the magnitude of the effect, the space between the source and screen was filled with carbon dioxide at diminished pressure and a known pressure of hydrogen was added. The pressure of the carbon dioxide and of hydrogen were adjusted so that the total absorption of α particles in the mixed gas should be equal to that of the air. In this way it was found that the curve of absorption of H atoms produced under these conditions was somewhat steeper than curve C of fig. 1. As a consequence, the amount of hydrogen mixed with carbon dioxide required to produce a number of scintillations equal to that of air, increased with the increase of absorption. For example, the effect in air was equal to about 4 cm. of hydrogen at 12 cm. absorption, and about 8 cm. at 19 cm. absorption. For a mean value of the absorption, the effect was equal to about 6 cm. of hydrogen. This increased absorption of H atoms under similar conditions indicated either that (1) the swift atoms from air had a somewhat greater range than the H atoms, or (2) that the atoms from air were projected more in the line of flight of the α particles.

While the maximum range of the scintillations from air using radium C as a source of α rays appeared to be about the same, viz. 28 cm., as for H atoms produced from hydrogen, it was difficult to fix the end of the range with certainty on account of the smallness of the number and the weakness of the scintillations. Some special experiments were made to test whether, under favourable conditions, any scintillations due to nitrogen could be observed beyond 28 cm. of air absorption. For this purpose a strong source (about 60 mg. Ra activity) was brought within 2·5 cm. of the zinc sulphide screen, the space between containing dry air. On still further reducing the distance, the screen became too bright to detect very feeble scintillations. No certain evidence of scintillations was found beyond a range of 28 cm. It would therefore appear that (2) above is the more probable explanation.

In a previous paper (III.) we have seen that the number of swift atoms of nitrogen or oxygen produced per unit path by collision with α particles is about the same as the corresponding number of H atoms in hydrogen. Since the number of long-range scintillations in air is equivalent to that produced under similar conditions in a column of hydrogen at 6 cm.

pressure, we may consequently conclude that only one long-range atom is produced for every 12 close collisions giving rise to a swift nitrogen atom of maximum range 9 cm

It is of interest to give data showing the number of long-range scintillations produced in nitrogen at atmospheric pressure under definite conditions. For a column of nitrogen 3·3 cm. long, and for a total absorption of 19 cm. of air from the source, the number due to nitrogen per milligram of activity is ·6 per minute on a screen of 3·14 sq. mm. area.

Both as regards range and brightness of scintillations, the long-range atoms from nitrogen closely resemble H atoms, and in all probability are hydrogen atoms. In order, however, to settle this important point definitely, it is necessary to determine the deflexion of these atoms in a magnetic field. Some preliminary experiments have been made by a method similar to that employed in measuring the velocity of the H atom (see paper II.). The main difficulty is to obtain a sufficiently large deflexion of the stream of atoms and yet have a sufficient number of scintillations per minute for counting. The α rays from a strong source passed through dry air between two parallel horizontal plates 3 cm. long and 1·6 mm. apart, and the number of scintillations on the screen placed near the end of the plates was observed for different strengths of the magnetic field. Under these conditions, when the scintillations arise from the whole length of the column of air between the plates, the strongest magnetic field available reduced the number of scintillations by only 30 per cent. When the air was replaced by a mixture of carbon dioxide and hydrogen of the same stopping power for α rays, about an equal reduction was noted. As far as the experiment goes, this is an indication that the scintillations are due to H atoms ; but the actual number of scintillations and the amount of reduction was too small to place much reliance on the result. In order to settle this question definitely, it will probably prove necessary to employ a solid nitrogen compound, free from hydrogen, as a source, and to use much stronger sources of α rays. In such experiments, it will be of importance to discriminate between the deflexions due to H atoms and possible atoms of atomic weight 2. From the calculations given in paper III., it is seen that a collision of an α particle with a free atom of mass 2 should give rise to an atom of range about 32 cm. in air, and of initial energy about ·89 of that of the H atom produced under similar conditions. The deflexion of the pencil of these rays in a magnetic field should be about ·6 of that shown by a corresponding pencil of H atoms.

Phil. Mag. S. 6. Vol. 37. No. 222. *June* 1919.   2 S

586   *Collision of α Particles with Light Atoms.*

Discussion of results.

From the results so far obtained it is difficult to avoid the conclusion that the long-range atoms arising from collision of α particles with nitrogen are not nitrogen atoms but probably atoms of hydrogen, or atoms of mass 2. If this be the case, we must conclude that the nitrogen atom is disintegrated under the intense forces developed in a close collision with a swift α particle, and that the hydrogen atom which is liberated formed a constituent part of the nitrogen nucleus. We have drawn attention in paper III. to the rather surprising observation that the range of the nitrogen atoms in air is about the same as the oxygen atoms, although we should expect a difference of about 19 per cent. If in collisions which give rise to swift nitrogen atoms, the hydrogen is at the same time disrupted, such a difference might be accounted for, for the energy is then shared between two systems.

It is of interest to note, that while the majority of the light atoms, as is well known, have atomic weights represented by $4n$ or $4n+3$ where n is a whole number, nitrogen is the only atom which is expressed by $4n+2$. We should anticipate from radioactive data that the nitrogen nucleus consists of three helium nuclei each of atomic mass 4 and either two hydrogen nuclei or one of mass 2. If the H nuclei were outriders of the main system of mass 12, the number of close collisions with the bound H nuclei would be less than if the latter were free, for the α particle in a collision comes under the combined field of the H nucleus and of the central mass. Under such conditions, it is to be expected that the α particle would only occasionally approach close enough to the H nucleus to give it the maximum velocity, although in many cases it may give it sufficient energy to break its bond with the central mass. Such a point of view would explain why the number of swift H atoms from nitrogen is less than the corresponding number in free hydrogen and less also than the number of swift nitrogen atoms. The general results indicate that the H nuclei, which are released, are distant about twice the diameter of the electron (7×10^{-13} cm.) from the centre of the main atom. Without a knowledge of the laws of force at such small distances, it is difficult to estimate the energy required to free the H nucleus or to calculate the maximum velocity that can be given to the escaping H atom. It is not to be expected, *a priori*, that the velocity or range of the H atom released from the nitrogen atom should be identical with that due to a collision in free hydrogen.

Taking into account the great energy of motion of the α particle expelled from radium C, the close collision of such

587

an α particle with a light atom seems to be the most likely agency to promote the disruption of the latter ; for the forces on the nuclei arising from such collisions appear to be greater than can be produced by any other agency at present available. Considering the enormous intensity of the forces brought into play, it is not so much a matter of surprise that the nitrogen atom should suffer disintegration as that the α particle itself escapes disruption into its constituents. The results as a whole suggest that, if α particles—or similar projectiles—of still greater energy were available for experiment, we might expect to break down the nucleus structure of many of the lighter atoms.

I desire to express my thanks to Mr. William Kay for his invaluable assistance in counting scintillations.

University of Manchester,
     April 1919.

[429]

XXXVIII. *Energy Spectrum of Fragments from the Spontaneous Fission of Natural Uranium.*

By W. J. WHITEHOUSE and W. GALBRAITH *

[Received January 31, 1950.]

ABSTRACT.

A direct comparison is made between the energy spectrum of fission fragments from the spontaneous fission and from the thermal neutron induced fission in the same foils of natural uranium. The results show that, within the limits of the experiment, the two distributions are not significantly different from one another. Calculations are made which indicate the magnitude of the distortion of the energy spectrum to be expected, and the theory is found to agree satisfactorily with experiment.

As a subsidiary result to the main experiment the rate for the spontaneous fission of natural uranium is found to be $24\cdot2 \pm 1\cdot5$ per hour per gram.

§1. INTRODUCTION.

THE possibility of the spontaneous fission of the heavy elements was forecast by Bohr and Wheeler (1939) who estimated a mean life for Uranium of the order of 10^{22} years. The phenomenon was first observed experimentally by Flerov and Petrzhak (1940, 1941). They used a multi-plate ionization chamber with active foils of U_3O_8 and obtained a value for the half-life (based on the assumption that all the spontaneous fissions were due to the U^{238} isotope) of 10^{16}–10^{17} years. This corresponds to between 2·1 and 21 fissions per hour per gram of Uranium.

Shortly after this discovery, as a result of the war, the publication of experimental results on this subject was seriously interrupted, and a review of the literature is further complicated for the present authors by the difficulty in obtaining translations of Russian papers. There have, however, been a number of studies of the neutrons emitted in spontaneous fission (Maurer and Pose 1943, Pose 1943, Chatterjee 1945, Scharff-Goldhaber and Klaiber 1946). Pose obtained a nominal half-life for uranium, with respect to neutron emission, of $3\cdot1 \times 10^{15}$ years, corresponding to the emission of about 65 neutrons per hour per gram.

N. A. Perfilov (1947) measured the spontaneous fission rates of Uranium and Thorium using the method of track counting in photographic emulsions which had been exposed to the fragments at a depth of 50 metres below the surface of the earth. An oxidizing agent was employed to remove

---

* Communicated by the Authors.

the tracks of alpha-particles from the emulsion. The value obtained for the half-life of Uranium with respect to spontaneous fission was $(1 \cdot 3 \pm 0 \cdot 2) \times 10^{16}$ years, $(16 \pm 2$ fissions per hour per gram). Perfilov also measured the ranges of the residual tracks, after oxidation, of the fragments from spontaneous fission, and from slow neutron-induced fission in Uranium, and in each case obtained a distribution with a single peak at about $1 \cdot 0$ cm. air equivalent. As he points out the fission tracks are much reduced in length by the oxidation. For this reason his results are not of any use in determining the amount of asymmetry in spontaneous fission.

A figure of 20 spontaneous fissions per hour per gram has been given for the spontaneous fission rate of U^{238} (*The Science and Engineering of Nuclear Power* 1947), but without any account of the method of measurement. It is evident, therefore, that the rate for Uranium is nearer the upper of the original limits given by Flerov and Petrzak.

It was felt that a knowledge of the distribution in energy of the fragments from spontaneous fission of uranium, as distinct from the rate at which the phenomenon occurs, would be useful in the formulation of theories of fission and it was with this object in view that the present work was undertaken.

The experiment consists essentially of a direct comparison of the energy distributions of the fragments from the spontaneous fission of uranium and from the slow neutron induced fission of U^{235}, which forms a convenient standard of comparison, under identical conditions. The latter distribution has been investigated by a number of workers (Kanner and Barschall 1940 ; Flammersfeld, Jensen and Gentner 1943 ; Jentschke 1943) but the experiments of Fowler and Rosen (1947), Deutsch and Ramsey (1945) and Brunton and Hanna (1949, 1950) * are particularly suitable for comparison with the present results, since all of them used gridded ionization chambers similar to ours.†

Brunton and Hanna give in their paper a summary of the results obtained by themselves and five previous experimenters on the distribution for U^{235}. In each case two rather broad peaks are observed. That for the light group of fragments has a most probable energy of about 92 MeV., and a width at half maximum of 12–15 MeV., while that for the heavy group of fragments has a most probable value of about 60 MeV. and a width at half-maximum of 20–25 MeV.‡ The two peaks overlap slightly, and the height of the minimum between them is approximately 1/7 to 1/10 of the higher peak (see fig. 5, curve I.).

---

* The results of the latter paper were kindly communicated to us before publication.

† In the last two experiments the energies of the two fragments were measured simultaneously, but results were also obtained when only one fragment was observed, and it is these which are used for comparison.

‡ Brunton and Hanna point out that there is a correction to be applied in their experiment for energy loss in the foil and collimator.

of Fragments from the Spontaneous Fission of Natural Uranium   431

The difficulties associated with a similar measurement of the energies of spontaneous fission fragments arise from the very slow rate of fission. We have used a large cylindrical ionization chamber with uranium foils of total area 675 square centimetres on the inner surface of the outer electrode. Even so, in order to obtain a counting rate of about 1·4 fissions per hour, we have had to use foils of approximately 0·1 mg./cm.^{2} thickness. This distorts the observed energy distribution considerably, in a manner which will be discussed in § 3.1. Another difficulty associated with the slow counting rate is that of maintaining constant conditions in the ionization chamber itself, and in the electronic amplifiers and recording apparatus. We have overcome the difficulty associated with the chamber by using pure argon as the gas filling instead of the more usual argon–carbon dioxide mixture, and purifying it continuously, throughout the course of the experiment, by convective flow over hot calcium turnings. Our purifier worked on the same principle as that mentioned by Rossi and Staub (1949) but was simpler in design. This practice results in a longer collection time for the electrons, but does not appear to have led to any noticeable errors. The electronic apparatus was calibrated daily with a signal generator and we have good reason to believe that the error caused by random variations in the sensitivity of the chamber, amplifier and recording apparatus was not greater than ±3 per cent during the experiment. Other possible sources of error will be discussed in § 3.

In order to compare the two energy distributions under identical conditions, we simply removed part of the cadmiun shielding at different times (see § 2.2) and placed a neutron source near the chamber to induce fissions in the U^{235} present in the foils. In this way we measured approximately 1000 neutron induced fissions in three groups over a period of a week, the number of spontaneous fissions recorded during a week being about 200. This procedure was repeated during subsequent weeks until in the end some 5000 neutron induced and 1000 spontaneous fissions were obtained.

§ 2. APPARATUS

2.1. *The Ionization Chamber.*

A cylindrical form for the ionization chamber was chosen, as it eased a number of mechanical problems, particularly the construction of the grid (see fig. 1). The outer electrode is 25 cm. in diameter and 20 cm. high and is supported on porcelain insulators. The uranium foils are carried on the inner surface of this electrode and are approximately 10 cm. across, ensuring that no ionization track left the active volume of the chamber. The inner electrode consists of a cylinder 20 cm. high and 11·5 cm. diameter, and is supported on insulators consisting of perspex rings. The grid which has diameter of 15 cm. consists of 64 beryllium copper wires (24 S.W.G.) of length 25 cm. supported on half-rings of brass attached to the same insulator system as the inner electrode. It is easily demountable

2 I 2

432   W. J. Whitehouse *and* W. Galbraith *on the Energy Spectrum*

and quite robust. There are two guard rings to prevent transient discharges along the insulators between the grid and the inner electrode. The system of electrical connections is shown in fig. 1.

The distance between the outer electrode and the grid is about 5 cm., which is considerably longer than the track length in argon at one atmosphere of either the fission fragments or the alpha particles from uranium. The distance from the grid to the inner electrode is about 1·7 cm.

In a chamber designed to measure spontaneous fission pulses the background either from gamma rays or alpha contamination is unimportant, and no attempt has been made to minimize it.

The theory of a planar grid chamber has been given by Bunemann, Cranshaw and Harvey (1949). Their formulæ have been used to calculate the dimensions of our grid, using the following transformations between electrically equivalent cylindrical and plane chambers.

Fig. 1.

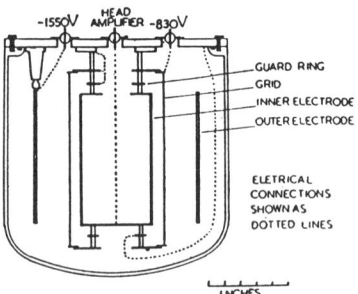

Diagram of ionization chamber.

If R_P is the radius of the inner electrode,

   R_G is the radius of the grid,

   R_A is the radius of the outer electrode,

then the dimensions of the electrically equivalent plane chamber are given by (see fig. 2):

$p^1 = R_G \log_e R_G/R_P$ where p^1 is the distance from the anode to the central plane of the grid.

$a^1 = R_G \log_e R_A/R_G$ where a^1 is the distance from the central plane of grid to the outer electrode.

When this system of transformation is used the physical dimensions of the grid are the same in both chambers.

of Fragments from the Spontaneous Fission of Natural Uranium   433

One important quantity to be calculated is the charge induced on the collecting electrode by the positive ions which, after each fission, are left in the neighbourhood of the outer electrode. The dimensions of the grid were so chosen that this was, theoretically, less than 2 per cent of the charge carried to the collecting electrode by the electrons. This rather low efficiency of the grid is tolerable because the accuracy of the measurements of the energies required in this experiment is much lower than that usually required in the measurement of alpha particle energies.

The other important condition to be satisfied is that none of the electrons should be collected on the grid wires. With our dimensions this requires that the voltages on the outer electrode (V_A), grid (V_G) and collector (V_P) should obey the relation

$$\frac{V_P - V_G}{V_G - V_A} \geqslant 0{\cdot}88. \qquad \cdots \cdots \cdots \quad (1)$$

The chamber was filled with argon (99·8 per cent pure), to a pressure of about 1·25 atmospheres, and this was continuously purified by convective flow over calcium turnings at about 400° C. The volume of

Fig. 2.

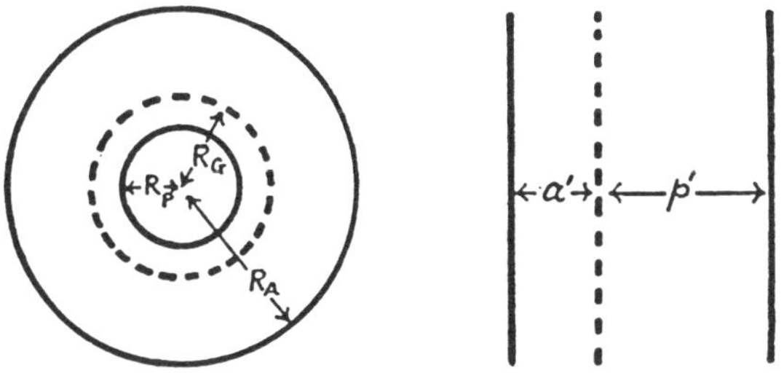

Electrically equivalent cylindrical and plane ionization chambers.

the chamber is approximately 30 litres, and the rate of flow of gas was about 40 c.c. per second, so that the chamber gas was frequently renewed. After the purifier had been switched on, the conditions in the chamber, as judged from the energy and resolution of the uranium alpha-peaks, improved for about fifteen hours, and then remained constant, within our limits of accuracy, for the three months during which preliminary work and the actual counting were carried out*.

The voltages for the outer electrode and the grid were supplied by dry batteries, through simple resistance-capacity filters.

---

* We were unfortunate enough to have three failures in the power supply, early in the experiment, which resulted in some loss in gas pressure, due to the contraction of the seals of the purifier on cooling down. Apart from this there is no reason to suppose that the chamber and purifier could not have been operated continuously for a year.

2.2. *The Neutron Shield for the Chamber.*

The chamber was surrounded by a cadmium shield (0·5 mm. thickness) around which was a paraffin shield of approximately 25 cm. thickness. A hole was cut in one of the paraffin blocks in which a neutron source could be placed. By removing a section of the cadmium shield and inserting the neutron source a satisfactory rate of neutron induced fissions could be obtained.

2.3. *The Uranium Foils.*

Eight uranium foils each 9·2 cm. square were prepared by evaporation under vacuum on to platinum. No small scale irregularities were seen when the foils were viewed under a low power microscope. The

Fig. 3.

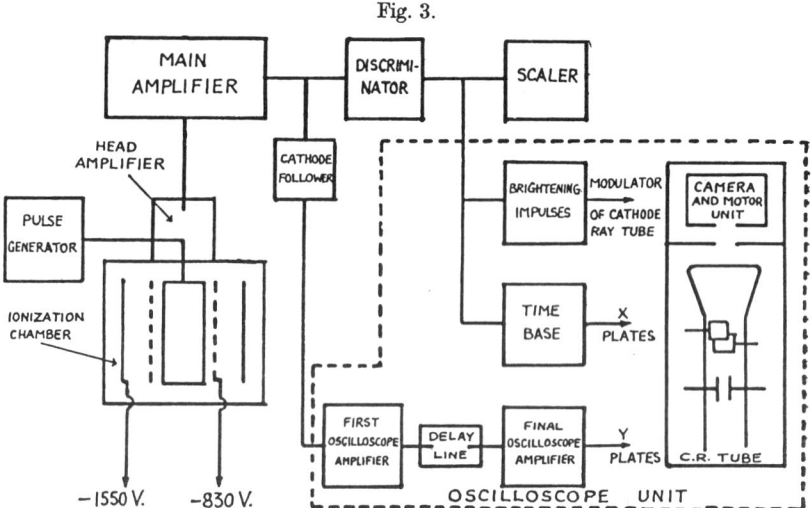

Block diagram of electronic apparatus.

uniformity over any one foil was checked by counting the alpha-particles from sections of the foils and was constant to within ±15 per cent. The total weight of the deposit, which is presumably oxidized, and of which the chemical composition is unknown, is 71 mg., which corresponds to an average surface density of 0·1 mg./cm². The separate foils were found, by weighing, to have average surface density per foil between 0·09 and 0·12 mg./cm². The total weight of uranium present, estimated by alpha-counting, is 57±3 mg. indicating either that the films are completely oxidized or that there is some impurity present. It is fortunately not important in this experiment to have a very exact knowledge of the weight of uranium in the foils.

2.4. *Amplifiers, Recorder, and other Electronic Apparatus.*

A block diagram of the apparatus is given in fig. 3. The head amplifier consists of one so-called " ring of three " circuit and the main amplifier of two such circuits. The circuits contained no electrolytic condensers, as these had been found to generate spurious pulses, and the grid bias voltages for the valves were derived from a resistance chain in the negative H.T. line. This modification increased the hum level of the amplifier, but not sufficiently to prevent an accurate determination of the position of the U^{234} alpha-peak. At the gain setting used for measuring the fission pulses the hum is negligible. The band width of the amplifier is controlled by " differentiating " and " integrating " condenser–resistance networks with time constants which can be varied in steps from $0.08 \,\mu\text{S}$ to $8 \,\mu\text{S}$. The gain is controlled by attenuators made up from high stability components.

The discriminator was a standard instrument (A.E.R.E. Type 1028). It was set at a level corresponding to an energy of 20 MeV.

The fission pulses were recorded by photographing an oscilloscope trace on 16 mm. film, and the times at which they occurred recorded by photographing a clock. All resistors in the oscilloscope amplifiers were high stability types. The fission pulses were passed through a 6μ second delay line in order to allow the time base and brightening circuits of the oscilloscope to operate. After each pulse the film was automatically moved on one frame. The interval between successive spontaneous fission pulses was sometimes several hours, in which case some fogging of the film occurred due mainly to the residual light from the filament of the cathode ray tube. This, however, was not serious.

The traces of the fission pulses were projected on to a screen for measurement.

The pulse generator, which was not a precision instrument, was coupled to the head amplifier through a small capacitor of about $1 \,\mu\mu\text{F}$. It was used for intercalibrations between the discriminator setting and the cathode ray tube deflections, and also for checks on the linearity of the main linear amplifiers, and the amplifiers associated with the cathode ray tube.

It was essential to have a mains supply free from fluctuations which would have generated spurious pulses in the amplifiers. All the apparatus, except the heater for the gas purifier, was therefore supplied from a separate motor driven single phase alternator.

§3. Errors.

It is instructive to consider what limitations are imposed on the accuracy of the results by the conditions of the experiment.

3.1. *Energy loss of Fragments—Distortion of Energy Distribution Curve.*
  (a) *Foil thickness.*

The distribution of the paths which the fission fragments traverse inside the foil can be shown to have the form given in fig. 4 provided that the thickness of the foil is small compared with the maximum range of the fragment in the foil. If T is the path length in the foil and X_0 is the thickness of the foil, and $t=T/X_0$, the distribution is a straight horizontal line with $N(t)=\frac{1}{2}$ for values of $t \leqslant 1$, and falls off for $t > 1$ according to the formula

$$N(t)\,dt = \frac{1}{2}\frac{dt}{t^2}. \qquad \ldots \ldots \ldots \quad (2)$$

The numerical constant $\frac{1}{2}$ is introduced to make the integral of $N(t)$ from $t=0$ to $t=\infty$ equal to unity, which corresponds with the fact that for every fission one fragment will emerge from the upper surface of the foil.

Fig. 4.

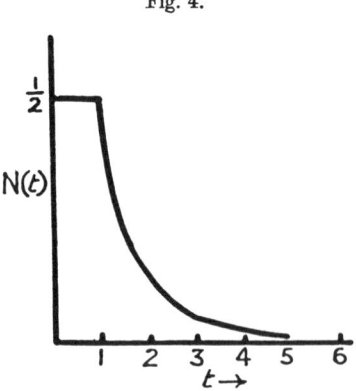

$N(t)$

Distribution of particles originating in a foil as a function of their length of path in the foil.

The energy lost by the fragment near the beginning of its path can be calculated from the approximate formula of Bohr (1941, 1948) according to which the velocity of the fragment is

$$v = v_0(1-T/R_F), \qquad \ldots \ldots \ldots \quad (3)$$

where v_0 is its initial velocity, T is the distance it has travelled, and R_F is an extrapolated fission range which is nearly the same as the actual range of the fragment. We have therefore

$$\epsilon = E_0\left(\frac{2T}{R_F} - \frac{T^2}{R_F^2}\right). \qquad \ldots \ldots \ldots \quad (4)$$

Where ϵ is the energy lost by the fragment in traversing a path T ; E_0 is the initial energy of the fragment.

The value of R_F can be calculated from the range of an alpha-particle with the same initial velocity as the fragment (Bohr 1941).

of Fragments from the Spontaneous Fission of Natural Uranium   437

We have calculated values of R_F in U_3O_8 for the light and heavy fission fragments using a relative stopping power of 4 mg./cm.^{2} U_3O_8 equivalent to 1 cm. of air (Bøggild *et al.* 1947) and have found the initial energy losses ($d\epsilon/dT$ for T=0) of the two fragments, in terms of the path expressed in mg./cm.^{2}. For the light fragment this is 19 MeV. per unit of path and for the heavy it is 17 MeV. This agrees reasonably well with the measurements of West (1948) who found that the fragments lose 6·5 MeV. per mm. of nitrogen, corresponding to 16 MeV./mg./cm.^{2} of U_3O_8.

In order to estimate, roughly, the distortion to be expected in our curve, we have taken the following round figures (*a*) an initial energy loss $d\epsilon/dT$ of 20 MeV. for each mg./cm^2 of film for all the fragments

Fig. 5.

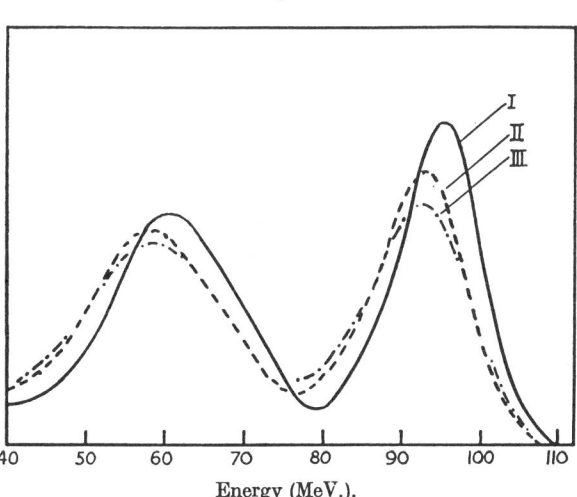

Energy (MeV.).

Corrections to be applied to the energy distribution of fission fragments.
 Curve I. Energy distribution of fission fragments from thermal neutron induced fission of U^{235} (after Brunton and Hanna).
 Curve II. Curve after correction is made for energy loss in foil.
 Curve III. Curve after correction both for energy loss and gain variation.

whether heavy or light, (*b*) a film of thickness 0·1 mg./cm^2. Thus all the fragments are assumed to lose 2 MeV. if they traverse the whole depth of the film in a direction normal to its surface.

By combining equations (2), (3) and (4) and taking terms as far as the fourth power it is possible to show that the number of particles losing an energy η is given by

$$\left. \begin{array}{l} n(\eta)=\tfrac{1}{2} \text{ for } \eta \leqslant 1, \\ n(\eta)\,d\eta=\tfrac{1}{2}\,d\eta/\eta^2\{1-1/16(\epsilon_0/E_0)^2\eta^2\} \text{ for } \eta>1 \end{array} \right\} \quad \cdots \quad (5)$$

438   W. J. Whitehouse *and* W. Galbraith *on the Energy Spectrum*

Where $\eta = \epsilon/\epsilon_0$ and ϵ_0 ($=2$ MeV) is the energy lost in traversing the film in a direction normal to its surface. Thus to a close degree of approximation the expressions for the distribution according to path lengths and to energy loss are identical in form.

When this correction is applied, curve II. of fig. 5 is obtained.

It has become a common practice in measurements of this type to collimate the particles with a perforated metal sheet, thus eliminating those which lose a large amount of energy in passing obliquely through the film. The disadvantage of this practice in an experiment such as this, where the counting rate is low, is that the collimator stops a large proportion of the particles, and necessitates the use of a much thicker source. Calculations have been carried out (Howlett and Whitehouse 1950) which show that in this experiment the use of a collimator would have increased the energy lost by the particles in the film, and a plain foil has therefore been used.

(b) *Energy lost by Particles striking the Electrode.*

In a cylindrical chamber such as ours some of the fragments lose energy at the ends of their paths by striking the curved surface of the outer electrode. This affects less than 5 per cent of the fragments, and we have estimated that the average loss in energy from this cause is less than $\frac{1}{4}$ MeV., whereas the average energy loss in the film itself is about 4 MeV. We therefore have not applied any correction for this effect.

(c) *Effect of Variations in the gain of the Amplifier and Recording Apparatus.*

Curve III. of fig. 5 shows the effect which might be expected from a random variation of the sensitivity of the apparatus. In order to simplify the computation we have assumed that the measurements for all the fragments regardless of their initial energy are subject to a random error with a standard deviation of ± 3 MeV.

This figure has been chosen as the most probable value of the spread in our results as a result of the combination of the drift of the apparatus over a period of time (§ 3.2) and of the differential sensitivity to pulses of different rise time (§ 3.3).

Owing to the rough nature of the corrections described above we do not invite detailed comparison between our results and the theoretically distorted curves of fig. 5. In fig. 6, however, we plot our experimental results for U^{235} along with curve III. of fig. 5. Our approximate energy calibration shows our results for the energies of the two peaks to be within experimental error of those given in the introduction, but in drawing fig. 6, we have adjusted the curves so that the total number of fissions is the same in each, and the positions of the minima between the peaks coincide.

It will be seen that the two main sources of error in the experiment do provide an adequate explanation for the difference between the shape of our curve and that of Brunton and Hanna.

of Fragments from the Spontaneous Fission of Natural Uranium   439·

3.2. *The Drift in Gain.*

A daily check was made of the linearity of the amplifiers and their overall gain, by means of the signal generator and the zero error of the discriminator was determined at the same time. It was decided to use the discriminator voltage as a standard and to refer all the pulse heights to it.

We have two independent checks of the constancy of the apparatus :

(*a*) the height of the U^{234} alpha-peak. This was measured every second day using, of course, a different gain setting on the amplifier, and found to be nearly constant on the discriminator scale. The greatest deviation from the mean value, over three months was ±2·2 per cent.

Fig. 6.

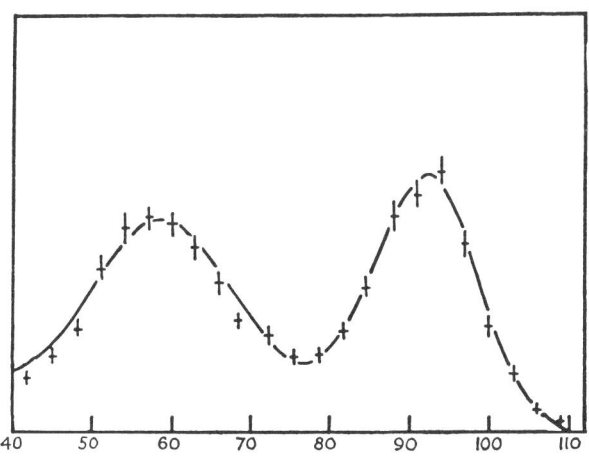

Approximate energy scale (MeV.).

Comparison of present results for fission fragments from thermal neutron induced fission of U^{235} with corrected curve.

(*b*) It is also possible to maintain a general check on the steadiness of the apparatus from the average of the heights of the fission pulses.

If we regard the high energy peak alone as a statistical population, and apply simple sampling theory (neglecting the difference between the shape of the observed peak and the normal error curve) we can assume that the mean energy of *n* fissions will have had a standard deviation of approximately $\sigma/\sqrt{(n)}$, where σ is the standard deviation of the whole population. The value of σ is of the order of 10–15 MeV ; consequently the mean energy of any hundred consecutive fission pulses in the high energy group should have a standard deviation of about 1 to 1·5 MeV. If the error due to slow drift in the apparatus is much greater than this, it will show up in the mean energies of groups of fragments. Exactly similar considerations apply to the mean energies of the fissions in the low energy group.

We have calculated the mean energies of the high and low energy fragments in groups of five-hundred for neutron induced fissions and of two hundred for spontaneous fissions. We find in both cases that the standard deviation of the mean of the groups is about 2 MeV., indicating that the error due to random slow drift could not have been larger than this.

3.3. *The Electron Collection time and the Resolving time of the Amplifier.*

Owing to the comparatively slow rate of electron drift in pure argon, the large dimensions of the chamber, and the moderate collecting voltages applied, it was considered necessary to check that the resolving time of the amplifier ($8 \mu S$) was sufficient to allow complete collection of the electron pulses. This was done by experiments on the U^{234} alpha-peaks and on the energy distribution of fission fragments from neutron induced fission of U^{235}.

(a) It was found that it was necessary to reduce the time-constant of the amplifier from eight to five microseconds before there was any observable diminution of the α-pulse height.

(b) The outer electrode potential could be reduced from -1550 volts to -830 volts, and the grid voltage from -830 volts to -450 volts, without any diminution in the height of the α-pulse.

(c) With the reduced electrode voltages given above, the energy distribution of some 240 neutron induced fission pulses from U^{235} showed no significant difference from that obtained with the chosen operation voltages. We concluded then that the resolving time of the amplifier was adequate.

Another small effect of the electron collection time, quite unconnected with the resolving time of the main amplifier, has been noticed.

By direct calibration with voltage pulses fed into the head amplifier we have found that the amplifier pulse height measured at the discriminator is proportional to the original voltage, and is not noticeably affected by the time of rise of the pulse within quite wide limits.

The heights of the traces on the oscilloscope do, however, decrease by about 3 per cent when the rise time is reduced from $3 \mu s$. to a fraction of a microsecond, an effect which we attribute to the delay line. Since the collection time of electrons in our chamber depends upon the angle of the track, a slight spread in the measured energy values of fission fragments will be introduced, in addition to that caused by random drift in the gain of the apparatus.

3.4. *The background Rate of Real or Spurious Pulses.*

The neutron induced fission rate, with the Ra-α-Be source in position, was about one hundred per hour, while the spontaneous fission rate

of Fragments from the Spontaneous Fission of Natural Uranium   441

was about 1·4 per hour. It follows that the background of spontaneous
fissions during a period of measurement on neutron-induced fissions
was too small to affect the observed distribution of pulse sizes.

By placing the neutron source outside the paraffin and comparing the
count rates with and without the cadmium we were able to show that the
percentage of spontaneous fissions which could be ascribed to neutrons
penetrating the shield is probably negligible.   Even if an unfavourable
interpretation is placed upon the statistics of the results, the percentage
would not be greater than of the order of 2 per cent.

Wheeler (1949) has pointed out that the capture of a negative μ meson
by a uranium atom will in almost every case lead to the fission of the
uranium nucleus, but that the number of fissions so induced will be
negligible in comparison with the spontaneous rate.

The background of spurious counts was very low in comparison to the
observed spontaneous fission rate.   When the foils were removed from
the chamber only two counts were recorded in 160 hours, and both these
were too small to have been confused with fission pulses.

3.5. *Coincidences of Alpha-particles and Fission Pulses.*

Examination of the photographs of the trace of the cathode ray tube
reveals the alpha-particles as small pulses along the time base.   It is
possible to make a fairly close estimate of how close the alpha-particle
and the main fission pulse would have to be before it became impossible to
distinguish them from each other.   In this way we estimate that less
than 1 per cent of the fission pulses are in effective coincidence with
alpha-particles.

The chance of sufficient alpha-particles coinciding with each other to
give an alpha build-up pulse which might be mistaken for a fission is
negligible.

§ 4. RESULTS.

The results of the experiment are presented in fig. 7.   The full line is
the energy distribution curve for neutron-induced fissions, which, as has
been explained, is consistent with the results of other observers, providing
that allowance is made for absorption in the foil, and for some random
variation in the recording apparatus.   About 4700 neutron-induced
fissions were used in the preparation of this curve, and the standard errors
of the points are shown.

The energy distribution of 960 spontaneous fissions is shown by the
histogram.   The errors indicated are the square roots of the numbers of
events in each cell of the histogram.   The curve and the histogram have
been adjusted to have the same area.

The energy scale on the diagram, derived from the height of the U^{234}
alpha peak, is approximate and, in view of the errors involved in its
derivation, we would not expect it to be accurate to better than 5 per cent.
However, it suffices to show that our observed values for the most

442   W. J. Whitehouse *and* W. Galbraith *on the Energy Spectrum*

probable energies of the fission fragments are not inconsistent with those
of other workers.   An accurate absolute energy determination was not
attempted, since the present experiment consists of a direct comparison
between the distributions for the two types of fission.

There is no significant difference between the observed distributions for
neutron-induced fissions in U^{235} and for spontaneous fissions, according
to the standard χ^2 test for goodness of fit (0·05 criterion)*.

It is too much to hope, with the small number of spontaneous fissions
observed, that detailed differences in the shape of the two distribution
curves would be revealed, but it might be thought possible to observe

Fig. 7.

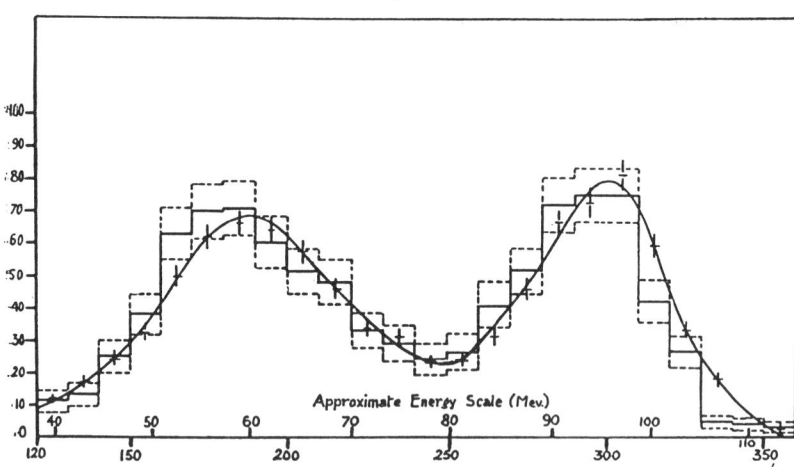

Arbitrary scale proportional to energy.
Comparison of thermal neutron induced fission pulses of U^{235}, and spontaneous
fission pulses of natural uranium:
    Full line curve—U^{235} fission with thermal neutrons.
    Histogram—spontaneous fission.

some difference in the positions of the peaks representing the most
probable energies for the fragments.   We have carried out a simple test
of this point by moving a tracing of the continuous curve along the energy
axis, over the histogram, and carrying out χ^2 tests for different positions.
We find that we get the best fit if we assume that the whole distribution
for the spontaneous fissions is about 1·5 MeV. below that for the neutron-
induced fissions.   The χ^2 test shows, however, that the two curves are
consistent for any downward displacement of the spontaneous fission
distribution between 0 and 3 MeV.

---

* This means that the probability of the difference between the two distributions
being due to chance is greater than 0·05.

of Fragments from the Spontaneous Fission of Natural Uranium   443

The present experiment is not particularly well adapted to an accurate measurement of the spontaneous fission rate but a combination of the observed frequency of the spontaneous fissions ($1\cdot38\pm0\cdot04$ per hour) and the mass of uranium in the foils (57 ± 3 mg.) gives a figure of $24\cdot2\pm1\cdot5$ per gram of natural uranium per hour.

ACKNOWLEDGMENTS.

This work is published by permission of the Director of the Atomic Energy Research Establishment. The authors wish to express their thanks to Messrs. Wells and Godfrey who designed and constructed the oscilloscope unit, to Messrs. Kandiah and Gillespie for the design and modification of the head and main amplifier circuits, to Mr. Robins and his group for the preparation of the uranium foils, and to Dr. E. Bretscher, Head of the Nuclear Physics Division at A.E.R.E.

REFERENCES.

BØGGILD et al., 1947, *Phys. Rev.*, **71**, 281.
BOHR, N., 1941, *Phys. Rev.*, **59**, 270; 1948, *Det. Kgl. Danske Vidensk Selsk., Mat. Fys. Medd*, 18, No. 8.
BOHR and WHEELER, 1939, *Phys. Rev.*, **56**, 426.
BRUNTON and HANNA, 1949, *Phys. Rev.*, **75**, 990.; *Can. Journ. Res.* A. In press.
BUNEMANN, CRANSHAW and HARVEY, 1949, *Can. Journ. Res.*, A 27, 191.
CHATTERJEE, 1945, *Indian Journal of Physics*, **19**, 211.
DEUTSCH and RAMSEY, 1945, LADC. 257.
FLAMMERSFELD, JENSEN and GENTNER, 1943, *Zeit f. Physik*, **120**, 450.
FLEROV and PETRZAK, 1940, *J. Phys. U.S.S.R.*, No. 4–5, **3**, 275; 1941, *Ibid.*, No. 3, **4**, 283.
FOWLER and ROSEN, 1947, *Phys. Rev.*, **72**, 926.
HOWLETT and WHITEHOUSE, 1950, A.E.R.E. Report N/R. 473.
JENTSCHKE, 1943, *Zeit. f. Physik*, **120**, 165.
KANNER and BARSCHALL, 1940, *Phys. Rev.*, **57**, 372.
MAURER and POSE, 1943, *Zeit. f. Physik*, **121**, 285.
PERFILOV, 1947, *J. Expt. Theor. Phys. U.S.S.R.*, **17**, 476; *J. Phys. U.S.S.R.*, No. 3, **11**.
POSE, 1943, *Zeit. f. Physik*, **121**, 293.
ROSSI and STAUB, 1949, *Ionization Chambers and Counters*, McGraw-Hill, p. 109.
SCHARFF-GOLDHABER and KLAIBER, 1946, *Phys. Rev.*, **70**, 229.
Science and Engineering of Nuclear Power, 1947, Addison Wesley Press, pp. 80 and 358.
WHEELER, 1949, *Rev. Mod. Phys.*, **21**, 133.
WEST, D., 1948, *Can. J. Res.* A, **26**, 115.

Part Five

Quanta

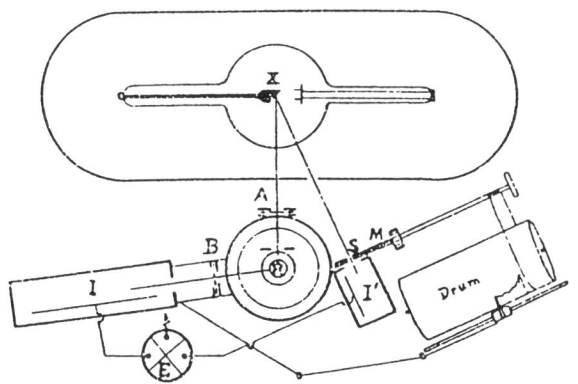

Part Five
Quanta

The year 1900 marked not only the dawn of a new century but a watershed in the history of the physical sciences. On 14 December of that year Max Planck submitted a paper to *Verh. Deutsch Phys. Ges.* (1900 **2** 237) in which the fundamental constant h (Planck's constant) appeared for the first time. That paper heralded the birth of quantum physics. 'Today I have made a discovery as important as that of Newton', Planck said to his son with uncharacteristic immodesty. Subsequent events were to reveal the true significance of this remark.

Niels Bohr's application of quantum concepts to describe the behaviour of electrons in atoms has already been described in Part Three of this volume. Here we backtrack to the beginnings of the new quantum theory which lie in Planck's explanation of the spectral dependence of thermal radiation emitted by a black body. Not content with formulating an empirical equation which perfectly fitted the spectra obtained experimentally for different temperatures, Planck laboured to discover its underlying meaning. Applying statistical thermodynamics to the problem, he arrived at the 'inescapable conclusion' that the molecules (Planck actually used the term resonators) comprising the source of the radiation, gained or lost energy in small but nevertheless finite amounts — in fact in integer multiples of $h\nu$ where ν is the frequency of oscillation of the molecules or, if the cavity walls are in equilibrium with the radiation, the frequency of the radiation itself.

Planck (1858–1947) was a professor of the famous Friedrich Wilhelm University in Berlin — having succeeded Kirchhoff who died in 1887 — when he made his momentous proposal at the age of 42. Although the success of his radiation law was recognized almost immediately, the wider significance of energy quanta was not appreciated for several years. In 1905, Albert Einstein, then 24 and an essentially unknown junior clerk in the Berne Patent Office, broadened the quantum hypothesis in a way that even Planck found difficult to accept. Einstein's proposal was that it is not only the material oscillators of a blackbody cavity that are quantized in energy but that electromagnetic radiation itself is composed of indivisible packets of energy. These light quanta (named *photons* by C. N. Lewis in 1926) carry an energy $h\nu$ and a momentum $h\nu/c$, and are emitted and absorbed as discrete units. In many respects they behave just like material particles.

As might be expected, such a radical proposal ran into strong opposition in the academic world. Had not a century of work by Young, Fresnel, Huygens and others demonstrated conclusively that, by its properties of interference and diffraction, light is a wave? The suggestion that it might have a particulate nature flew in the face of all the evidence and seemed to be little more than a return to Newton's corpuscular theory. Even Einstein's explanation of the photoelectric effect (the ejection of electrons from a metal surface by ultraviolet light) in terms of the absorption of quanta of radiation, did little to establish belief in the concept. When Millikan established in 1915 that the value of h deduced using Einstein's photoelectric equations was exceedingly close to that obtained from Planck's radiation formula, he (Millikan) felt it necessary to write '... [the results] were correctly and exactly predicted nine years ago by a form of quantum theory which has now been pretty generally abandoned'.

Undaunted by the general lack of acceptance of his ideas, Einstein applied the quantum concept to the vibrations of atoms in a crystal, thereby explaining the fall in specific heat with decreasing temperature of diamond and other substances, an idea taken up by Debye and Stern. There were others who warmed to the ideas; Johannes Stark used the quantum hypothesis to account for certain features of spectral lines and Nernst introduced quanta into chemistry. Also, as we have seen, Bohr proposed the quantization of the angular momentum of electrons in orbits around nuclei and Sommerfeld extended this model to more complicated atoms containing many electrons.

It was, however, work by Arthur H. Compton on the scattering of X-rays in 1923 which finally turned the tide in favour of the quantum nature of electromagnetic radiation. Following pre-war work at the Westinghouse Corporation in Pittsburgh on X-ray absorption and scattering, Compton spent a year with Rutherford and J. J. Thomson at the Cavendish Laboratory before returning to the United States in 1920 as Wayman Crow Professor of Physics at Washington University. There he made the important discovery that, when X-rays are scattered by solids, they undergo a definite shift of wavelength which depends systematically on the scattering angle. The results could not be reconciled with classical theories based on the assumption of a wave-like nature for the X-rays but, by treating the scattering process as a collision between a photon, having an energy $h\nu$ and a momentum $h\nu/c$, and a stationary electron, Compton calculated that the wavelength of the radiation increased by just the amount he had observed. For the discovery that bears his name, Compton shared the Nobel Prize for Physics in 1927. His equations and experimental results are given in one of his classic papers reproduced in this Part.

With Compton's work, resistance to the quantum nature of light began to evaporate but there still remained the difficult problem of reconciling the particle and wave pictures of light. It was not a question of deciding between the two viewpoints, for both were required — which one depending on the type of experiment or observation. A puzzling type of duality emerged, first recognized by Einstein who, as early as 1909, had prophesized 'It is my opinion...that the next phase of development of theoretical physics will bring us a theory of light that can be interpreted as a kind of fusion of the wave and emission [particle] theories.' The wave–particle duality paradox in fact became the dominant theme of quantum theory during the 1920s.

The next surprising development came just one year after Compton's paper was published and, in terms of radical ideas, must rank amongst the most startling in the history of science. Prince Louis-Victor de Broglie, a French aristocrat, proposed in 1924 that wave–particle duality applied not only to photons but to matter in general. In particular he put forward the revolutionary idea that electrons could behave like waves. Furthermore the frequency and wavelength associated with these waves was given by precisely the same formulae that Einstein had proposed and Compton had used for photons. Thus an electron of total energy E has associated with it a wave of frequency ν given by $E = h\nu$ and its momentum p is related to the wavelength by $p = h/\lambda$. (In the case of a photon $\lambda = c/\nu$ with c the velocity of light; for an electron $\lambda = u/\nu$ with u the *phase* velocity.) It is unlikely that de Broglie's ideas would have attracted much attention were it not for the unmitigated support that Einstein, now famous, gave to them. Three years later, Davisson and Germer, and independently G. P. Thomson, demonstrated the wave-like nature of electrons by showing that they could produce diffraction patterns and be reflected from single crystals just like X-rays. De Broglie's classic paper is reproduced in this Part.

1914   **28**   *New Paths of Physical Knowledge. By Dr Max Planck, Professor of Theoretical Physics*

Max Planck, the founder of quantum physics, did not publish any of his original scientific papers in the *Philosophical Magazine*, choosing instead to submit his work to German journals in his native language. In the present communication, which is a translation of an Address he gave on commencing the Rectorate of the Friedrich-Wilhelm University in Berlin on October 15th 1913, he expresses his views on the revolutionary new ideas in physics which were born in the first decade of the present century.

In 1900 Planck had directed himself to the task of attempting to explain the wavelength dependence of the energy of electromagnetic radiation emitted from a black body held at a given temperature. The concept of a 'black body' (i.e. one that absorbs and emits all wavelengths) had been introduced by Kirchhoff in 1860. The spectrum of radiation from such a body was known experimentally to depend only on its temperature and to peak at a wavelength inversely proportional to that temperature — an effect easily observed (qualitatively at least) by the changing colour of a heated metal. Lord Rayleigh provided the first theoretical treatment of the problem. He used classical physics to describe the emission in terms of allowed standing wave patterns in a black body cavity, each of which carried an energy kT where k is the Boltzmann constant and T the temperature (see *Science in the Making*, Volume 2). Although Rayleigh's formula fitted the experimentally determined curve exactly at long wavelengths, no peak in the emission was predicted and at short wavelengths the theoretical curve headed for infinity in what has been dramatically called the 'ultraviolet catastrophe'.

Planck's bold assumption was that each mode of vibration carried discrete amounts of energy of magnitude $h\nu$ or hc/λ where ν is the frequency, λ the wavelength, c the velocity of light and h is Planck's constant. Development of this *quantum hypothesis* led to a complete explanation of the form of the black body radiation curve, a success which was to initiate further revolutionary concepts. Einstein, in particular, proposed in 1905 that the bundles of energy were preserved in the emitted radiation so that light, and indeed all electromagnetic radiation, consisted of such quanta or, as they are now called, *photons*.

In this University Address, Planck refers to three 'theorems' of physics which had been 'proved untenable' in the light of developments during the prior decade and a half. These he lists as The Invariability of Chemical Atoms, the Mutual Independence of Space and Time, and the Continuity of all Dynamical Effects. It was the last of these which the quantum hypothesis had dispelled. He cites radioactivity as evidence for the breakdown of the first and the theory of special relativity (as proposed by Einstein in 1905) for the demolition of the second. In contrast, the principles of conservation of energy and momentum, the principle of least action and the laws of thermodynamics remained unscathed by the 'new physics' — and indeed this is still true today.

1923   **46**   *Absorption Measurements of the Change of Wave*
Length accompanying the Scattering of X-rays. By
Arthur H. Compton, Wayman Crow Professor of
Physics, Washington University, Saint Louis

The Compton effect is the change of wavelength experienced by electromagnetic radiation when it is scattered by electrons. The phenomenon holds a special place in the history of physics by providing a direct verification of the quantum theory of radiation, as first proposed by Planck in 1900 and subsequently given a firm foundation by Einstein in 1905. In spite of Einstein's success in explaining the photoelectric effect, the quantum hypothesis was not widely accepted until Compton applied it so effectively to account for experimental data from X-ray scattering from solids. In the same year that Compton's first paper (*Physical Review* 1923 **22**) on this topic appeared, Debye in Germany published a similar analysis.

In the present paper, Compton's main hypotheses and resulting formulae are first summarized. Treating the X-rays as a stream of particles (photons) each of energy $E = h\nu$ (ν being the frequency of the radiation and h Planck's constant) and carrying a momentum $p = h\nu/c$, Compton considered the scattering in terms of billiard-ball-like collisions in which energy is transferred to a free (i.e. unbound) and stationary electron. The electron recoils and the photon suffers a loss of energy, or increase in wavelength. Relativistic mechanics is used in the equations expressing conservation of energy and momentum and manipulation of these equations yields expressions for the shift of wavelength as a function of the angle through which the X-rays are scattered.

As shown in Figure 1 of the paper, curves of X-ray intensity versus scattering angle for carbon show excellent agreement with theoretical expectations. However, they also illustrate that not all the X-rays are scattered, as evidenced by the 'unmodified' lines of substantial strength, for which Compton offers explanations later in the paper.

The new experiments reported here were designed to test the theoretical predictions for scattering from heavier elements and also using longer primary wavelengths. Careful measurements of the change in absorption coefficient[1] were carried out by a method that is described in detail, and the wavelength shifts deduced as a function of scattering angle. For light metals and short wavelengths the results agreed with theory but, for heavy metals and long wavelengths, considerable departures from the theoretical predictions were observed. Compton attributes this to a greater proportion of X-rays being unmodified, i.e. not undergoing a shift of wavelength.

Two explanations are proposed for the existence of the unmodified ray. The first is that some of the electrons are too firmly bound to their nuclei to experience recoil — a situation more likely in heavier elements. The second is that two or more electrons on adjacent atoms scatter the X-rays *simultaneously*. Since the change in wavelength is proportional to the reciprocal of the mass of the scattering body (see equations (5) and (6) in the paper), collective scattering of this kind should lead to a smaller shift. Compton invokes the latter hypothesis to explain the absence of a wavelength shift 'when X-rays are regularly reflected from a crystal [i.e. Bragg reflection] or when light is reflected by the free electrons of a metallic mirror'.

In 1929, Klein and Nishina improved on Compton's theory by using Dirac's relativistic equation to include the effect of electron spin in a calculation of the scattering cross-section.

1924 **47** *A Tentative Theory of Light Quanta. By Louis de Broglie*

This remarkable paper, based on de Broglie's thesis for a University of Paris doctorate, contains far more than is suggested by the title and is noteworthy in several respects.

In the first place, it seeks to reconcile what was, at the time, seemingly irreconcilable evidence that light — indeed electromagnetic radiation of all kinds — sometimes behaves like a wave (as demonstrated by interference and diffraction) and sometimes like a particle (as in the photoelectric effect and Compton scattering). Here de Broglie formulates a relationship between these two apparently contradictory viewpoints by proposing that light quanta have associated with them a frequency ν which can be obtained by equating Einstein's relativistic formula for the total energy of a particle of rest mass m_0 to Planck's quantum energy $h\nu$

$$h\nu = m_0 c^2 / (1 - \beta^2)^{\frac{1}{2}} \qquad \text{where } \beta = v/c$$

According to this formula, different velocities v produce different frequencies. De Broglie postulates that 'the atoms of light' (the name 'photon' was introduced two years later) have velocities *nearly* equal to Einstein's limiting velocity c, i.e. $\beta \sim 1$, and concludes therefore that 'm_0 should be at most of the order of 10^{-50} grams'. (Of course, if $v = c$, the rest mass of the photon is zero.)

In Part II of the paper, de Broglie shows that use of relativistic expressions for the energy (W) and momentum (G) of a particle yields, for $v = c$, the fundamental relationship, $G = W/c$. (In more familiar notation, $p = E/c$, which, with $E = h\nu$, gives $p = h\nu/c = h\lambda$ — the Compton expression for the momentum of a photon in terms of its wavelength.) He then derives Wien's limiting form for the frequency distribution of black body radiation by regarding the radiation as a 'gas' of light quanta. In Part VIII he develops these ideas further, obtaining the full Planck expression for the radiant energy density. (This theory, in essence, represents a change from Boltzmann classical statistics to Bose–Einstein quantum statistics.)

The most revolutionary suggestion in the paper, however, appears in Part III. Driven by an intuitive feeling that if waves can sometimes behave as particles then particles should sometimes behave as waves, de Broglie proceeds to analyse what form such waves might take. The 'internal' frequency ν_0 associated with a particle at rest is given by $h\nu_0 = m_0 c^2$, i.e. $\nu_0 = m_0 c^2/h$, but for a particle moving with velocity $v(= \beta c)$ this becomes $\nu_0 \gamma$ (where γ denotes $(1 - \beta)^{-\frac{1}{2}}$ in the paper) just as used above for photons. There is, however, another frequency seen by a fixed observer. This is ν_0/γ, i.e. an *increase* in ν_0 resulting from dilation of the time intervals of the internal clock. The waves associated with the two frequencies remain in phase and travel with a velocity equal to c^2/v. He writes 'We are then inclined to admit that any moving body may be accompanied by a wave and that it is impossible to disjoin motion of body and propagation of wave'.

De Broglie is careful to point out that the 'phase' velocity c^2/v, being greater than c, is *not* the velocity at which energy is transported. Energy propagation occurs at the 'group' velocity which he shows is equal to that of the particle.

In Part IV de Broglie considers the situation of a particle moving in a curved path. What of the wave in this case? He acknowledges that the problem is difficult: 'Perhaps a new electromagnetism will give us the laws of this complicated propagation', he writes, 'but it seems that we know beforehand the final result: *The rays of the phase wave are identical with the paths which are dynamically possible*'. He then goes on to give a beautiful 'explanation' of the stability of electrons in Bohr-like orbits around atomic nuclei, which is that stable motion results 'if the phase wave is tuned to the length of the path' — in simple terms when an integral number of whole wavelengths exists around the closed orbit, thereby ensuring that the electron stays in phase with itself.

This condition is, in fact, equivalent to Bohr's postulate that the angular momentum in the orbits is quantized, for if the de Broglie condition is written $2\pi r = n\lambda$ for an orbit of radius r, we have for the angular momentum $mvr = mvn\lambda/2\pi$ and, using the de Broglie wavelength $\lambda = h/p = h/mv$, we find $mvr = nh/2\pi$.

At the time this historic paper was written, there was no direct experimental evidence for de Broglie's 'matter waves'. This first came in 1927 when Davisson and Germer, and almost simultaneously G. P. Thomson, discovered that electrons could be diffracted by crystals in a manner entirely analogous to X-rays and that their wavelength changed with velocity precisely according to the de Broglie relation, $\lambda = h/mv$. The complementary aspects of matter and waves were then demonstrated in a convincing and unequivocal way.

Notes

1. Strictly speaking X-rays are not *absorbed* in Compton scattering but are instead scattered out of the line of measurement. A 'photoelectric effect', in which X-rays *are* absorbed, accompanied by the ejection of electrons out of core levels, occurs for materials of high atomic number.

60   Prof. Max Planck : *New Paths*

Now $m=(p-\beta-1)n-\alpha$ and $\alpha<n-1$. For, if $\alpha=n-1$ the integration in (4) could be effected without separation into partial fractions. Therefore $m-1$ contains n as a factor $p-\beta-2$ times.

We then obtain

$$\frac{1}{x^m(x^n+1)} = -\frac{1}{n}\sum_{\kappa=1}^{n}\frac{r_\kappa^{-m+1}}{x-r_\kappa} + \sum_{\kappa=0}^{p-\beta-2}\frac{(-1)^\kappa}{x^{(p-\beta-\kappa-1)n-\alpha}}. \quad (18)$$

And

$$\int_1^x \frac{dx}{x^m(x^n+1)} = -\frac{1}{n}\sum_{\kappa=0}^{\left[\frac{n-2}{2}\right]}\left[\cos\frac{2\kappa+1}{n}\pi(m-1)\left\{\log\left(x^2-2x\cos\frac{2\kappa+1}{n}\pi+1\right)\right.\right.$$

$$\left.-2\log\left(2\sin\frac{2\kappa+1}{2n}\pi\right)\right\}$$

$$+2\sin\frac{2\kappa+1}{n}\pi(m-1)\left\{\tan^{-1}\frac{x-\cos\frac{2\kappa+1}{n}\pi}{\sin\frac{2\kappa+1}{n}\pi}-\frac{2\kappa+1}{2n}\pi\right\}\right]$$

$$+\frac{(-1)^m[1-(-1)^n]}{2n}\log\frac{1}{2}(x+1)+\sum_{\kappa=0}^{p-\beta-2}\frac{(-1)^{\kappa-1}}{(p-\beta-\kappa-1)n-\alpha-1}$$

$$\times\frac{1}{x^{(p-\beta-\kappa-1)n-\alpha-1}}-\frac{1}{2}[1-(-1)^{p-\beta}]. \quad . \quad . \quad . \quad . \quad . \quad . \quad (19)$$

The required Integral (1) is obtained by combining (18) or (19) with (7) (due regard being paid to the limits).

University of Pennsylvania,
    Philadelphia, Pa., U.S.A.

IX. *New Paths of Physical Knowledge*: being the Address delivered on commencing the Rectorate of the *Friedrich-Wilhelm University, Berlin,* on October 15th, 1913. *By* Dr. MAX PLANCK, *Professor of Theoretical Physics* *.

The Rector began his Oration thus :—

HONOURED ASSEMBLY, ESTEEMED COLLEAGUES, DEAR COMRADES :

Called to the head of the Administration by the confidence of the accredited representatives of our Corporation, I have undertaken as my first official duty the task of greeting the members and friends of our ALMA MATER at the commencement of the new

* Communicated after revision by Sir Oliver Lodge, on the basis of a translation by Dr. Fournier d'Albe.

Session in an Address concerning the beginning of the Course of Studies, as the Statutes prescribe.

Rather special are the feelings with which on this occasion we, Teachers and Taught, regard the problems facing us in the new Term ; for while the year just passed appears bathed in a festive glory, illumined and warmed over its whole course by the thought of national ideas, of the heavy sacrifices brought and of glorious resulting victories,—the last and greatest of which is to be celebrated in these days by the whole German people,—the coming Term appears, on the other hand, likely to bear an everyday character and to be devoted to regular work.

The best we can derive from the Memorial festivities of the past year is the fervent wish that our successors may at some future time look up to us as we look up to those men who, one hundred years ago, fought and suffered in word and deed for the Fatherland. Let no one reject such a wish as entirely baseless, on the ground that such high aims cannot be contemplated to-day. For, in the first place, we must not forget that the forces which were then gloriously displayed derived their real strength from the quiet everyday work of the simpler times preceding, which may not have been so conscious of their high import but may have been all the more intense and creative ; and in the second place, none of us can know with what standard coming generations may approach the estimate of our present-day performances. But what we can assert with certainty, under all circumstances, is that our generation can only hope to be judged honourably by posterity if it endeavours to solve its own problems according to its lights, in strict fulfilment of its duties : each one in the place to which his calling and his fortune have led him.

So then may I be allowed to-day to lay before you a section of the special work going on within the Science I represent, by a survey of the progressive development of Physical Knowledge and an endeavour to sketch those new paths which it has trodden since the beginning of this century.

NEVER, probably, has experimental physical investigation experienced so strenuous an advance as during the last generation, and never probably has the perception of its significance for human culture penetrated into wider circles than it does today. The Waves of Wireless Telegraphy, Electrons, Röntgen Rays, the Phenomena of Radio-activity, appeal more or less to the interest of everyone. But if we face the larger question in what respect have these new and brilliant discoveries influenced and advanced our understanding of Nature and her laws, the outlook, at first glance, does not appear at all correspondingly brilliant. On the contrary, whoever endeavours to judge of the state of present Physical theories from a higher point of view, may readily get the impression that theoretical investigation has been rather confused by so many new experimental discoveries, which have been for the most part entirely unexpected, and that it is now in a profitless period of blind groping, which contrasts strongly with the clear calm and security marking the theoretical epoch just passed,—an epoch which, with some justification, can be described as the Classical Epoch. Everywhere old and firmly-rooted conceptions are attacked, universally recognized theories are discarded and are replaced by new hypotheses,—some of them of a boldness which makes almost intolerable demands on the intelligence even of the scientifically educated, and, in any case, does not appear calculated to strengthen confidence in a steady and effective progress of Science. So the present science of Theoretical Physics may give the impression of an edifice, venerable indeed but fragile, in which one part after another commences to crumble away, and whose foundations even are threatened.

Yet nothing would be more incorrect than such an idea. It is true that in the main structure of Physical Theories fundamental changes are taking place. But on closer inspection, we see that this is not a work of destruction but a work of completion and amplification, that certain blocks of the edifice are only removed from their place in order to find a better and securer place elsewhere, and that the real foundations of theory are still as firm and safe as they have ever been. I shall endeavour to prove in some detail this assertion.

First a general consideration :—The initial impulse towards revision and modification of a physical theory comes almost aways from the discovery of one or more facts which do not fit into the present framework of the theory. Facts always furnish the Archimedian point from which even the most ponderous theory may be lifted off its hinges. In that sense nothing is more interesting to the real Theoretician than a fact which directly contradicts a hitherto universally accepted theory, for it is just here that his real work begins.

Now what are we to do in a case like this? Only one thing is certain ; something must be changed in the accepted theory, and in such a manner that it agrees with the new fact, but it is often a difficult and complicated question at what point of the theory the correction is to be applied, for one fact is insufficient to furnish a theory. A theory, indeed, consists as a rule of a whole series of theorems connected with each other. It may be compared to a complicated organism, whose separate parts fit together so intimately that any interference at one place is felt in various other places, sometimes far removed. Wherefore, since every conclusion of theory results from the cooperation of several theorems, it follows that, as a rule, several theorems may be made responsible for each failure of the theory, and there are generally several possibilities of finding the way out. Usually, the question is eventually reduced to a conflict between two or three propositions which hitherto have found a place in the theory, but of which one must be abandoned in face of the new fact. The conflict lasts often for years or decades ; and its final decision not only means the destruction of the defeated theorem, but also quite naturally—and this is specially important—a corresponding confirmation and elevation of the victorious constituent theorems which survive.

And now we must note the extremely important and remarkable result that in all this war and conflict it is just the great Physical Principles which have held the field,—such as the Principle of the Conservation of Energy, the Principle of the Conservation of Momentum, the Principle of Least Action, and the chief laws of Thermodynamics. Their importance has thus been considerably increased ; while, on the other hand, the theorems which have succumbed in the fight are those on which theoretical developments were based tacitly, either because they seemed so self-evident that it was not, as a rule, considered necessary to mention them, or because they were forgotten. In general, then, one may assert that the most recent development of theoretical physics is marked by the victory of the great Physical Principles over certain deeply-rooted and yet merely habitual assumptions and conceptions.

To illustrate these statements, I may adduce some of those theorems which have hitherto been used without any hesitation as the self-evident foundations of any theory, but which, in the light of new facts, have proved untenable, or extremely doubtful, in face of the general principles of Physics. I mention three : The Invariability of Chemical Atoms; The Mutual Independence of Space and Time ; and The Continuity of all Dynamical Effects.

Of course it is not my intention here to quote all the important arguments which tell against the Invariability of Chemical Atoms. I shall only mention the single fact which brought about an inevitable conflict between this assumption —formerly always regarded as self-evident—and a general physical principle. The *fact* is the constant evolution of

64   Prof. Max Planck : *New Paths*

heat by every compound of radium ; and the physical *principle* is that of the Conservation of Energy. The conflict ended finally in the complete victory of that principle, although voices were heard endeavouring to throw doubt upon its complete validity.

A radium salt enclosed in a sufficiently thick envelope of lead constantly evolves heat amounting for 1 gram of radium to 135 calories per hour. It therefore constantly remains warmer than its surroundings, just like a heated stove. The Principle of Conservation of Energy asserts that this heat cannot be evolved from nothing, but must have its cause in some other change representing its equivalent. In the case of the stove, it is the constant process of combustion, but as no chemical process is going on in the case of the Radium compound, we must assume a change in the Radium atom itself ; and this hypothesis, which from the point of view of previous chemical science is bold and unprecedented, has been corroborated in every direction.

From a purely formal point of view, there is, no doubt, a certain contradiction in the conception of a changeable atom, since originally atoms were defined as the unchangeable constituents of all matter. Accordingly one might feel bound to reserve the term " atom " for the really unchangeable elements, such perhaps as Electrons and Hydrogen. But apart from the fact that we may perhaps never be able to establish the existence of invariable elements in the absolute sense, such a change in terminology would produce a wild confusion in literature. Indeed the modern chemical atoms have long ceased to be the atoms of Democritus, they can be numerically and accurately specified by a much sharper definition, it is only they that are meant when we speak of a Transmutation of the Atom, and any misunderstanding in the direction indicated seems clearly excluded.

A thesis not less self-evident than the Invariability of Atoms was, until recently, the Mutual Independence of Space and Time. The question whether two occurrences taking place at different points are simultaneous or are not, had a definite physical meaning, without the need of any enquiry about the observer who measured the time. To-day, that is altered. For a fact—so far invariably corroborated by the most delicate optical and electrodynamic experiments—which is briefly and not quite clearly described as the Relativity of all Motion, has brought that simple conception into conflict with the so-called principle of the Constancy of the Velocity of Light established by the electrodynamics of Maxwell and Lorentz. This principle asserts that the velocity of the propagation of light in open space is independent of the motion of the source of light. If, therefore, we assume that Relativity is experimentally established, we must sacrifice either the principle of the Constancy of the Velocity of Light or the Mutual Independence of Space and Time.

For let us consider a simple example. Let a time signal be given out by Wireless Telegraphy from a central station,—say the Eiffel Tower,—as provided by the International Time Service already projected. Then all stations in the vicinity which are at the same distance from the central station receive the signal at the same time, and can set their clocks accordingly. But this kind of time regulation becomes theoretically faulty, if accepting the relativity of all motion we transfer our standpoint from the earth to the sun, whence we must regard the earth as moving. For, according to the principle of the Constancy of the Velocity of Light, it is clear that those stations which, seen from the central station, lie in the direction of the earth's motion, receive the signal later than those lying in the opposite direction, because the former stations are moving on in advance of the light waves which they have to receive, and must be overtaken by them, whereas the latter stations travel to meet the waves. Thus the principle of the Constancy of the Velocity of Light renders impossible an absolute determination of time which shall be independent of the motion of the observer. The two are incompatible. So far as the conflict has proceeded, the principle of the Constancy of the Velocity of Light has been decidedly victorious, and in spite of many doubts which have latterly been raised, it is not at all probable that any abandonment of that position will occur.

The third of the above theories concerns the Continuity of all Dynamical Effects. This was formerly taken for granted as the basis of all physical theories, and, in close correspondence with Aristotle, was condensed into the well-known dogma—*Natura non facit saltus.* But even in this venerable stronghold of Physical Science present-day investigation has made a considerable breach. This time it is the principles of Thermodynamics with which that theorem has been brought into collision by new facts, and unless all signs are misleading, the days of its validity are numbered. Nature does indeed seem to make jumps—and very extraordinary ones. As an illustration, let me make an instructive comparison :—

Let us imagine a sheet of water in which strong winds have produced high waves. Even after the total cessation

66   Prof. Max Planck : *New Paths*

of the wind, the waves will be maintained for some time and will pass from one shore to the other. But there will be a certain characteristic change in them. During their impact on the shore, or on other solid obstacles, the energy of motion of the longer and coarser waves is converted to an ever greater extent into the energy of motion of shorter and slighter waves ; and this process will continue until at last the waves have become so small and their motion so slight that they are quite lost to view. That is the familiar transmutation of visible motion into heat, of molar into molecular, of ordered into disordered motion ; for in ordered motion many neighbouring molecules have a common velocity, whilst in disordered motion every molecule has its separate and separately directed velocity.

This process of disintegration or subdivision does not proceed indefinitely, but finds its natural limit in the size of the atoms. For the motion of a single atom by itself is always an ordered one, since the separate parts of an atom all move with the same common velocity. The larger the atoms, the less can the total energy of motion be subdivided. So far, everything is perfectly clear, and the Classical Theory is in excellent agreement with experience.

But now let us take another and quite analogous process, not dealing with water waves but with waves of light and heat. Let us assume that rays emitted by a brightly glowing body are collected by suitable mirrors into a completely enclosed hollow space, and that they are continually thrown to and fro between the reflecting walls of that space. Here also there will be a gradual transmutation of the energy of radiation from longer waves to shorter waves, from ordered radiation to disordered radiation. The longer and coarser waves correspond to the infra-red rays, and the shorter and slighter waves correspond to the ultra-violet rays of the spectrum. Hence, according to the Classical Theory, we must expect the total energy of radiation to concentrate itself upon the ultra-violet portion of the spectrum ; or, in other words, we must expect the infra-red and the visible rays to disappear gradually and convert themselves ultimately into invisible ultra-violet or chemical rays.

But of such a phenomenon no trace can be discovered in Nature. The conversion sooner or later attains a perfectly definite and assignable limit, and after that, the radiation-conditions remain stable in every respect.

In order to reconcile this fact with the Classical Theory the most varied experiments have already been made, but the result has always been that the contradiction went too deep into the roots of the Theory to leave them unhurt. So again nothing remains but to re-examine the foundations of the Theory. And again we must admit that the principles of Thermodynamics have shown themselves to be unshakable. For the only method so far found to promise a complete solution of the riddle depends directly upon the two laws of Thermodynamics; though it combines with them a new and peculiar hypothesis, which, if we utilize the two illustrations above mentioned, can be expressed somewhat as follows :—

In the case of the Water waves, the disintegration of the energy of motion is limited by the fact that the atoms hold the energy together, in a way, each atom representing a certain finite material Quantum which can only move as a whole. In the same sort of way certain processes must be at work in the case of light and heat rays, although they are quite of an immaterial nature, which shall hold together the energy of radiation in definite finite Quanta, and shall unite it the more strongly the shorter the waves and the quicker therefore the frequency of the oscillations.

In what way we are to conceive the nature of quanta of a purely dynamical nature, we cannot yet say for certain. Possibly such quanta might be accounted for if each source of radiation can only emit energy when that energy attains at least a certain minimum value ; just as a rubber pipe, into which air is gradually compressed, bursts and scatters its contents only when the elastic energy in it attains a certain quantity.

In any case, the hypothesis of Quanta has led to the idea that there are changes in Nature which do not occur continuously but in an explosive manner. I need hardly remind you that this view has become much more conceivable since the discovery and investigation of Radio-Active Phenomena. Besides, all difficulties connected with detailed explanation are at present overshadowed by the circumstance that the Quantum Hypothesis has yielded results which are in closer agreement with radiation-measurements than are all previous theories.

Moreover, if it is a good sign for a new hypothesis that it is found applicable to regions for which it was not originally devised, then the Quantum Hypothesis can surely claim a favourable testimony. I shall only refer to a quite particularly striking point. Since we have succeeded in liquefying Air, Hydrogen, and Helium, a new field of activity has been opened for experimental investigation in the region of low temperatures, and in this region a number of new and in some

68   Prof. Max Planck : *New Paths*

ways highly surprising results have been obtained. To heat a piece of copper from $-250°$ to $-249°$, *i. e.* by one degree, we do not require the same quantity of heat as for heating it from $0°$ to $1°$, but a quantity about thirty times less. If we took the original temperature of the copper still lower, the corresponding quantity of heat would turn out many times smaller, without any assignable limit. This fact not only runs counter to our habitual ideas, but also is out of harmony with the demands of the Classical Theory. For although for more than 100 years we have learnt to distinguish between temperature and quantity of heat, yet we were led by the Kinetic Theory of Matter to suppose that these two quantities, even if not strictly proportional to each other, preserved at all events a sensibly parallel course.

· The Quantum Hypothesis has entirely cleared up this difficulty, and in addition has yielded another result of high importance, viz. that the forces controlling the thermal oscillations in a solid body are of the same kind as those which control its elastic oscillations. With the help of the Quantum Hypothesis, therefore, we can now calculate quantitatively the thermal energy of a monatomic substance at various temperatures, from its elastic properties,—a performance which was far beyond the reach of the Classical Theory. Hence arise a number of further questions which appear very strange at first sight,—for instance, whether perhaps the vibrations of a tuning-fork are not absolutely continuous but are broken up into quanta. It is true that in acoustic vibrations the energy quanta will be extremely small, on account of their relatively low frequency. Thus, in the middle *a*, they would amount to only 3 quadrillionths of a unit of work in absolute mechanical measure. It would be just as little necessary to alter the Theory of Elasticity on that account as on account of the quite analogous circumstance that it treats *matter* as perfectly continuous, whereas it is really constituted atomistically, *i. e.* according to quanta. But fundamentally the revolutionary aspect of the new conception must be clear to everybody ; and although the nature of dynamical quanta still remains somewhat puzzling, yet, in view of the facts now known, it is difficult to doubt their existence, in some form or other. For whatever we can measure must exist.

Thus in the light of recent investigation, the Physical representation of the Universe exhibits an ever more intimate correspondence between its various features, and also manifests a certain peculiar structure whose refinement was hidden to the less trained eye and therefore remained concealed. But ever the question arises : What is the significance of this progress in fundamental conceptions for the satisfaction of our thirst for knowledge ? Do we approach one step nearer to a real knowledge of Nature by the refining of our world image ? To this fundamental question let us devote a brief consideration. It is not as if anything essentially new could be said in this region, already traversed by manifold and endless speculation, but that while on this point modern views are often diametrically opposed, yet everyone who takes a deep interest in the real aims of Science must necessarily take up some position.

Thirty-five years ago, Hermann von Helmholtz in this very place expounded the view that our perceptions never give us an image of, but at most a message from, the external world. For every attempt fails to demonstrate any kind of similarity between the nature of the external impression and the nature of the corresponding sensation ; all conceptions which we make for ourselves of the external world only reflect our own sensations in the last resort. Is there any sense, therefore, in opposing to our consciousness an independent "intrinsic Nature" ? Are not indeed all so-called "Laws of Nature" essentially but more or less effective rules by means of which we summarize the temporal course of our sensations as accurately and conveniently as possible? If that were so, then not only common-sense but exact Science would have been fundamentally at fault from the beginning. For it is impossible to deny that the whole evolution of Physical knowledge up to now has aimed towards the completest fundamental division between the happenings of external Nature and the processes of human perception. The way out of this embarrassing difficulty is seen as soon as we go one step further along this line of thought. Let us suppose that a Physical representation of the Universe had been found which fulfils all our demands, and therefore one that can completely and accurately represent all laws of Nature empirically known ; still that that image even remotely resembles "real" Nature, can in no way be proven. But this assertion has another side to it, which is generally too little emphasized : for, in exactly the same sense, the much bolder assertion that the proposed image represents real Nature in all points with absolute fidelity cannot be in any way refuted. For the first step in such a disproof would be the ability to assert anything with certainty concerning real Nature, and that, as everybody agrees, is absolutely excluded.

70   *New Paths of Physical Knowledge.*

We see that an immense gulf yawns here, into which no Science can ever penetrate. The filling of this gulf is a function not of pure reason, but of practical reason,—it is a matter of common-sense.

Just as a given cosmic scheme cannot be scientifically established, so we may also be assured that it will survive every attack so long as it agrees with itself and with the facts of experience. But we must not fall into the error of supposing that it is possible to advance, even in the exactest of all Sciences, without the help of any world-image, *i. e.* without any unprovable hypotheses. Even in Physics, the phrase holds good that "There is no Salvation without Faith,"—at least a faith in a certain reality outside ourselves. It is this confident faith which guides the advancing creative impulse, this it is which gives the necessary support to the groping imagination, this which alone can raise the spirit depressed by failure and inspire it to new efforts. An observer who does not allow himself to be led in his work by any hypothesis, however cautious and provisional, renounces beforehand all deeper understanding of his own results. Whoever rejects faith in the reality of atoms and electrons, or the electromagnetic nature of Light-waves, or the identity of Heat and Motion, can never be found guilty of a logical or empirical contradiction, but he will find it difficult from his standpoint to advance Physical knowledge.

It is true that faith alone does nothing. As the history of all Science shows, it is liable also to lead astray and to issue in narrowness and fanaticism. If it is to be a reliable guide, it must constantly be tested by the laws of thought and by experience which in the last resort can only be furnished by conscientious and often laborious self-denying solitary work. There is no Prince of Science who is not willing, in case of necessity, to do menial work, whether in the laboratory, the library, in the open air, or at the writing-desk. It is just this hard struggle which ripens and purifies the cosmic view. Only he who has in his own body gone through the process can fully realize its meaning and importance.

———————

And the Rector of the University concluded his Address as follows :—

Thus I address myself, finally, particularly to you, dear Comrades, who are about to cross the threshhold of a new term of studies. The gates of our University are open, shortly you will fill the Lecture Halls, and many a grain of seed will be sown afresh, many a plant will approach fruition, fed and nurtured

71

by the treasures of infinitely manifold mental work transmitted to you by your teachers. But do not believe that all that is offered to you from the Chair is the last word in wisdom. So long as there is progress in Science, so long it will be subject to temporary error. He who has got so far that he never errs, has ceased to work. If, therefore, doubts and difficulties occur to you in your studies, do not consider them something unpleasant or forbidden which must be shaken off or suppressed, but seek out their meaning carefully, and go to your Teachers (who are your guides), trust in their riper experience, and cling to the hope of attaining a gradually increasing understanding of dark and difficult questions by means of conscientious and sustained effort; and so secure the fullest and truest scientific advancement.

And if your honest efforts, verified by many tests, decisively indicate to you new paths differing from the old, then—follow your own conviction beyond any other. That is and must remain your highest, your most precious possession; for just as training for scientific independence is the highest aim of academic instruction, so does a scientific conviction acquired by honest work give a firm anchorage for holding fast to a moral conception of the universe in face of all the vicissitudes of life.

The noblest among all the moral fruits of science, and that which is peculiarly its own, is Truthfulness: that truthfulness which leads through the sense of personal responsibility to inner Freedom, and whose estimation in our present public and private life should be much higher than it is. In whatever degree our younger generation takes part in the fight to win for Truth an ever fuller recognition, to that extent it may feel at one with those heroes who, a hundred years ago, sealed the genuineness of their love for the Fatherland with their hearts' blood. With such memories and such thoughts, let us enter upon the work of the new Session.

[897]

XCV. *Absorption Measurements of the Change of Wave-Length accompanying the Scattering of X-Rays.* By ARTHUR H. COMPTON, *Wayman Crow Professor of Physics, Washington University, Saint Louis* *.

IN some recent papers † the writer has described spectroscopic experiments which have shown that when the characteristic X-rays from molybdenum are scattered by graphite, the wave-length of the X-rays is increased. While these spectroscopic investigations have been made for only two wave-lengths, ·708 and ·630 Å., a quantum theory of the phenomenon has been developed ‡ which predicts that a similar change in wave-length should occur whatever the wave-length of the primary beam. Absorption measurements on scattered γ-rays have indicated a change in wave-length of about the theoretical amount §, but interferometer measurements on light scattered by paraffin have failed to show any effect of this character ‖. Apparently, therefore, the change in wave-length due to scattering depends in some way upon the wave-length of the primary radiation used. The present experiments, in which the change in wave-length was measured by an absorption method, have as their primary object to test the theory over a wider range of wave-lengths, and for a greater variety of scattering materials than could be done conveniently by the spectroscopic method.

The quantum theory of this change in wave-length is based upon the hypothesis that each quantum of primary X-rays is scattered by an individual electron. If the frequency of the incident quantum is ν_0, its energy is $h\nu_0$, and its momentum is $h\nu_0/c$, where c is the velocity of light. Due to the change in direction of the quantum on scattering, its momentum is altered, resulting in a recoil of the scattering electron. Equating the momentum of recoil of the electron to the change in momentum of the quantum, we have,

$$\left\{ \frac{m\beta c}{\sqrt{(1-\beta^2)}} \right\}^2 = \left(\frac{h\nu_0}{c}\right)^2 + \left(\frac{h\nu_\theta}{c}\right)^2 + 2\frac{h\nu_0}{c}\frac{h\nu_\theta}{c}\cos\theta. \quad . \quad (1)$$

* Communicated by the Author.
† A. H. Compton, Bulletin National Research Council, xx. p. 16 (Oct. 1922); Paper before American Physical Society, April 28, 1923, Phys. Rev. June 1923; and Phys. Rev. xxii. (1923).
‡ A. H. Compton, Bull. N. R. C. xx. p. 18 (Oct. 1922); Paper before Am. Phys. Soc., Dec. 1, 1923; Phys. Rev. xxi. p. 207 (Dec. 1923) & xxi. p. 483 (May 1923). P. Debye, *Phys. Zeitschr.* xxviii. p. 161 (April 15, 1921).
§ A. H. Compton, Phil. Mag. (May 1921); Phys. Rev. xxii. (1923).
‖ P. A. Ross, Science, lvii. p. 614 (1923).

898   Prof. Compton : *Absorption Measurements of Change*

Here βc is the velocity with which the electron recoils, and ν_θ is the frequency of the rays scattered at the angle θ. But the energy of the scattered quantum is less than that of the incident quantum because of the energy spent in setting the scattering electron in motion. Thus,

$$h\nu_0 - h\nu_\theta = mc^2 \left\{ \frac{1}{\sqrt{(1-\beta^2)}} - 1 \right\} \quad \ldots \quad (2)$$

Combining these two equations we find,

$$\nu_\theta = \nu_0 / 1 + \alpha(1 - \cos\theta), \quad \ldots \ldots \quad (3)$$

where

$$\alpha = h\nu_0 / mc^2 = h/mc\lambda_0, \quad \ldots \ldots \quad (4)$$

or

$$\delta\lambda = \lambda_\theta - \lambda_0 = \gamma(1 - \cos\theta), \quad \ldots \ldots \quad (5)$$

where

$$\gamma = h/mc = 0{\cdot}0242 \times 10^{-8} \text{ cm.}^* \quad \ldots \quad (6)$$

Typical results of the spectrum measurements of this change in wave-length are shown in fig. 1, which represents, for slits of two different widths, the spectra of the Kα ray from molybdenum : (A) the primary ray, (B) as scattered by graphite at 45°, (C) as scattered at 90°, and (D) as scattered at 135° †. The line P is drawn in each case at the position of the primary line, and the line T at the theoretical position of the scattered line as given by equation (5). It will be noticed that, within a comparatively small probable error, the wave-length of one component of the scattered beam is exactly that predicted by this quantum theory. There remains, however, a part of the scattered beam which is unchanged in wave-length.

Experimental Method.

For the measurement of the difference in absorption co-efficient between the primary and the scattered ray, a balance method was employed, as is shown diagrammatically in fig. 2. Two beams of X-rays from the target X of a Coolidge tube came through separate windows in the lead box. One of them was scattered by a radiator R into an ionization chamber I, and the other went directly through a slit of variable width S into a second ionization chamber I'.

---

* Since the mass of a quantum is $h\nu/c^2 = h/\lambda c$, the mass of a quantum of radiation of wave-length γ is $h/(h/mc)c = m$; *i. e.* a quantum of radiation of wave-length γ has a mass equal to that of the electron. This fact was pointed out to me by Dr. Eldridge through Prof. A. Sommerfeld.

† *Cf.* A. H. Compton, Phys. Rev. xxii. (1923).

of Wave-Length accompanying Scattering of X-Rays. 899

One ionization chamber was kept at a positive and the other at a negative potential, so that with equal ionization currents

Fig. 1.

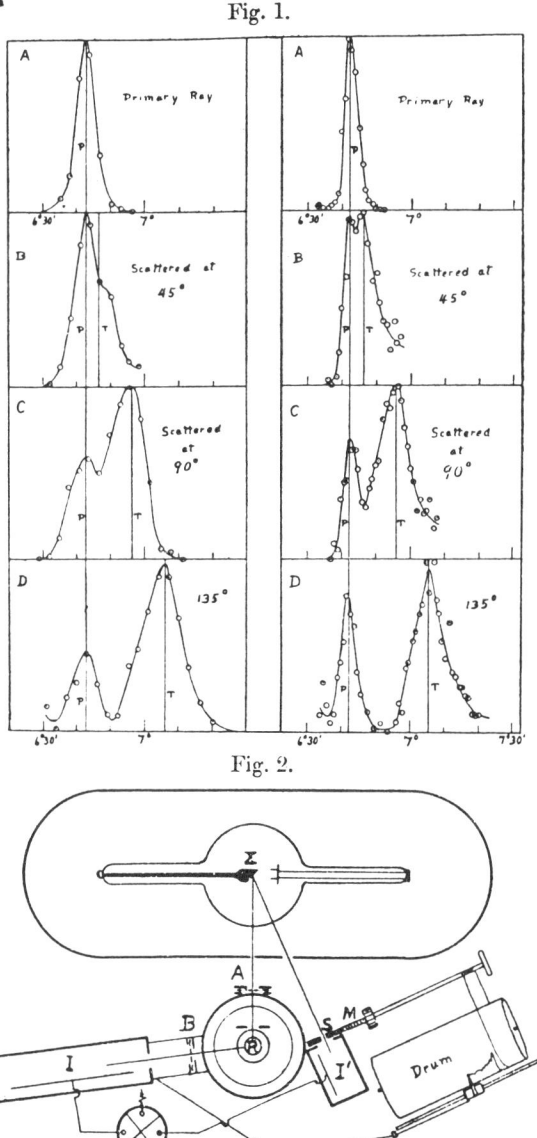

Fig. 2.

in each there was no deflexion on the electrometer E. The variable slit S was opened by a micrometer screw M.

3 M 2

900   Prof. Compton : *Absorption Measurements of Change*

The shaft driving this screw had wrapped about it a metal cord which was wrapped also about a drum in such a manner that the drum was rotated when the slit was opened by the micrometer screw. A recording pen moving along the drum was actuated by a metal cord fastened to the movable ionization chamber. Thus each position of the pen corresponded to a particular angle of the ionization chamber.

The chamber I was rotated through a range of angles of about 75° by a motor driving a worm gear. As the chamber was moving, the micrometer screw was turned so as to keep the electrometer at its zero position. In this manner a record was obtained on a sheet of paper placed around the drum showing the width of the slit for every angle of the

Fig. 3.

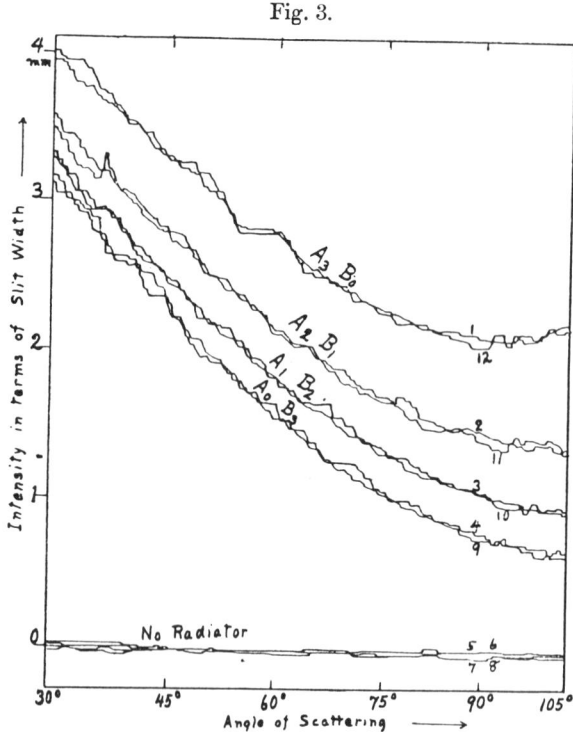

The scattering of hard X-rays by paraffin. In curve A_2B_1, 2 mm. of copper are in the path of the primary beam and 1 mm. in the path of the scattered beam, and similarly for the other curves. The fact that the intensity is greatest with the absorbing screen in the path of the primary beam shows that the wave-length of the X-rays is increased by scattering. The curves were made in the order of their numbers.

of Wave-Length accompanying Scattering of X-Rays.   901

chamber I. A record of this type, copied in indian ink for reproduction, is shown in fig. 3, which represents the scattering of hard X-rays by a block of paraffin. In this figure the ordinates represent the width of the slit-opening, and the abscissæ represent the angle of the ionization chamber.

To interpret the graphs thus obtained, it was necessary to determine how the ionization in the chamber I′ varied with the width of the slit. This was done by means of a sector disk made of lead which cut off a known and adjustable fraction of the primary X-ray beam. A calibration curve taken in this manner is shown in fig. 4, where the intensity

Fig. 4.

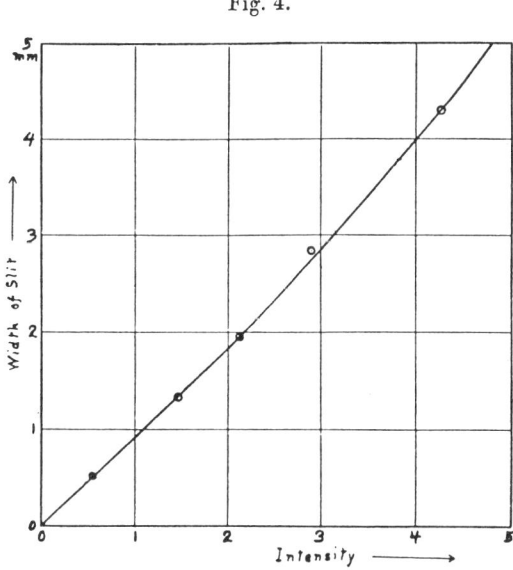

of the ionization is plotted against the width of the slit in millimetres. It will be seen that for slit-widths less than 2 mm. the calibration curve is sensibly a straight line, and that the departure from such a line is not large even for the greater widths.

The X-rays scattered into the chamber I were filtered through an absorption screen placed at A or B. It was necessary to place a similar absorption screen in the path of the rays entering I′ in order that slight variations in the voltage applied to the X-ray tube should not destroy the balance between the two ionization currents. Of course,

902   Prof. Compton : *Absorption Measurements of Change*

the beam entering the chamber I was much less intense than that entering the chamber 1'. This difference in intensity was balanced by making the chamber I much larger and filling it with methyl-iodide vapour.

The change in absorption coefficient, corresponding to the change in wave-length of the scattered beam, was measured by observing the relative intensity of the scattered beam when an absorbing screen was transferred from position A to position B. If I_p is the intensity of the primary beam whose wave-length is λ and whose absorption coefficient in the absorbing screen is μ, and if S is the fraction of the energy of this beam which is scattered into the ionization chamber when no absorption screens are present, then the intensity of the scattered beam when a screen of thickness x is placed at A is $I_A = I_p S e^{-\mu x}$. If the absorption coefficient of the scattered ray is μ', the intensity of the scattered ray when the screen is placed at B is $I_B = I_p S e^{-\mu' x}$. Thus

$$\frac{I_A}{I_B} = e^{(\mu' - \mu)x}, \quad \ldots \ldots \quad (7)$$

whence

$$\delta\mu = \mu' - \mu = \frac{1}{x}\log(I_A/I_B). \quad \ldots \quad (8)$$

It is clear from equation (7) that if μ' were equal to μ, I_A would be equal to I_B. If the absorption coefficient of each component of a complex beam of X-rays were unchanged by scattering, its intensity should therefore be unaltered by moving the absorbing screen from A to B. Hence the intensity of the whole beam should also remain unchanged. Thus any change observed in the intensity of the scattered ray when the absorption screen is shifted from position A to position B represents a change in the absorption coefficient of the component rays when the X-rays are scattered *. The fact, which is shown clearly by fig. 3, that the intensity of the rays is greater when the screen is at A therefore means that μ' is greater than μ †, that is, that each component of the X-ray beam is softened during the scattering process.

* This conclusion is strictly true only in case the fraction of the rays which is scattered from the absorber placed at A onto the radiator is the same as the fraction scattered from the absorber when placed at B into the ionization chamber. The apparatus was so designed that this condition was at least very nearly fulfilled. However, even if uncorrected, the error from this source would have been very small.

† It is by no means always true, when heterogeneous X-rays are used, that the scattered ray is softer than the primary ray. For if the radiator is of considerable thickness, the more penetrating parts of the primary beam are scattered by the whole radiator while the softer components

of Wave-Length accompanying Scattering of X-Rays. 903

Knowing the absorption coefficient of the primary beam, and the change in absorption coefficient due to scattering, one can determine the wave-length of both the primary and the scattered beam. For this purpose I have employed the absorption date for different wave-lengths given by Hewlett, Richtmyer, and Duane *.

Experimental Results.

The results of these absorption measurements are collected in Table I., under the head of observer 2, and the corresponding wave-lengths are exhibited in Table II. In each case the scattering material was in the form of a cylinder, 2 cm. in diameter, with walls of such thickness that more than half of the X-rays were transmitted. The wave-length changes are very consistent with each other in every case except that of the secondary radiation from lead for the effective primary wave-length 0·14 Å. In this case the change in wave-length, especially at the small angles, is considerably greater than the theory predicts. The difficulty obviously lies in the fact that the fluorescent K radiation from lead is being excited in large amounts, and that this secondary radiation is always softer than the ray which excites it. The measurements of the scattering by the other elements for wave-lengths 0·12 and 0·13 Å. are also slightly affected by the fluorescent K radiation from the lead slits, the tendency being to make the observed change in wave-length greater at small angles and less at large angles with the primary beam, just as in the case of the scattering by lead. These fluorescent rays are not excited appreciably when the effective wave-length of the primary beam is

are absorbed before they have penetrated very deeply. Because of this filtering process, it usually happens that the scattered ray is more penetrating than the primary. One cannot help but feel that this process may account in part for the small magnitude of the change in absorption observed by Barkla and Miss Sale (Phil. Mag. xlv. p. 758, 1923), even though they took the precaution of using thin sheets of paper as radiators. Changes in absorption due to scattering similar to those described here have been observed for γ-rays by Eve, Phil. Mag. viii. p. 669 (1904); R. D. Kleeman, Phil. Mag. xv. p. 638 (1908); Madsen, Phil. Mag. xvii. p. 423 (1909); D. C. H. Florance, Phil. Mag. xx. p. 921 (1910), xxvii. p. 225 (1914); J. A. Gray, Phil. Mag. xxvi. p. 611 (1913); A. H. Compton, Phil. Mag. xli. p. 749 (1921); *et al.* For X-rays the change has been observed by Sadler and Mesham, Phil. Mag. xxiv. p. 138 (1912); J. Laub, *Ann. de Phys.* xlvi. p. 785 (1915); J. A. Gray, Frank. Inst. Jour. p. 643, Nov. (1920); A. H. Compton, Phys. Rev. xviii. p. 96 (1921); Nature, cviii. p. 366 (1921); and J. A. Crowther, Phil. Mag. xlii. p. 719 (1921).

* These data are collected in Bulletin N. R. C. no. xx. p. 32 (1922).

904   Prof. Compton : *Absorption Measurements of Change*

TABLE I.

Mass Absorption Coefficients of Primary and Scattered X-Rays.

Radiator	Absorber	μ/ρ Primary	$\delta(\mu/\rho)$, Scattered Ray.								Observer.
			30°.	45°.	60°.	75°.	90°.	103°.	120°.	135°.	
Paraffin	Pb	·073	…	·027	…	…	…	…	…	·71	1
Aluminium	Pb	·073	…	·027	…	…	…	…	…	·43	1
Iron	Pb	·073	…	·037	…	…	…	…	…	·43	1
Tin	Pb	·073	…	·017	…	…	…	…	…	·25	1
Lead	Pb	·073	…	(?)	…	…	…	…	…	·08	1
Paraffin	Cu	·45	·043	·08	·12	·20	·245	·32	…	…	2
Graphite	Cu	·46	·08	·10	·12	·19	·15	·33	…	…	2
Aluminium	Cu	·45	·05	·08	·13	·19	·24	·30	…	…	2
Copper	Cu	·59	·04	·07	·10	·15	·19	·26	…	…	2
Lead	Cu	·64	·21	·27	·31	·32	·33	·34	…	…	2
Paraffin	Cu	·76	·065	·11	·17	·26	·34	·42	·50	·59	2
Graphite	Cu	·76	…	·105	·16	·23	·36	·46	·52	·59	2
Aluminium	Cu	·76	…	·10	·14	·21	·295	·35	·42	·53	2
Copper	Cu	·76	…	·06	·15	·20	·23	·32	·39	…	2
Tin	Cu	1·03	…	·07	·12	·17	·23	·36	…	…	2
Lead	Cu	1·05	…	·10	·14	·14	·19	·20	…	…	2
Paraffin	Cu	1·74	·09	·16	·29	·405	·52	·63	…	…	2
Paraffin	Cu	2·84	·15	·25	·33	·51	·60	·73	…	…	2
Paraffin	Cu	5·25	·17	·29	·50	·77	1·00	1·16	…	…	2
Paper	Al	·70–·90	…	…	…	…	·22	…	…	…	3
Paper	Al	1·15–1·83	…	…	…	…	·22	…	…	…	3
Paper	Al	1·90–4·42	…	…	…	…	·18	…	…	…	3
Paper	Al	4·8–10·3	…	…	…	…	·07	…	…	…	3

greater than 0·15 Å.   The fluorescent K-rays from the copper absorbing screen would also have been a source of difficulty had not the differential absorption coefficients $\delta(\mu/\rho)$ been measured after the scattered rays had been

TABLE II.

Change in Wave-Length Accompanying the Scattering of X-rays.

Radiator.	λ, A.U., Primary.	\(δλ\) for Scattered Ray, A.U.								Theory.	Observer.
		30°.	45°.	60°.	75°.	90°.	105°.	120°.	135°.		
Free electron	All.	·003	·007	·012	·018	·024	·030	·036	·041	·041	1
Paraffin	·024		·006						·051		1
Aluminium......	·024		·006						·039		1
Iron............	·024		·007			·021			·029		1
Tin	·024		·004						·029		1
Lead	·024								·013		1
Paraffin	·12	·005	·010	·013	·020	·024	·031				2
Graphite......	·12	·010	·013	·015	·021	·027	·033				2
Aluminium......	·12	·006	·010	·015	·020	·025	·030				2
Copper	·13	·004	·007	·010	·014	·017	·022				2
Lead	·14	·016	·021	·024	·025	·025	·026				2
Paraffin	·15	·004	·008	·012	·018	·023	·027	·033	·039		2
Graphite......	·15		·007	·012	·017	·025	·031	·034	·038		2
Aluminium......	·15		·007	·010	·016	·021	·025	·028	·035		2
Copper	·15		·005	·011	·015	·017	·023	·027			2
Tin	·17		·004	·007	·010	·013	·020				2
Lead	·17		·005	·008	·008	·011	·012				2
Paraffin	·21	·004	·008	·014	·018	·023	·027				2
Paraffin	·25	·005	·008	·012	·017	·020	·024				2
Paraffin	·32	·003	·006	·010	·015	·020	·023				2
Paper	·30–·33					·047					3
Paper	·41–·50					·023					3
Paper	·50–·69					·015					3
Graphite......	·63					·020					4
Graphite......	·71		·003			·018			·031		4
Paper	·70–·90					·003					3

filtered through at least ·5 mm. of copper. This precaution was sufficient also to eliminate the effect of the K-rays from tin when it was used as radiator.

For sake of completeness, I have included in Tables I.

906   Prof. Compton : *Absorption Measurements of Change*

and II. the results of some earlier experiments on γ-rays [*]
(Observer 1), Barkla and Sale's recent experiments on the
change in absorption coefficient of soft X-rays scattered by
paper [†] (Observer 3), and my spectrum measurements on the
change of wave-length of molybdenum K-rays [‡] (Observer 4).
In the case of the γ-rays, the wave-lengths are calculated
from the absorption coefficients according to the equation
$\mu/\rho = \tau/\rho + \sigma/\rho$, where [§] $\tau/\rho = 1\cdot64 \times 10^3 \lambda^3$ and [||] $\sigma/\rho =$
$\cdot151/(1 + \cdot0485/\lambda)$. In Barkla and Sale's work, it did not
seem possible to reproduce the results in successive series of
experiments. I have accordingly averaged their results
obtained for certain arbitrarily chosen ranges of wave-
lengths, and have estimated the wave-lengths from Hewlett's
data for the absorption of different wave-lengths in aluminium. The wave-length changes estimated from the spectrum
measurements are the weighted mean values of the modified
and the unmodified rays.

From Table II. it is apparent that in order that the
scattered ray shall be changed in wave-length by the amount
predicted by equation (5), X-rays of very short wave-length
and radiators of low atomic number must be employed.
These facts are exhibited in figs. 5 and 6. In fig. 5 is plotted
the change in wave-length observed when X-rays of widely
differing wave-lengths are scattered by paraffin [¶]. For both
wave-lengths 0·024 and 0·15 Å., the observed change is very
nearly that predicted by the theory (as represented by the
solid curve); but the change for λ=0·32 Å. is slightly less,
and that for λ=0·71 is still less than the theoretical value.
Similarly in fig. 6, whereas the change in wave-length of the
rays scattered by carbon is within experimental error that
demanded by theory, the wave-length change for the heavier
elements becomes less and less as the atomic number becomes
greater. The difference here shown between the rays

* The value of μ/ρ for the primary γ-rays in lead is that of M. Ishino,
Phil. Mag. xxxiii. p. 140 (1917), and for the scattered rays is from
A. H. Compton, Phil. Mag. xli. p. 760 (1921).

† C. G. Barkla & Rhoda Sale, Phil. Mag. xlv. p. 748 (1923).

‡ For λ=0·63 (MoKβ-line, *cf.* A. H. Compton, Phys. Rev. xxi.
p. 495 (1923). For λ=0·71 (MoKα-line), *cf.* fig. 1 of this paper.

§ A. W. Huil & Marion Rice, Phys. Rev. viii. p. 836 (1916).

|| The mass absorption due to scattering, according to the writer's
quantum theory (Phys. Rev. xxi. p. 493, 1923), is $\sigma/\rho = \sigma_0/\rho(1 + 2a)$,
where σ_0/ρ is the mass scattering calculated on the classical theory, and
has the value 0·151.

¶ In the measurements on γ-rays (λ = ·024 Å.), as plotted in fig. 5, the
mean wave-length change for paraffin, aluminium, and iron is used, in
order to reduce the probable experimental error. All of these elements
may be considered as of low atomic number when γ-rays are employed.

of Wave-Length accompanying Scattering of X-Rays. 907

Fig. 5.

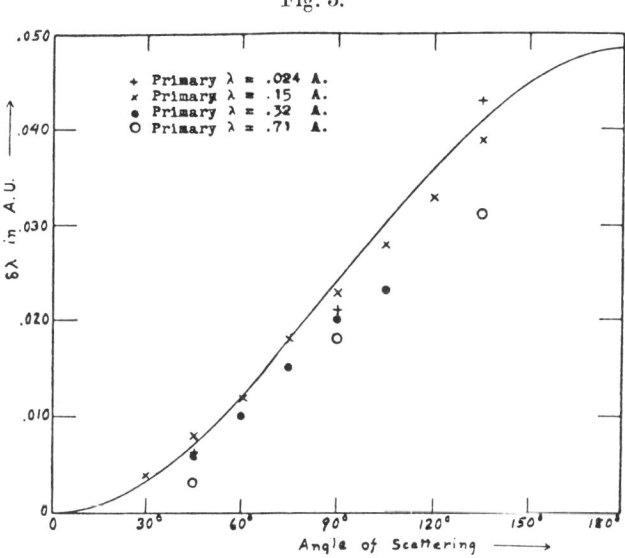

The change in wave-length accompanying the scattering of X-rays by
paraffin, when different primary wave-lengths are employed.

Fig. 6.

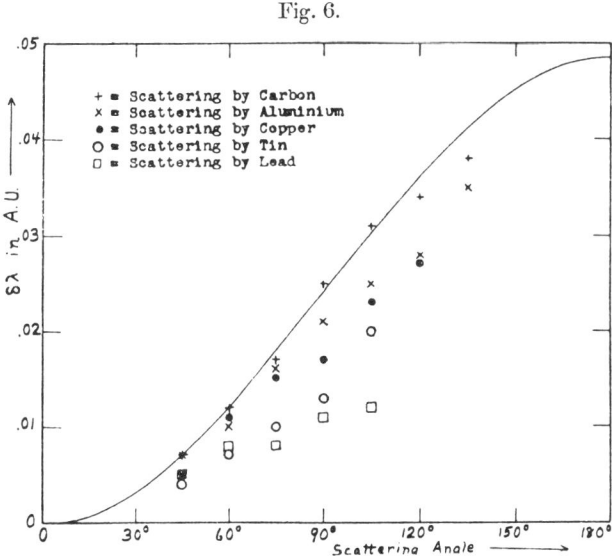

The change in wave-length when X-rays of wave-length ·15 ·17 Å. are
scattered by different radiators, showing a smaller change for the
heavier elements.

908   Prof. Compton: *Absorption Measurements of Change*

scattered by different elements when hard X-rays are used appears also when hard γ-rays are employed, as is shown in Table II. In this case, however, the difference between the different elements does not become apparent for elements lighter than iron.

I do not feel that these absorption experiments are sufficiently accurate to make from them any reliable estimate of the homogeneity of the scattered X-rays. The spectrum measurements, however, such as those shown in fig. 1, indicate clearly the existence of both a *modified ray*, whose wave-length is changed by the theoretical amount, and an *unmodified* ray of unchanged wave-length. It would seem possible to explain all of the present results on the assumption that only these two rays exist in the scattered beam, but that the energy distribution between the two rays varies with the wave-length, the angle of scattering, and the atomic number of the radiator. According to this view, for short wave-lengths and low atomic numbers nearly all of the energy lies in the modified ray, while for long waves and high atomic numbers the unmodified ray has the greater energy.

These experiments therefore suggest that for such comparatively great wave-lengths as those used in optics, when the usual materials are used as radiators, the unmodified ray should predominate, and the effective change in wave-length due to scattering should be very much less than that which occurs when X-rays are scattered. This is in accord with the negative result of Ross's experiment, in which he attempted to detect a change in wave-length when light was scattered by paraffin.

The Limb-Effect.

If the electrons were really free, which would correspond to atoms of zero atomic number, the present experiments suggest that the change in wave-length predicted by the theory should occur for even very long waves. J. Q. Stewart has recently presented an argument which suggests strongly that there exists about the sun a comparatively dense atmosphere of free electrons *. If this is the case, we should expect, in addition to the spectrum lines transmitted directly through this atmosphere †, to find some scattered light from

---

* J. Q. Stewart, Nature, cxi. p. 186 (1923); Phys. Rev. xxii. (1923).

† Of course, the solar lines are absorption, rather than emission lines. The change in wave-length should occur, however, in exactly the same manner. For the continuous background on either side of a dark line should be shifted toward the red, which would shift the centre of gravity of the dark line itself.

of Wave-Length accompanying Scattering of X-Rays.   909

the atmosphere which would be of greater wave-length than the direct ray. Since the thickness of the atmosphere traversed is greater near the limb, the amount of scattering, and hence the effective increase in wave-length, should be greater at the limb. We might thus expect the mean wave-length of a spectrum line from the sun's limb to be slightly greater than that of the same line from the middle of the photosphere. The difference should probably be less than 0.024 Å., since even at the limb the direct ray would probably be responsible for a large part of the spectrum line.

An effect of exactly this character is found in the solar spectrum, and is known as the "limb effect." Dr. C. E. St. John writes me that the wave-lengths from the limb are greater than those from the centre by from 0.004 to about 0.010 Å. in passing from the violet to the red. This neglects the very strong lines, which show no change in wave-length, and which presumably originate above the denser part of the electron atmosphere. The observed limb effect is thus of the right sign and of the right order of magnitude. At first sight it seems difficult to account for the fact that the red lines are shifted more than the violet. It is very possible, however, that the violet light is the more rapidly absorbed by the sun's atmosphere, so that the violet light reaching us traverses a thinner stratum of electrons. Our ignorance of the relative amount of the primary and the scattered light in the solar lines makes it impossible at present to give this explanation of the limb effect a quantitative test; but qualitatively it seems to be satisfactory.

Possible Origin of the Unmodified Ray.

Two different hypotheses suggest themselves to account for the presence of the unmodified ray. The first is that if the momentum of the light quantum is insufficient to impart to the scattering electron enough kinetic energy to eject it from the atom, the electron is held so firmly that it cannot recoil. Since in this case no energy is lost by recoil, the frequency of the scattered ray is the same as that of the incident ray. According to this view, many of the electrons in the heavier elements would be so tightly bound that they could recoil only from quanta possessing great energy, whereas for low energy or long wave-length quanta, even in the lighter elements some of the electrons would not recoil, thus giving rise to unmodified scattering. This is in accord with experiment. Quantitatively, however, the hypothesis is not so satisfactory. Thus the kinetic energy of an electron recoiling with the impulse imparted by a molybdenum $K\alpha$-ray when deflected through 135° is greater than the critical ionizing energy (280 volts) of the K electrons in carbon. We should therefore expect that no unmodified ray should appear when these rays are scattered at 135° by carbon. The spectra exhibited in fig. 1, however, show that the unmodified ray is present under these conditions.

The second hypothesis is based upon the view that when interference occurs, two or more electrons must scatter the same quantum. The theory upon which equation (5) is based, however, supposed that each quantum is scattered by a single electron. The change in wave-length is proportional to $1/m$, where m is the mass of the body which scatters the ray. If, then, the ray is scattered simultaneously by two electrons, the change in wave-length should be 1/2 the maximum value, and if interference occurs between the rays scattered by a large number of electrons, as in the case of crystal reflexion, the change in wave-length should be negligible. According to the wave-theory, partial interference should always occur when more than one electron is traversed by an electromagnetic wave. Experimentally, however, we have no evidence that the rays scattered by small groups of electrons, such as those in an atom of low atomic number, interfere with each other except when the phases of the rays scattered by the individual electrons are nearly identical. This is the condition, for example, under which excess scattering of X-rays occurs. There is accordingly some justification for the assumption that an electron scatters independently only when removed from other electrons by a distance greater than some fraction of a wave-length of the incident ray. If the electrons are closer than this, they will cooperate in their scattering, and in view of their large total mass, no appreciable change in wave-length will result. This hypothesis therefore leads also to an unmodified ray which possesses greater relative energy as the wave-length of the primary ray and the atomic number of the radiator are increased.

On the latter hypothesis, there should be no change of wave-length when X-rays are regularly reflected from a crystal, or when light is reflected by the free electrons of a metallic mirror, whereas according to the former hypothesis such a change might have been anticipated. The fact that experiment seems not to show any wave-length change in these cases * is a point in favour of the view that the

* The test on light reflected by a mirror has been made by P. A. Ross (*loc. cit.*); that on the wave-length of reflected X-rays is being made in this laboratory.

911

unmodified ray results from scattering by groups instead of by single electrons.

Summary.

The present absorption measurements on hard X-rays, when combined with the writer's earlier measurements on γ-rays and his spectrum measurements on soft X-rays, show that over the range of primary wave-lengths from 0·7 to 0·024 Å., there occurs a change in wave-length during the scattering process.

For light elements and short wave-lengths the effective wave-length change is very near the theoretical value $\delta\lambda = 0·024\ (1 - \cos\theta)$, but is less for long wave-length X-rays and for radiators composed of heavy elements.

It appears probable that in each case the scattered ray consists of two portions, an unmodified ray for which $\delta\lambda = 0$, and a modified ray for which $\delta\lambda = 0·024\ (1 - \cos\theta)$. The effective wave-length change then depends upon the distribution of energy between these two rays.

Two different hypotheses are suggested to account for the existence of the unmodified ray.

The limb effect, or difference in wave-length of solar lines between the centre and the limb of the photosphere, receives a satisfactory qualitative explanation on the view that it is due to a change in wave-length as the light is scattered by an electron atmosphere around the sun.

Washington University,
     Saint Louis, U.S.A.
   June 23, 1923.

446   M. L. de Broglie *on a*

continued increase of pressure. The greatest pressure for which a minimum value of pV on an isothermal occurs is $7 \cdot 23 p_c$, the corresponding temperature being $2 \cdot 06 T_c$.

(5) Inversion in the porous-plug experiment is indicated, but this can only take place if the temperature is below $5 \cdot 6 T_c$. The greatest pressure which will allow of inversion is $24 \cdot 36 p_c$, the corresponding temperature being $2 \cdot 825 T_c$.

(6) The specific heat K_v at constant volume is found to vary with the volume (with the temperature constant). As the volume increases from a small value, K_v at first rises, attains a maximum value, and then diminishes. For temperatures in the neighbourhood of $2 T_c$, the specific volume giving the maximum value of K_v is about the critical volume.

The chief difficulty in giving the proposed equation of state more than an empirical status lies in the fact that the pressure-correction term ϕ of (2) involves p as well as V and T.

XXXV. *A Tentative Theory of Light Quanta.* By LOUIS DE BROGLIE [*].

I. *The Light Quantum.*

THE experimental evidence accumulated in recent years seems to be quite conclusive in favour of the actual reality of light quanta. The photoelectric effect, which is the chief mechanism of energy exchange between radiation and matter, seems with increasing probability to be always governed by Einstein's photoelectric law. Experiments on the photographic actions, the recent results of A. H. Compton on the change in wave-length of scattered X-rays, would be very difficult to explain without using the notion of the light quantum. On the theoretical side Bohr's theory, which is supported by so many experimental proofs, is grounded on the postulate that atoms can only emit or absorb radiant energy of frequency ν by finite amounts equal to $h\nu$, and Einstein's theory of energy fluctuations in the black radiation leads us necessarily to the same ideas.

I shall in the present paper assume the real existence of light quanta, and try to see how it would be possible to reconcile with it the strong experimental evidence on which was based the wave theory.

For the sake of simplicity, it is a very natural assumption to admit that all light quanta are identical and that only

[*] Communicated by R. H. Fowler, M.A.

Tentative Theory of Light Quanta.   **447**

their velocities are different. We shall then assume that the "mass at rest" of every light quantum has a given value m_0: since the atoms of light have velocities very nearly equal to the Einstein's limiting velocity c, they must have an extremely small mass (not infinitely small in a mathematical sense); the frequency of the corresponding radiation must be related to the whole energy of a quantum by the relation

$$h\nu = \frac{m_0 c^2}{\sqrt{1-\beta^2}}, \quad \left(\beta = \frac{v}{c}\right);$$

but, since $1-\beta^2$ is very small, we can write

$$\beta = \frac{v}{c} = 1 - \frac{1}{2}\frac{m_0^2 c^4}{h^2 \nu^2}.$$

The light quanta would have velocities of slightly different values, but such that they cannot be discriminated from c by any experimental means. It then seems that m_0 should be at most of the order of 10^{-50} gr.

Naturally, the light quantum must have an internal binary symmetry corresponding to the symmetry of an electromagnetic wave and defined by some axis of polarization. We shall refer again later to this remark.

II. *The Black Radiation as a Gas of Light Quanta.*

Let us consider a gas made up by the light quanta we have described above. At a given temperature (not too near to the absolute zero) almost all these atoms of light would have velocities $v=\beta c$ very nearly equal to c. The whole energy of one of these atoms is

$$W = \frac{m_0 c^2}{\sqrt{1-\beta^2}},$$

and its momentum is

$$G = \frac{m_0 v}{\sqrt{1-\beta^2}};$$

so we have approximately

$$G = \frac{W}{c}.$$

The pressure of such a gas on the walls of the enclosure is easily seen to be

$$p = \frac{n}{6} \cdot 2Gc = \tfrac{1}{3}nW,$$

if n is the number of light quanta in an element of volume.

This expression is the same as the one given by the electro-magnetic theory, whilst without using the Relativity formulæ we should have found a result twice as great as this.

Now the question arises, Can we use for the quanta gas Maxwell's energy partition law? In Einstein's dynamics, Liouville's theorem, on which is based all Statistical Dynamics, is still valid; we can then use for the elementary cell of extension-in-phase a value proportional to $dx\,dy\,dz\,dp\,dq\,dr$, if x, y, z are rectangular coordinates and p, q, r the corresponding momenta. In consequence of the canonical distribution law, the number of atoms whose representative points are in the element $dx\,dy\,dz\,dp\,dq\,dr$ must be proportional to

$$e^{-\frac{W}{kT}}dx\,dy\,dz\,dp\,dq\,dr = e^{-\frac{W}{T}}.4\pi G^2\,dG\,dv,$$

if dv is the element of volume and G the momentum. But, since $G = \dfrac{W}{c}$, this number is also given by

$$C^t \times e^{-\frac{W}{kT}}w^2\,dw\,dv.$$

Each quantum has a total energy $h\nu$; then the whole energy contained in the volume dv and carried by light quanta of energy $h\nu$ is

$$C^t e^{-\frac{h\nu}{kT}}\nu^3\,dv\,dv.$$

This is obviously Wien's limiting form of the radiation law. Two years ago * I was able to show that, by using the hypotheses made by Planck that the element of extension in phase was $\dfrac{1}{h_3}dx\,dy\,dz\,dp\,dq\,dr$, it was possible to find for the radiant energy density the value

$$u_\nu\,d\nu = \frac{8\pi h}{c^3}\nu^3\,e^{-\frac{h\nu}{kT}}d\nu.$$

This was an encouraging result, but not quite complete. The assumption of finite elements of extension in phase seemed to have a somewhat arbitrary and mysterious character. Moreover, Wien's law is only a limiting form of the actual radiation law, and I was obliged to suppose some kind of quanta aggregation for explaining the other terms of the series.

It seems that these difficulties are now removed, but we shall first of all explain many other ideas; we shall later on return to the "black radiation" gas.

* See *Journal de Physique*, November 1922.

Tentative Theory of Light Quanta.   **449**

III. *An important Theorem on the Motion of Bodies.*

Let us consider a moving body whose "mass at rest" is m_0; it moves with regard to a given observer with velocity $v = \beta c$ $(\beta < 1)$. In consequence of the principle of energy inertia, it must contain an internal energy equal to $m_0 c^2$. Moreover, the quantum relation suggests the ascription of this internal energy to a periodical phenomenon whose frequency is $\nu_0 = \dfrac{1}{h} m_0 c^2$. For the fixed observer, the whole energy is $\dfrac{m_0 c^2}{\sqrt{1 - \beta^2}}$ and the corresponding frequency is

$$\nu = \frac{1}{h} \frac{m_0 c^2}{\sqrt{1 - \beta^2}}.$$

But if the fixed observer is looking at the internal periodical phenomenon, he will see its frequency lowered and equal to $\nu_1 = \nu_0 \sqrt{1 - \beta^2}$, that is to say this phenomenon seems for him to vary as $\sin 2\pi \nu_1 t$. The frequency ν_1 is widely different from the frequency ν; but they are related by an important theorem which gives us the physical interpretation of ν.

Let us suppose that, at time 0, the moving body coincides in space with a wave whose frequency ν has the value given above and which spreads with velocity $\dfrac{c}{\beta} = \dfrac{c^2}{v}$. This wave, however, cannot carry energy, according to Einstein's ideas.

Our theorem is the following:—" *If, at the beginning, the internal phenomenon of the moving body is in phase with the wave, this harmony of phase will always persist.*" In fact, at time t, the moving body is at a distance from the origin $x = vt$ and its internal phenomenon is proportional to $\sin 2\pi \nu_1 \dfrac{x}{v}$; at the same place the wave is given by $\sin 2\pi \nu \left(t - \dfrac{\beta x}{c} \right) = \sin 2\pi \nu \, x \left(\dfrac{1}{v} - \dfrac{\beta}{c} \right)$. The two sines will be equal; the harmony of phase will again occur if the following condition is realized:

$$\nu_1 = \nu (1 - \beta^2),$$

a condition clearly satisfied by the definitions of ν and ν_1.

This important result is implicitly contained in Lorentz's time transformation. If τ is the local time of an observer carried along with the moving body, he will define the

Phil. Mag. S. 6. Vol. 47. No. 278. *Feb.* 1924.   2 G

450   M. L. de Broglie *on a*

internal phenomenon by the function $\sin 2\pi\nu_0\tau$. According to
Lorentz's transformation, the fixed observer must describe the
same phenomenon by the function $\sin 2\pi\nu_0 \dfrac{1}{\sqrt{1-\beta^2}}\left(t-\dfrac{\beta x}{c}\right)$,
which can be interpreted as the representation of a wave
of frequency $\dfrac{\nu_0}{\sqrt{1-\beta^2}}$ spreading along the x axis with
velocity $\dfrac{c}{\beta}$.

We are then inclined to admit that any moving body may
be accompanied by a wave and that it is impossible to disjoin
motion of body and propagation of wave.

This idea can also be expressed in another way. A group
of waves whose frequencies are very nearly equal has a
"group velocity" U, which has been studied by the late
Lord Rayleigh, and which in the usual theory is the velocity
of "energy propagation." This group velocity is linked
with the "phase velocity" V by the relation

$$\frac{1}{U} = \frac{d\left(\dfrac{\nu}{V}\right)}{d\nu}.$$

If ν is equal to $\dfrac{1}{h}\dfrac{m_0 c^2}{\sqrt{1-\beta^2}}$ and V to $\dfrac{c}{\beta}$, we find $U=\beta c$—
that is to say, "*The velocity of the moving body is the energy
velocity of a group of waves having frequencies* $\nu=\dfrac{1}{h}\cdot\dfrac{m_0 c^2}{\sqrt{1-\beta^2}}$
and velocities $\dfrac{c}{\beta}$ *corresponding to very slightly different
values of* β."

IV. *Dynamics and Geometrical Optics.*

Trying to extend the former ideas to the case of variable
velocity is a rather difficult but very suggestive problem.
If in any medium a moving body describes a curved path.
we say that there is a field of force ; at each point the
potential energy may be calculated, and the body when
crossing this point has a velocity determined by the constant
value of its whole energy. Now, it seems natural to suppose
that the phase wave must have at any point a velocity and a
frequency fixed by the value *which β would have if the
body were there.* During its propagation the phase wave
has a constant frequency ν and a constantly variable
velocity V.

Tentative Theory of Light Quanta.   451

Perhaps a new electromagnetism will give us the laws of this complicated propagation, but it seems that we know beforehand the final result : " *The rays of the phase wave are identical with the paths which are dynamically possible.*" In fact, the paths of the rays can be computed as in a medium of variable dispersion by means of Fermat's principle, which may be written here (λ wave-length, ds element of path) :

$$\delta \int \frac{ds}{\lambda} = \delta \int \frac{v\,ds}{V} = \delta \int \frac{m_0\beta c}{\sqrt{1-\beta^2}}\,ds = 0.$$

The principle of least action in its Maupertuisian form gives the dynamical paths by the equation

$$\delta \int m_0 c^2 \left(\frac{1}{\sqrt{1-\beta^2}} - \sqrt{1-\beta^2} \right) dt$$
$$= \delta \int \frac{m_0\beta^2 c^2}{\sqrt{1-\beta^2}}\,dt = \delta \int \frac{m_0\beta c}{\sqrt{1-\beta^2}}\,ds = 0,$$

a result which justifies the above statement.

It is now so simple a matter to show that the theorem of phase harmony is always valid that it seems not necessary to develop the proof.

The present theory suggests an interesting explanation of Bohr's stability conditions. At time 0 the electron is in a point A of its trajectory. The phase wave starting at this instant from A will describe all the path and meet again the electron in A'. It seems quite necessary that the phase wave shall find the electron in phase with itself. This is to say : "The motion can only be stable if the phase wave is tuned with the length of the path." The tune relation is then :

$$\int \frac{ds}{\lambda} = \int_0^T \frac{m_0\beta^2 c^2}{h\sqrt{1-\beta^2}}\,dt = n$$

(n whole number ; T revolution period).

Now, we can write the stability condition of the quantum theory in a general form given by Einstein which degenerates into the manifold Sommerfeld's conditions for quasi-periodical cases in consequence of the infinite number of the pseudo-periods. Let us call $p_x\,p_y\,p_z$ the momenta ; then Einstein's general condition is

$$\int (p_x\,dx + p_y\,dy + p_z\,dz) = nh \quad (n \text{ whole number}),$$

2 G 2

or also

$$\int_0^T \frac{m_0}{\sqrt{1-\beta^2}} (v_x{}^2 + v_y{}^2 + v_z{}^2)\, dt = \int_0^T \frac{m_0}{\sqrt{1-\beta^2}} \beta^2 c^2 dt = nh,$$

which is precisely the result obtained above.

V. *The Propagation of Light Quanta and the Coherence Problem.*

We will now make use of our results for studying the propagation of free light quanta whose velocities are always slightly lower than *c*. We can say: "The atom of light whose whole energy is equal to *hν* is the seat of an internal periodical phenomenon which, for a fixed observer, has at each point of space the same phase as a wave spreading in the same direction with velocity very nearly equal to *c* (very little greater)." The light quantum is in some manner a part of the wave, but for explaining interferences and other phenomena of the wave optics it is necessary to see how several light quanta can be parts of the *same* wave. This is the coherence problem.

In the light quanta theory, it seems necessary to make the following hypothesis : "When a phase wave crosses an excited atom, this atom has a certain probability of emitting a light quantum determined at each instant by the intensity of the wave." Perhaps this hypothesis will seem arbitary, but I think that any theory of coherence must adopt some postulate of this kind.

The emissions of γ-rays by radioactive substances are known to be quite independent, but this cannot be considered as an objection against our view because the "mean life" of any known radioactive atom is always so much greater than the period of the γ-rays.

Thus when an atom emits a light quantum, a spherical phase wave is simultaneously emitted and, crossing over the neighbouring atoms of the point source, will excite other emissions. The non-material phase wave will carry many little drops of energy which slide slowly upon it and whose internal phenomena are coherent.

VI. *Diffraction by a Screen Edge and the Inertia Principle.*

The corpuscular theory of light meets here a great difficulty. It is known since Newton that the light rays passing within a short distance of a screen edge are no longer straight but that some penetrate into the geometrical shadow. Newton ascribed this deviation to the action of some forces which would be exerted by the edge upon the light corpuscle. It seems to me that this phenomenon is perhaps worthy of a more general explanation. Since an intimate connexion seems to exist between motion of bodies and propagation of waves, and since the rays of the phase wave may now be considered as the paths (the possible paths) of the energy quanta, we are inclined to give up the inertia principle and to say : "A moving body must always follow the same ray of its phase wave." In the continuous spreading of the wave, the form of the surfaces of equal phase will change continuously and the body will always follow the common perpendicular to two infinitely near surfaces.

When Fermat's principle is no longer valid for computing the ray path, the principle of least action is no longer valid for computing the body path. I think that these ideas may be considered as a kind of synthesis of optics and dynamics.

We must still specify some points. The ray which now assumes in our ideas an important physical significance may be defined as above by the *continuous* spreading of a small part of the phase wave : it cannot be defined at each point by the geometrical sum for all the waves of the vector which is called in electromagnetic theory "radiant or Poynting's vector." Let us consider a sort of Wiener's experiment. We send a train of plane waves on a perfectly reflecting plane mirror in the normal direction ; standing waves are set up ; the reflecting mirror is a nodal plane for the electric vector, the plane at a distance $\frac{1}{4}\lambda$ from the mirror is a nodal plane for the magnetic vector, the plane at a distance $\frac{1}{2}\lambda$ of the mirror is again a nodal plane for the electric vector, and so on. In each nodal plane the radiant vector is null. May we say that these planes are not crossed by any energy ? Evidently not, we can only say that the interference states in these planes are always the same. In every interference case we should find similar intricacies. In the wave theory, the propagation of energy has a somewhat fictitious character, but, in return, the exact calculation of the interference fringes is easily made ; we shall try in the next paragraph to see why it is so.

VII. *A new Explanation of Interference Fringes.*

Consider how we can detect the presence of light at a point in space—by direct perception of the scattered light, by photographic tests, by calorific effects, and perhaps by some other means.   It seems that all these means can, in fact, be reduced to photoelectric actions and scattering.   Now, when a light quantum crosses a material atom, it has a certain probability of being absorbed or scattered, which can depend on external agencies.   If then a theory succeeds in determining these probabilities without taking account of the actual motion of energy, it may be able to predict correctly the average reaction between radiation and matter at each place.   Following the electromagnetic theory (and Bohr's correspondence principle is consistent with this view) I have been inclined to suppose that, for a material atom, the probability of absorbing or scattering a light quantum is determined by the geometrical sum of one of the defining vectors of the phase waves crossing upon it.   The last hypothesis is, in fact, quite analogous to that which is admitted by electromagnetic theory when it links intensity of disclosable light with the intensity of the resultant electric vector. Thus, in the Wiener's experiment, the photographic action only occurs in the nodal planes of the electric vector; according to the electromagnetic theory, the luminous magnetic energy is not disclosable.

Let us now consider Young's interference experiment. Some atoms of light pass through the holes and diffract along the ray of the neighbouring part of their phase waves. In the space behind the wall, their capacity of photoelectric action will vary from point to point according to the interference state of the two phase waves which have crossed the two holes.   We shall then see interference fringes, however small may be the number of diffracted quanta, however feeble may be the incident light intensity.   The light quanta do cross all the dark and bright fringes; only their ability to act on matter is constantly changing.   This kind of explanation, which seems to remove at the same time the objections against light quanta and against the energy propagation through dark fringes, may be generalized for all interference and diffraction phenomena.

VIII. *The Quanta and the Dynamical Theory of Gases.*

For the sake of calculating the entropy constants and the so-called "chemical constants," Planck and Nernst have been obliged to introduce the quantum idea into the theory of gases.   As explained above, Planck chooses an element of extension in phase equal to

$$\frac{1}{h^3} dx\, dy\, dz\, dp\, dq\, dr \quad \text{or} \quad \frac{4\pi}{h^3} m_0^{3\cdot2} \sqrt{2w}\, dw\, dx\, dy\, dz.$$

Tentative Theory of Light Quanta.   455

We shall now try to justify this assumption.

Each atom of velocity βc may be considered as linked with a group of waves whose phase velocity is $V = \dfrac{c}{\beta}$, frequency $\dfrac{1}{h}\dfrac{m_0 c^2}{\sqrt{1-\beta^2}}$, and group velocity $U = \beta c$. The state of the gas can only be stable if the waves corresponding to all the atoms make up a system of standing waves. Using a well-known method given by Jeans, we find for the number of waves per unit of volume whose frequencies are included in the interval $\nu, \nu + d\nu$ * :

$$n_\nu d\nu = \frac{4\pi}{U V^2}\nu^2\, d\nu = \frac{4\pi}{c^3}\beta\nu^2\, d\nu.$$

If w is the kinetic energy of an atom and ν the corresponding frequency, then :

$$h\nu = \frac{m_0 c^2}{\sqrt{1-\beta^2}} = w + m_0 c^2 = m_0 c^2(1+\alpha),$$

where $\alpha = \dfrac{w}{m_0 c^2}$.

It is now very easy to find that $n_\nu d\nu$ is given by the equation :

$$n_\nu d\nu = \frac{4\pi}{h^3} m_0^2 c(1+\alpha)\sqrt{\alpha(2+\alpha)}\, dw.$$

Each phase wave can carry with it one, two, or more atoms, so that, according to the canonical law, the number of atoms whose energy is $h \cdot \nu$, will be proportional to :

$$\frac{4\pi}{h^3} m_0^2 c(1+\alpha)\sqrt{\alpha(2+\alpha)}\, dw\, dx\, dy\, dz \sum_1^\infty e^{-\frac{nh\nu}{kT}}.$$

Let us first consider a material gas whose atoms have relatively great mass and relatively small velocities. We can then neglect all the terms of the series except the first, and we can also put $1+\alpha=1$. The number of atoms whose kinetic energy is w will be, neglecting a constant factor,

$$\frac{4\pi}{h^3} m_0^{3/2}\sqrt{2w}\, dw\, dx\, dy\, dz\, e^{-\frac{w}{kT}},$$

a result which justifies Planck's method and leads to the usual form of Maxwell's law.

In the case of the light quanta gas α is always great, and, moreover, we must use all the series. In consequence of the

* Léon Brillouin, *Théorie des Quanta*, p. 38. Paris: A. Blanchard.

internal binary symmetry of the light quantum, we must introduce a factor 2, and we find that the radiant energy density is proportional to :

$$\frac{8\pi}{h^3 c^3} w^3 \sum_1^\infty e^{-\frac{nh\nu}{kT}} dw = \frac{8\pi h}{c^2} \frac{\nu^3}{e^{\frac{h\nu}{kT}} - 1} d\nu.$$

A method developed in the *Journal de Physique*, of November 1922, shows that the proportionality factor is unity, so that we obtain the actual radiation law.

IX. *Open Questions.*

The conceptions stated in this paper, if they are received, will necessitate a wide modification of the electromagnetic theory. The so-called "electric and magnetic energies" must be only a kind of average value, all the real energy of the fields being probably of a corpuscular fine-grained structure. The building up of a new electromagnetism seems to be a very difficult task, but we have one directive idea : according to the correspondence principle and to the above statements, the defining vectors of the old electromagnetic theory would give the probability of the reaction between matter and the fine-grained energy.

The new electromagnetism will give the solution of many problems. The laws of propagation of waves given by Maxwell's theory will probably be valid for the energyless light phase waves, and the scattering of the radiant energy will be explained by the resulting curvature of the rays (viz., light quanta paths). There seems to be a great analogy between scattering of radiation and scattering of corpuscles ; lowering of particles' velocities by crossing through a screen may also be similar to the lowering of X-ray frequencies by scattering, which recently has been computed and experimentally checked by A. H. Compton.

Explaining optical dispersion will be more difficult. The classical theories (including the electron theory) gives only an average view of this phenomenon, which is produced by complex elementary reactions between radiation and atoms ; we shall certainly be obliged here also to distinguish accurately the real motion of energy from the propagation of the resulting interference state. The kind of "resonance" shown by the variations of refractive index seems no longer to be irreconcilable with discontinuity of light.

Many other questions remain open : What is the mechanism

of Bragg's absorption ? What .occurs when an atom passes from a stable state to another, and how does it eject a single quantum ? How may we introduce the granular structure of energy into our conceptions of elastic. waves and into Debye's theory of specific heats?

Finally, we must remark that the quantum relation remains still a kind of postulate defining the constant h whose actual significance is not at all cleared up ; but it seems that the quantum enigma is now reduced to this unique point.

Summary.

In the present paper it is assumed that the light is essentially made up of light quanta, all having the same extraordinarily small mass. It is shown mathematically that the Lorentz-Einstein transformation joined with the quantum relation leads us necessarily to associate motion of body and propagation of wave, and that this idea gives a physical interpretation of Bohr's analytical stability conditions. Diffraction seems to be consistent with an extension of the Newtonian Dynamics. It is then possible to save both the corpuscular and the undulatory characters of light, and, by means of hypotheses suggested by the electromagnetic theory and the correspondence principle, to give a plausible explanation of coherence and interference fringes. Finally, it is shown why quanta must take a part in the dynamical theory of gases and how Planck's law is the limiting form of Maxwell's law for a light quanta gas.

Many of these ideas may be criticized and perhaps reformed, but it seems that now little doubt should remain of the real existence of light quanta. Moreover, if our opinions are received, as they are grounded on the relativity of time, all the enormous experimental evidence of the "quantum" will turn in favour of Einstein's conceptions.

1 October, 1923.

Note.—Since I have written this paper, I have been able to give to the results contained in the fourth section a somewhat different, but much more general form.

The principle of least action for a material point can be expressed in the space-time notation by the equation :—

$$\delta \int \sum_{1}^{4} J_i \, dx^i = 0,$$

if the J_i are the covariant components of a four-dimensional

458

vector whose time component is the energy of the point divided by c and space components are the components of its momentum.

Similarly, in studying the propagation of waves, we have to write :—

$$\delta \int \sum_{1}^{4} O_i \, dx^i = 0,$$

if the O_i are the covariant components of a four-dimensional vector whose time component is the frequency divided by c and space components the components of a vector drawn along the ray and equal to $\dfrac{\nu}{V} = \dfrac{1}{\lambda}$ (V phase velocity). Now, the quantum relation says that $J_4 = h\,O_4$. More generally, I suggest putting $\overrightarrow{J} = h\,\overrightarrow{O}$. From this statement, the identity of the two principles of Fermat and Maupertius follows immediately, and it is possible to deduce rigorously the velocity of the phase wave in any electromagnetic field.

Part Six

The Solid State

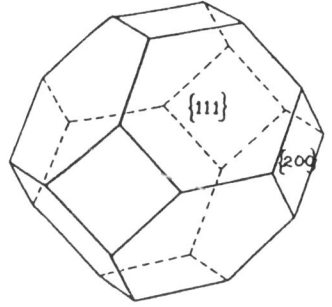

Part Six
The Solid State

The history of solid state physics appears at first sight to lack the dramatic milestones characterizing the development of atomic and nuclear physics as highlighted by the papers reproduced in the earlier Parts of this volume. It is certainly the case that major breakthroughs in our understanding of condensed matter were spread over a much longer period of time and that significant achievements were made by a much larger cohort of scientists than the relatively few credited with unravelling the secrets of the atom. Perhaps the passage of time will change the perspective, and history will identify more clearly than seems possible at present the major landmarks and key figures in the story.

The triumphs of quantum mechanics are nowhere better seen than in the application of its principles to account for the behaviour of electrons and atoms in solids. Classical treatments of electrical and thermal conductivities, notably by Drude and Lorentz in the first decade of this century, soon ran into trouble when attempting to explain certain experimental facts, for example the temperature dependence of the specific heats of metals. This situation was partially rectified by Einstein's 1907 treatment of lattice vibrations in terms of quantized modes — the first application of Planck's quantum hypothesis to solids. However, it was not until 1926, when Pauli's Exclusion Principle and Fermi–Dirac statistics were applied to the 'electron gas' in a metal, that real progress was made. The subject was carried forward by Sommerfeld, Bethe, Peierls, Born, Pohl and Pauling, among others, and in 1928 Felix Bloch produced the first full quantum mechanical theory of electrons in metals using the Schrödinger equation. Bloch showed that in a periodic potential, such as would exist in a perfect single crystal at zero temperature, the electrons could move freely. Scattering, and hence electrical resistance, was seen to arise from atomic vibrations or defects which distort the lattice from its ideal form.

Prior to Bloch's work it was difficult to understand the existence of metals. Afterwards it was the existence of insulators that required explanation! The answer came in 1930 from A. H. Wilson, who pointed out that the electron energy levels of isolated atoms are broadened into bands as the atoms are squeezed together in a solid. If these bands do not overlap in energy, elements with an even valency can produce an insulating material. Furthermore, if the bands are separated by a small enough energy gap, semiconducting properties arise. It was to be another 10 years before germanium and silicon were obtained in a sufficiently pure state to realize that these were archetypical semiconductors; the potential for their use in solid state electronic devices was fulfilled in 1947 with the invention of the transistor by Bardeen, Shockley and Brattain at the Bell Telephone Laboratories.

As the middle of the century approached, papers published in the *Philosophical Magazine* reflected the growing interest in the physics of the solid state. Indeed, as is evident from the articles reproduced in *Science in the Making*, Volume 4, a deliberate policy was adopted by the Editors of the journal to encourage submissions in this field.

The first paper reproduced in this Part, however, takes us back to 1914, when the first seminal papers by W. L. Bragg on the determination of crystal structures began to appear. The diffraction law named after Bragg (who succeeded Rutherford twice, first to the Chair of Physics at Manchester and then in 1938 to the Cavendish Professorship) is now so well known and utilized in its original form that it is hard to believe it was formulated 84 years ago.

Nevill Mott (1905–1996) had a close relationship with the journal (see his Foreword to *Science in the Making*, Volume 1) not only as an author but also as its Editor from 1948 to 1970. His first paper in the *Philosophical Magazine*, reproduced here, appeared in 1936 when he was Professor of Theoretical Physics at Bristol University, a chair to which he had been appointed three years earlier aged 28. Mott's contributions to solid state physics spanned 70 years in total, beginning in the early days of quantum mechanics. His work on metals and alloys, illustrated here, was to continue throughout his life, but was supplemented after his appointment as Cavendish Professor in 1954 by his growing interest in semiconductors, metal–insulator transitions, amorphous (non-crystalline) materials and glasses, for which work he was awarded the Nobel Prize for Physics with Anderson and van Vleck in 1977.

The final paper in this part is included to represent the extent to which Brillouin zones had grown in importance, on account of the predictions that could be made from them of the influence of crystalline structure on electronic energy bands in solids. The concept of zones, introduced by Brillouin in 1930, has a direct connection with Bragg's law: electrons at the surfaces of Brillouin zones in momentum (or k) space have a de Broglie wavelength corresponding to that of the real-space lattice, and as a consequence suffer 'internal' Bragg reflection. Standing waves are set up and these give rise to a forbidden gap in the allowed spectrum of electron energy levels. One of the early demonstrations of the usefulness of these ideas was made by Hume-Rothery who was able to account for the electrical behaviour of metallic alloys as their composition was changed and their Fermi surfaces[1] interacted with Brillouin zone boundaries.

1914 **28** *The Crystalline Structure of Copper. By W. Lawrence Bragg, B.A., Allen Scholar of the University of Cambridge*

The first photographs of X-ray diffraction from crystals were published in 1912 by von Laue, Friedrich and Knipping in Germany. They were brought to the attention of William Lawrence Bragg by his father, William Henry Bragg, who had been struggling for more than a decade over the nature of X-rays before coming down fairly decisively in favour of a corpuscular theory. In curious parallel with another father–son-related witticism — namely that 'J. J. Thomson received the Nobel Prize for discovering that the electron was a particle while his son, G. P. Thomson, received it for showing that it was a wave' — W. L. Bragg demonstrated that the Laue photographs strongly supported the wave nature of X-rays.

Adapting an equation known to explain optical diffraction from ruled gratings, Lawrence Bragg formulated the famous equation which became known as Bragg's law, namely $n\lambda = 2d\sin\theta$ relating the wavelength λ of X-rays to the angle of incidence θ and the spacing d of atomic planes in a crystalline lattice. Before 1912, the actual arrangement of atoms was not known for any material, but a few years later the two Braggs had developed standard methods for determining crystal structures. They shared a Nobel Prize in 1915.

In this paper, Lawrence Bragg describes X-ray reflection results on copper. Application of his simple equation enabled him to determine the separation of the most densely populated crystal planes and to deduce that the atoms were arranged in a face-centred-cubic lattice.

These were the early days of X-ray crystallography. The technique was developed after the 1914–18 war and eventually led to the determination of much more complicated structures, including those of biological systems.

1936 **22** *Optical Constants of Copper–Nickel Alloys. By N. F. Mott, M.A., Professor of Theoretical Physics in the University of Bristol*

Between the two world wars, considerable progress had been made in the understanding of the electrical and magnetic properties of metals. In particular the electronic states associated with isolated atoms were known to be broadened into bands of allowed energies when the atoms were arranged as closely together as they are in the solid state.

This is the first paper published by Nevill Mott in the *Philosophical Magazine* — a journal to which he was later to be appointed Editor. Alloys of the type considered here were to figure in much of Mott's later work and, indeed, in the same year as this paper appeared, he published with Jones a book entitled *The Theory of Properties of Metals and Alloys*, which became a classic text.

The style of the paper is typical of Mott's writing, which continued unabated for another 60 years. Although a theoretician, he was always quick to spot an interesting experimental result. Here he interprets the optical properties of a series of copper–nickel alloys in terms of the degree of filling of the electronic *d*-band. Reference to 'positive holes' to describe missing electrons at the top of an otherwise filled band — a concept which was to be of particular significance for semiconductors — is of interest. These, Mott says, are responsible for the ferromagnetism of nickel. Addition of copper with its extra (1s) electrons results in a filling of the *d*-band, a transition to paramagnetic behaviour, and a fall in the optical absorption coefficient.

1941 **31** *Relationships between Lattice Types and Brillouin Zones. By Geoffrey Vincent Raynor*

The electronic band structure of solids obtains its most useful visualization in *k-space* in which the energies of allowed levels are drawn with respect to the momentum (*hk*) of the electrons (see for example C. Kittel, *Introduction to Solid State Physics*). In this representation, Brillouin zones (regular shaped figures, related to the crystal structure, which subdivide *k*-space) play an important role in visualizing electronic properties of solids. For instance, metals have Brillouin zones (or energy bands) partly filled with electrons, whereas semiconductors and insulators are characterized by zones that are either full or empty.

The importance of establishing the Brillouin zones for known crystallographic structures took on added impetus when Hume-Rothery invoked them to account for the properties and phase diagrams of certain alloys. In this paper, Hume-Rothery's colleague, Raynor, gives the Brillouin zones for several common structures and relates them to known properties of materials which crystallize in these forms.

An interesting feature of the paper, apart from its content, is the form of giving references to previously published and related work at the end of the paper — a practice which became common at about this time.

Notes

1. A term introduced by Jones and Zener in 1934 to describe the upper-most filled electron states.

[355]

XLII. *The Crystalline Structure of Copper.* By W. Lawrence Bragg, *B.A., Allen Scholar of the University of
Cambridge* *.

THE copper crystals used in this investigation were some
natural specimens, for which I am indebted to
Mr. Hutchinson, of the Mineralogical Laboratory at Cambridge. In their natural state these specimens are obviously
rough crystals, and have some faces of large dimensions
(1 cm. each way), but these faces are very much warped and
distorted. An attempt was made to obtain an X-ray reflexion
from various natural faces of such crystals, but it was not
successful. Apparently the outer surface of the crystal has
been so battered and distorted that little regular crystalline
arrangement is left. Any attempt to grind crystal faces
artificially also destroys the crystalline character of the
surface and so prevents the reflexion of X rays from the face.

It was observed, however, that when the crystal was placed
in nitric acid until the surface was eaten away to an extent
of perhaps $\frac{1}{4}$ millimetre, the faces were etched deeply into
numerous parallel facets, which all reflected the light simultaneously in the usual way. This suggested that, internally
the crystal structure was perfect, and showed further that in
some cases the whole specimen was composed of a single
crystal. Moreover, in this case the surface layers are not
pulled about, and so are capable of reflecting the X rays
falling on them. This method of obtaining a crystal surface
was suggested by some previous experiments on natural zinc
oxide, zincite. Zincite occurs very rarely as crystals, and
the specimens used had merely a platy structure. However,
by partly dissolving a block of the mineral in hydrochloric
acid, the etched mass showed indications of crystalline
structure sufficient to serve as a guide in the preparation of
various faces. The faces reflected the X rays and led to the
determination of the arrangement of zinc and oxygen atoms.

Copper crystallizes in the cubic system, holohedral class.
The natural crystals of copper used in the experiments were
mostly of one type, being composed of two individuals
twinned about the plane (111). The faces of the simplest
crystals approximated to those of the rhombic dodecahedron
{110}. The face first investigated was that parallel to the
twin plane (111) of the best of the crystals. The crystal
appeared like two triangular pyramids joined base to base,
and the apex of one pyramid was ground down on a carborundum wheel until a triangular face (111) was formed.

* Communicated by the Author.

2 A 2

Mr. W. L. Bragg *on the*

This was roughly polished, treated with nitric acid to dissolve away the outer layers, and then mounted in the usual way in the X-ray spectrometer *.

By assuming various arrangements for the copper atoms, we can calculate the spacings of various planes of the crystal, starting from the density of copper, 8·96, as a basis. For instance, if the copper atoms are arranged at the corners of cubes, so as to form a simple cubic lattice, we have the relation

$$(d_{(100)})^3 . 8·96 = 63·57 . 1·64 . 10^{-24},$$

since we know that

1. Mass of a copper atom $= 63·57$. (Mass of a hydrogen atom)
   $= 63·57 . 1·64 . 10^{-24}$ gram,

2. The unit cube of the structure contains one copper atom.

This gives the relation

$$d_{(100)} = 2·26 . 10^{-8} \text{ cm.}$$

The X rays from an anticathode of palladium, such as was used for the purposes of this experiment, have a wave-length of ·576 . 10^{-8} cm. This is reflected from the planes $d_{(100)}$ at an angle given by the equation

$$\lambda = 2 . d_{(100)} . \sin \theta.$$

Substituting the above values for λ and $d_{(100)}$, we find that $\theta = 7° 20'$.

Suppose, on the other hand, that the copper atoms were on a face-centred cubic lattice. This lattice has a point at each corner of a set of cubes, and one at the centre of each cube face. The volume $(d_{(100)})^3$ now contains only one-half a copper atom. From this it can be calculated that $d_{(100)} = 1·80 . 10^{-8}$ cm. So for the other faces; the results are summarized below.

	Spacing of planes.	Glancing angle of reflexion, Pd rays.
Face-centred lattice :		
	$d_{(100)} = 1·80 . 10^{-8}$ cm.	$\theta_{(100)} = 9° 13'$
	$d_{(110)} = 1·27 . 10^{-8}$ cm.	$\theta_{(110)} = 13° 2'$
	$d_{(111)} = 2·08 . 10^{-8}$ cm.	$\theta_{(111)} = 8° 0'$
Simple cubic lattice :		
	$d_{(100)} = 2·26 . 10^{-8}$ cm.	$\theta_{(100)} = 7° 20'$
	$d_{(110)} = 1·59 . 10^{-8}$ cm.	$\theta_{(110)} = 10° 22'$
	$d_{(111)} = 1·30 . 10^{-8}$ cm.	$\theta_{(111)} = 12° 50'$

* For a description of the instrument and its manipulation, see Proc. Roy. Soc. A. vol. lxxxviii. p. 428, and A. vol. lxxxix. p. 468.

The face (111) of the copper crystal, prepared as described above, was first investigated. The chamber was set at 25° 40′ (= 2 × 12° 50′). The crystal face, being adjusted so that the rays fell on it at a small glancing angle, was turned so that this angle assumed in turn a series of values between 6° and 20°, in order to find if, in some position, it reflected the X rays. This had to be done because the true orientation of the plane (111) in the crystal was not known with any exactness. At none of these angles was there a reflexion into the chamber, the simple cubic arrangement of the copper atoms being therefore ruled out.

On setting the ionization chamber at 16°, however, a marked effect was found. Fig. 1 shows the current in the chamber for a series of angles at which the crystal was set.

Fig. 1.

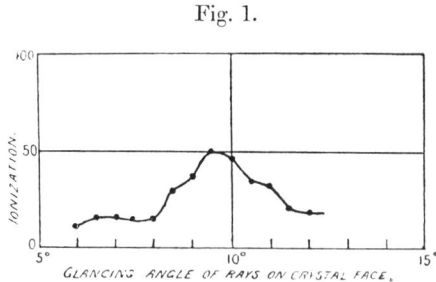

GLANCING ANGLE OF RAYS ON CRYSTAL FACE.

Between 8° and 11° 30′ there is a marked increase in the ionization current, which rises to a maximum at 9° 30′. Now if the crystal were perfect, the range of angles at which it reflected the X rays would be limited to some 30′ at most. The fact that the crystal reflects over such a wide range of angles shows that its planes are distorted to an extent of several degrees, instead of being strictly parallel. As the crystal is turned round, one set of planes after another comes into the reflecting position and causes an ionization current in the chamber. From the curve we deduce that when the crystal is set at 9° 30′ the area of its face so oriented as to reflect is larger than at any other angle.

A series of readings are now taken with the chamber at various angles " θ," and the crystal in each case at the angle

" $\dfrac{\theta}{2} + 1°\ 30′$," in order always to make use of this larger

reflecting area. The results are shown in the curve for the face

358   Mr. W. L. Bragg *on the*

(111) of fig. 2. Here we have two peaks close together in the curve, representing the two lines in the spectrum of palladium. The curve shows a decided first- and second-order spectrum, and even perhaps a third, though this last is somewhat doubtful.

Fig. 2.

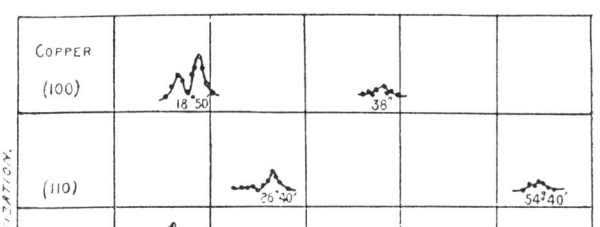

somewhat doubtful. It must be noted that the range of angles, over which the ionization chamber may be set so as to receive the reflected beam, is confined to 1°, though the crystal structure is so imperfect. If the crystal were perfect it would be scarcely smaller, and the reason for this is clear. Although there are a number of settings for the crystal which enable a set of planes somewhere on its distorted face to receive the incident X rays at a glancing angle of 8° and so to reflect them, the reflected rays all converge and are received by the chamber when set exactly at 16° *. Therefore, when the chamber and crystal are moved simultaneously, the reflexion is only found when the chamber is in the neighbourhood of 16°.

Fig. 2 also shows the curves for faces (110) and (100) of copper. The angles at which the spectra are found are as follows :—

$\theta_{(100)} = 9° 24'$ $\sin \theta_{(100)} = \cdot163$.   Calculated angle 9° 13'

$\theta_{(110)} = 13° 18'$ $\sin \theta_{(110)} = \cdot230$.   ,,   ,,   13° 2'

$\theta_{(111)} = 8° 0'$ $\sin \theta_{(111)} = \cdot139$.   ,,   ,,   8° 0'

$$\sin \theta_{(100)} : \sin \theta_{(110)} : \sin \theta_{(111)} = 1 : 1\cdot41 : \cdot85 = 1 : \sqrt{2} : \sqrt{\tfrac{3}{2}}.$$

This relation between the angles of reflexion from the

* *Cf.* W. H. Bragg, Phil. Mag. May 1914, p. 887.

three principal faces of the crystal is that which would exist
for a face-centred cubic lattice.　In such a lattice

$$d_{(100)}:d_{(110)}:d_{(111)} = 1:\frac{1}{\sqrt{2}}:\frac{2}{\sqrt{3}}.$$

We have already seen that the absolute values of the angles
at which reflexion occurs are those to be expected if the
copper atoms lie on a face-centred lattice.　It is further to
be observed that the 1st, 2nd, and 3rd order peaks reflected
from the faces of the crystals are in every case quite normal,
the first order being greater than the second, and that
greater than the third.　This, as has been shown in former
papers (Proc. Roy. Soc. A. vol. lxxxix. p. 472) implies a
regular arrangement of reflecting planes, equally spaced and
identical in nature.　Lastly, as a check, a search was made
for spectra at half the angles at which the first spectra of
fig. 2 occur, in the case of the planes (100) and (110).
This search gave a negative result.　Taking all this into
consideration, there can be little doubt that *the atoms
of a copper crystal are arranged on a face-centred cubic
lattice.*

The results are interesting in that they show that considerable accuracy of measurement can be obtained with the
X-ray spectrometer, even when the crystal itself is highly
irregular.　If the cubic symmetry of the copper crystal had
not rendered this unnecessary, it would have been possible
to measure the axial ratio of the crystal within 1 per cent. of
the truth, although the faces of the crystal were distorted by
many degrees.　Since the crystal is so irregular, only a
fraction of its surface reflects at any one angle, and therefore
the electroscope had to be very sensitive when measuring the
small ionization current.　This explains the very obvious
irregularities of the curves in fig. 2.　The dots in this figure
represent a set of readings.　All these curves were repeated
several times, some with different crystal faces.　Some curves
were more irregular than others, but all agreed closely in the
positions of the spectra.

I wish to take this opportunity of again thanking Mr. Hutchinson for his kind help, both in supplying material and in
aiding with his advice the preparation of the various crystal
faces.　I wish to thank Professor Sir J. J. Thomson for his
kind interest in the experiments.　I am indebted to the
Institut International de Physique Solvay for a grant with
the aid of which the apparatus used in these experiments was
purchased.

360

Summary.

It was found possible, by treating with acid prepared surfaces of a natural crystal of copper, to obtain crystal faces which could be used as reflectors in the X-ray spectrometer.

The results of the investigations, thus rendered feasible, showed that in a copper crystal the atoms are arranged on a face-centred cubic lattice. This is the close-packed lattice, to which attention has been drawn by Pope and Barlow. The crystal structure is the most simple of any as yet analysed.

The Cavendish Laboratory,
    July 16th, 1914.

[287]

XXIII. *Optical Constants of Copper-Nickel Alloys. By* N. F. Mott, *M.A., Professor of Theoretical Physics in the University of Bristol* *.

THE optical constants of the copper-nickel series of alloys have recently been measured by Lowery, Wilkinson, and Bor †. The purpose of this note is to give a theoretical interpretation of the results achieved and to relate them to the magnetic properties of these alloys.

The copper-nickel alloys form a face-centred cubic lattice throughout the whole range. A theoretical discussion of their magnetic properties has been given by the author in a recent paper ‡. In metallic nickel there exists a band of allowed energy levels corresponding to the $3d$ state of an electron in the free atom. This band of levels would be completely full if it contained ten electrons per atom; actually in the metal there are 0·6 " positive holes " per atom in the band, which are responsible for the ferromagnetism, and which give a saturation moment of 0·6 Bohr magnetons per atom. The copper atom contains one more electron than the nickel atom; when copper is alloyed with nickel the extra electron goes into the $3d$ band, so that the saturation moment decreases linearly with concentration; at a composition of 60 per cent. copper it disappears, so that the d band is then full.

The figure below shows the absorption coefficient of Cu–Ni measured by Lowery for a wave-length of 5780 Å.U. It will be seen that the absorption coefficient of the alloys increases rapidly with increasing concentration of nickel while the alloy remains paramagnetic, but at a composition near to 40 per cent. Ni, where the ferromagnetism begins, the curve flattens out, and the absorption coefficient then remains roughly constant.

The wave-length 5780 Å.U. lies on the low frequency side of the absorption band of copper The origin of this band will not concern us here; it may be due to the ionization of the $3d$ level of copper, or to the ejection of an s (conduction) electron to a higher state.

* Communicated by the Author.
† Phil. Mag. xx. p. 390 (1935).
‡ Proc. Phys. Soc. xlvii. p. 571 (1935).

288   Prof. N. F. Mott *on the Optical*

Only one explanation seems possible for the rapid rise in nk as nickel is added to copper.   It must be due to absorption resulting from the ejection of electrons from the $3d$ shells of the nickel atoms.   We should expect the energy $h\nu$ required for this process to be less than that required to eject an electron from the d shell of copper.   An estimate of it will be made below.

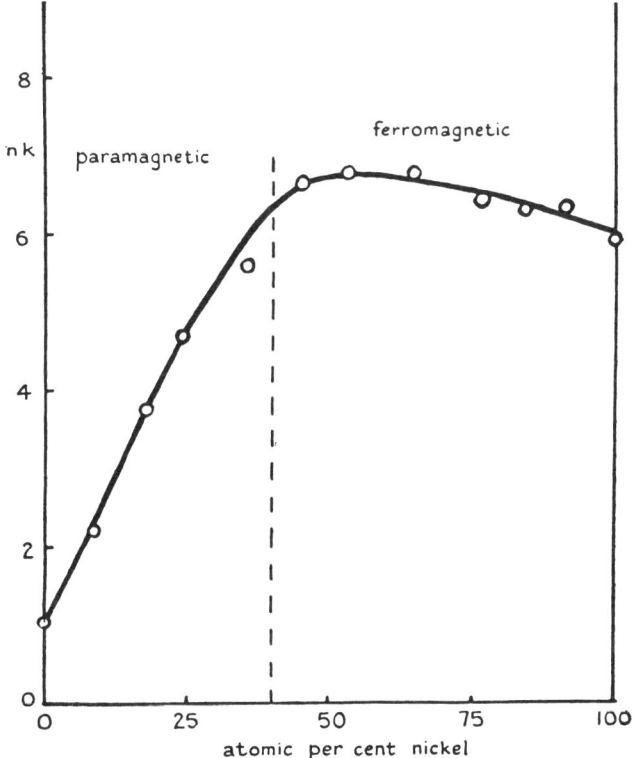

Absorption coefficient nk of Cu–Ni alloys.

We thus expect the absorption coefficient to rise linearly with increasing proportion of nickel.   We have now to explain why in fact it only rises linearly as long as the alloy is not ferromagnetic.   The explanation is as follows :—Up to a composition of 40 per cent. nickel the nickel ions contain, as we have seen, ten d electrons, forming, therefore, closed shells ; for greater compositions each nickel ion added gives one Bohr ·magneton to the magnetic moment, and must therefore contain only nine *d* electrons. There would thus be, among 100 atoms of all kinds, 40 nickel atoms with full ions, the remainder of the nickel ions containing a " positive hole." These positive holes would of course move from atom to atom through the lattice.

Now it is natural to suppose that the work required to remove an electron from a closed *d* shell is considerably less than that required to remove one from a *d* shell containing only nine electrons ; hence we conclude that only those ions which have no " positive hole " in them can contribute to the absorption for the wave-length considered. It follows that the absorption coefficient will only increase in the observed range.

The small increase in *nk* when copper is added to nickel may well be due to the high resistance of the copper-nickel alloys.

Finally, we shall attempt to estimate the low frequency limit of the absorption band. For nickel and for all alloys with up to 60 per cent. copper, the work required to remove an electron from the *d* band to the *s* band is *zero* (*cf.* Mott, *loc. cit.* fig. 2). Actually the poor reflecting power of nickel in the infra-red suggests that the absorption band does in fact extend to quite long wave-lengths *.

For alloys containing less than 40 per cent. of nickel, however, the highest occupied *s* state has higher energy than the 3*d* state of nickel. Since the number of *s* electrons is $1-x$ per atom, the energy of the former state is, according to the formula due to Sommerfeld

$$(1-x)^{\frac{2}{3}}E_F, \quad E_F = \frac{h^2}{8m}\left(\frac{3}{\pi\Omega_0}\right)^{\frac{2}{3}} = 7 \cdot 1 \text{ e.v.}$$

where Ω_0 is the atomic volume and $x : (1-x)$ the ratio of the number of nickel atoms to those of copper. For $x = 0 \cdot 4$ this will be just equal to the energy of the *d* shell. Therefore the work required to remove a *d* electron is

$$\{(1-x)^{\frac{2}{3}} - (0 \cdot 6)^{\frac{2}{3}}\}E_F,$$

* The poor reflecting power can also be accounted for by the Hagen-Rubens formula in terms of the measured electrical resistance ; but, as pointed out by Mott and Zener, Proc. Camb. Phil. Soc. xxx. p. 249 (1934), it is unlikely, on theoretical grounds, that this formula is a good approximation for wave-lengths less than 30 μ. A formula taking into account the inertia of the electrons, but neglecting photo-electric absorption, gives much higher reflecting power.

290

which will give the low frequency limit of the absorption band.   For instance, for an alloy containing 10 per cent. of nickel it is

$$\{(0 \cdot 9)^{\frac{2}{3}} - (0 \cdot 6)^{\frac{2}{3}}\} E_F = 1 \cdot 57 \text{ e.v.,}$$

corresponding to a wave-length of $0 \cdot 8\mu$.

[131]

XIV. *Relationships between Lattice Types
and Brillouin Zones.*

By Geoffrey Vincent Raynor *.

[Received July 29, 1940.]

In recent years the forms of the Brillouin Zones for the simple metallic types of crystal structure have been securely established, and the results have been extended to cover such complex examples of crystal structure as the γ-brass and β-manganese types. This has led in many cases to a satisfactory interpretation of some of the physical and electrical properties of metals and alloys [1].

Since a knowledge of the Brillouin Zones for the various structures which may be found in metals and intermetallic systems is becoming of increasing importance in any theoretical examination, it is of interest to discuss certain lattice relationships and the corresponding Brillouin Zone relationships. In this way it is possible to obtain at least some idea of the form of the zones for the more complex structures, if they can be related to more simple forms.

In the first place, we may consider crystal lattices which may be referred to cubic axes. The face-centred cubic structure forms a convenient basis for discussion. The appropriate Brillouin Zone for this lattice may readily be shown, by the usual methods, to be bounded by planes of the type {200} and {111} in k-space [2]. These are the only planes of low indices for which the structure factor does not vanish. The form of the zone for the face-centred cubic structure is shown in fig. 1.

The face-centred cubic lattice may, however, be regarded as a body centred tetragonal lattice with an axial ratio, c/a, equal to $\sqrt{2}$. Thus it follows that the form of the Brillouin Zone for a body-centred tetragonal lattice with parameters $a = \dfrac{\alpha}{\sqrt{2}}$ and $c = \alpha$ is the same as that for a face-centred cubic lattice of parameter α. The bounding planes of the zone will, of course, have a different nomenclature when referred to the tetragonal structure. The {002} planes of the face-centred cube remain the {002} planes of the tetragonal structure. The {200} and {020} planes of the face-centred cube become the {110} planes of the tetragon, while

---

* Communicated by Professor N. F. Mott, F.R.S.

the {111} planes of the cube become the {101} family of tetragonal planes. The form of the zone, together with the unit cell of the body-centred tetragonal lattice equivalent to a face-centred cube, is shown in fig. 2. For this particular structure with c/a equal to $\sqrt{2}$, the slopes of the faces of the zone correspond with the slopes of equivalent planes in the face-centred cube.

Fig. 1.

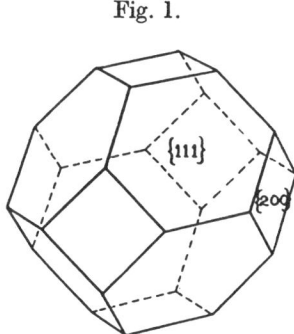

Brillouin Zone for face-centred cubic lattice.

Fig. 2.

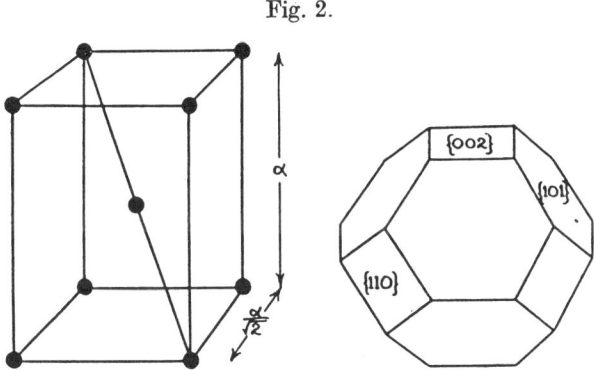

Unit cell of lattice, and Brillouin Zone for body-centred tetragonal structure of $c/a = \sqrt{2}$.

We may now consider the body-centred tetragonal structure with axial ratio $\sqrt{2}$ to be compressed along the c axis until it becomes a body-centred cube, of side $\dfrac{\alpha}{\sqrt{2}}$ for instance, and may consider the corresponding changes in the form of the Brillouin Zone. As compression of the lattice takes place, the inter-planar spacing of the {002} planes decreases, so that the {002} faces of the Brillouin Zone are displaced outwards from the origin of k-space. At the same time the planes of the type {101} become more horizontal in slope, while the planes of the type {101} and {110} tend to become, and eventually do become, exactly equivalent. We may thus visualize the changes in the form of the Brillouin Zone as shown in fig. 3. The derived zone for the body-centred cube is the same as that established by wave-mechanical calculations [3].

It is therefore clear that the form of Brillouin Zone for the face-centred cube is the same as that for the body-centred tetragonal lattice of axial ratio $\sqrt{2}$, while the zone for the body-centred cube is of the same form as that for a face-centred tetragonal of axial ratio $\dfrac{1}{\sqrt{2}}$. There are clearly two electron states per atom in each of these zones, as in the case of the zone for the face-centred cube.

Fig. 3.

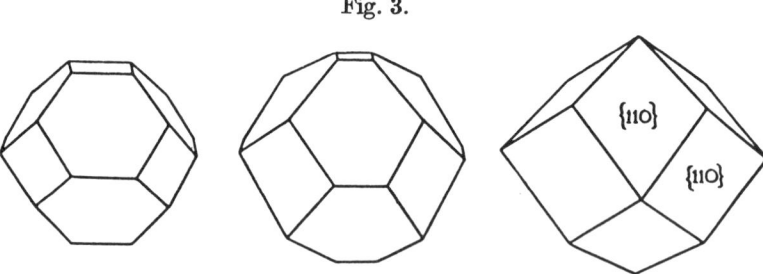

| Body-centred tetragonal lattice, $c/a = \sqrt{2}$. | Body-centred tetragonal lattice, $c/a < \sqrt{2}$ but >1. | Body-centred cube. |

The changes in the form of the Brillouin Zone in passing from a body-centred tetragonal lattice of axial ratio $\sqrt{2}$ to a body-centred cube have been traced. We may continue the process through the stage of the body-centred cube in order to obtain the form of the zone for the body-centred tetragonal structure of axial ratio less than unity. The {002} planes will clearly have disappeared from the zone structure; the slopes of the {110} type of plane in the lattice and in the zone retain their character. The {101} type of plane in the lattice, however, falls towards the horizontal, while the interplanar spacing decreases. In the zone, therefore, the {101} type of plane becomes more horizontal and at the same time is displaced outwards from the origin. The result of these changes is shown in fig. 4.

It is evident that the Brillouin Zones for the body-centred and face-centred tetragonal lattices are intimately related in form. They may be visualized as undergoing exactly the same variations in form and shape

134   Mr. G. V. Raynor *on the Relationships*

as the axial ratio varies, but with a lag between them. If the axial ratio for a body-centred tetragon is greater than $\sqrt{2}$ it is equivalent to a face-centred tetragon of axial ratio greater than 1, and the zones are the same in form. This conclusion is not surprising, since any face-centred tetragonal lattice, by simple rotation of the two equal axes through 45°, can be transformed into a body-centred tetragonal lattice with a unit cell of half the size.

The body-centred cubic structure is equivalent to a face-centred tetragonal lattice of axial ratio 0·707 ; this may be used to deduce the form of the Brillouin Zone for the white-tin structure. This structure belongs to the tetragonal system ; there are four atoms in the unit cell with the basis (0, 0, 0), ($\frac{1}{4}$, $\frac{1}{4}$, $\frac{1}{4}$), ($\frac{1}{2}$, 0, $\frac{1}{4}$), (0, $\frac{1}{2}$, $\frac{3}{4}$). It may be regarded as built up of two interpenetrating face-centred tetragonal lattices ; if the basis of one of these lattices is (0, 0, 0), (0, $\frac{1}{2}$, $\frac{1}{2}$), ($\frac{1}{2}$, 0, $\frac{1}{2}$), ($\frac{1}{2}$, $\frac{1}{2}$, 0),

Fig. 4.

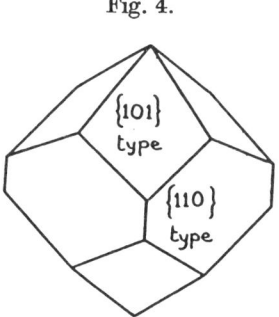

Brillouin Zone for body-centred tetragonal
structure of $c/a < 1$.

then that of the other is, relative to the first ($\frac{1}{4}$, $\frac{1}{4}$, $\frac{1}{4}$), ($\frac{1}{4}$, $\frac{3}{4}$, $\frac{3}{4}$), ($\frac{3}{4}$, $\frac{1}{4}$, $\frac{3}{4}$), ($\frac{3}{4}$, $\frac{3}{4}$, $\frac{1}{4}$). The Brillouin Zone for the white-tin structure may therefore be considered as similar to that for a face-centred tetragonal lattice of axial ratio less than $\sqrt{2}$, and therefore to that for a body-centred tetragonal lattice of axial ratio less than 1. The zone would at first sight be expected to take the form shown in fig. 4. The effect of the relative displacements of the two interpenetrating lattices, however, is to eliminate the X-ray reflexions from the {110} planes. That is to say, the structure factor for planes of this type becomes zero ; and since only planes for which the structure factor does not vanish can give rise to planes of energy-discontinuity in k-space, the {110} type of plane must be absent from the zone. The Brillouin Zone for white-tin is therefore bounded by the planes of the type {101}, and its form is shown in fig. 5.

This zone may be shown to contain 4·24 electron states per atom. White-tin has four valency electrons per atom and is a conductor of electricity. It is probable, therefore, that there is a small overlap across each of the eight faces of the zone ; this should be reflected in the properties of the metal and its solid solutions, and this will be discussed in later publications.

The diamond structure presents a problem similar to that of white-tin. The diamond lattice may be regarded as two interpenetrating face-centred cubic lattices, so that we might expect the form of the zone to be similar to that for the face-centred cube. The effect of the relative displacement of the two interpenetrating lattices, however, is to reduce the structure factor for the {200} planes to zero. The zone bounded by the {111} planes contains too few electron states per atom, and since the properties of diamond suggest that the structure corresponds with a completely filled zone, we have to find a system of planes enclosing a

Fig. 5.

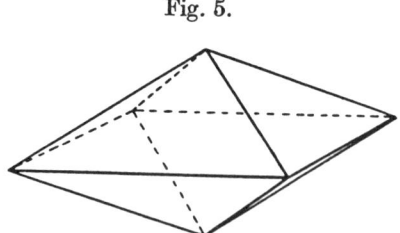

Brillouin Zone for the white tin structure.

zone of four states per atom. Such a zone is bounded by the {220} planes, so that the probable form of the zone for the diamond structure is the same as that for the body-centred cube.

The metal indium has a face-centred tetragonal structure of axial ratio 1·06, and the form of the corresponding zone should thus be intermediate between that for a face-centred cube and that for a body-centred cube. It will obviously be closely similar to the zone for the face-centred cube, but compressed vertically. Alternatively, it may be obtained from that of the body-centred cube by expanding the " c " axis of the lattice and compressing the " c " axis of the zone so that the {002} planes appear as bounding planes of the zone again. The zone for the indium structure will contain two electron states per atom, so that there is a considerable overlap across all faces of the zone. The overlap across the {002} planes will, however, exceed that across the four vertical faces owing to the shape of the zone, and this will lead to interesting solid solution effects, such as are met with in the system cadmium-indium [4].

136   Mr. G. V. Raynor *on the Relationships*

Other and more complex metallic structures may be treated in a similar way. As an instance, we may consider the mercury structure. This metal has a simple rhombohedral lattice. At −46° C. the lattice constant is 2·999 Å. with α=70° 31·7′. Although this cannot be referred directly to cubic axes, we may still obtain an idea of the form of the zone.

The face-centred cubic lattice may be considered as a simple rhombohedral lattice with α=60°, as shown in fig. 6. To put the basal plane of the rhombohedral lattice into the horizontal we must rotate the cube along

Fig. 6.

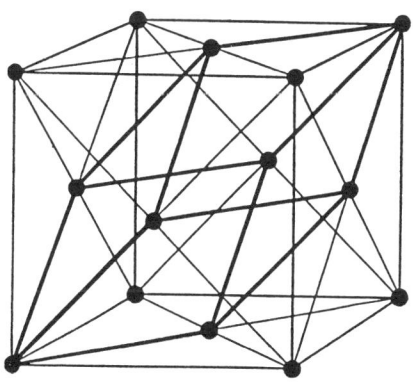

Fig. 7.

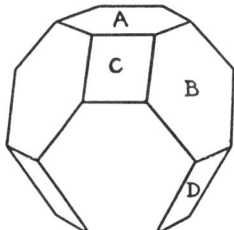

Zone for rhombohedral lattice, α=60°.

xy through an angle of 45° and then rotate the line xy about x in a vertical plane through an angle $\tan^{-1} \dfrac{1}{\sqrt{2}}$. Thus the Brillouin Zone for the rhombohedral lattice with α=60° is the same as that for the face-centred cube, taken through the rotations described above (fig. 7). For the face-centred cube, A and B are planes of the type {111}, and C and D are planes of the type {200}. Referred to the rhombohedral lattice, A is of the type

{001}, C of the type {101}, and B of the type {111}. At the particular angle $\alpha = 60°$, $\{111\}_{\text{f.c.c.}} = \{001\}_{\text{rhomb.}}$ and {111} rhomb.

If now the angle α is increased, the slope of the {001} pair of planes is unchanged, but they get farther apart in the lattice, and the zone is therefore compressed in the direction perpendicular to these planes. The {111} and {101} types of plane become more horizontal and the inter-planar spacing in the lattice decreases. In the zone therefore these boundary planes become more horizontal and are displaced outwards from the origin of k-space. Thus the zone for mercury, with $\alpha = 70° 31 \cdot 7'$, will take the form of fig. 8.

The number of electron states per atom in the rhombohedral zone is easily seen to be 2. The volume of the zone, when $\alpha = 60°$, is the same as that for the face-centred cube, since the planes and interplanar distances concerned are the same in both cases. Its value is $\dfrac{4}{a^3}$, where a is the cubic

Fig. 8.

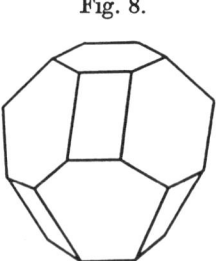

Zone for the mercury structure.

lattice constant. The volume of the unit cell of the rhombohedral lattice outlined in fig. 6 is $\dfrac{a^3}{4}$, where a is the lattice constant of the parent face-centred cubic lattice. Since the rhombohedral cell contains one atom, the volume per atom is $\dfrac{a^3}{4}$, so that the number of electron states $= 2$ per atom.

It is suggested that, by the application of processes similar to those described above, the form or the approximate form of the Brillouin Zones for complex structures may be obtained, even in cases where rigid application of the more formal methods might lead to difficulties.

Acknowledgments.

The author expresses his gratitude to Professor C. N. Hinshelwood, F.R.S., for his kindness in providing laboratory accommodation, and to

138   *Relationships between Lattice Types and Brillouin Zones.*

the Department of Scientific and Industrial Research for a Senior Research Award. Grateful acknowledgment is also made to Dr. W. Hume-Rothery, F.R.S., and Professor N. F. Mott, F.R.S., for helpful comment.

References.
(1) N. F. Mott and H. Jones, 'The Theory of the Properties of Metals and Alloys.' Oxford, 1936, chapter v.
(2) N. F. Mott and H. Jones, *ibid.* p. 156.
(3) N. F. Mott and H. Jones, *ibid.* p. 158.
(4) W. Hume-Rothery and G. V. Raynor.   (To be published.)

Theory of structure of atom

Suppose atom consists of + charge n at centre + − charge as electron distributed throughout sphere of radius b.

Force at P on electron $= Ne^2 \left\{ \dfrac{1}{r^2} - \dfrac{r}{b} \right.$

$$= Ne^2 \left\{ \frac{1}{r^2} - \frac{r}{b^3} \right\} =$$

Suppose charged particle e moves through atom so that is small but $\perp^r$ distance of

Deflecting force $\perp^r$ direction of motion
$$= Ne^2 \left\{ \frac{1}{r^2} - \frac{r}{b^3} \right\} \cos\theta$$

∴ Acc^l $\perp^r$ direction of motion $= d\alpha$

∴ Whole u acquired in passing through atom

$$u = \int d\alpha \cdot dt = Ne \int d\alpha \cdot \frac{ds}{V}$$

$$= \frac{Ne^2}{mV} \int_{c} \left(\frac{1}{r^2} - \frac{r}{b^3} \right) \frac{a}{r} \cdot \frac{r\, dr}{\sqrt{r^2 - a^2}}$$

$$= 2 \frac{Ne^2}{mV} \int_{0} \frac{da \left(b^3 - r^3 \right)}{r^2 \cdot b^3} \cdot \frac{dr}{\sqrt{r^2 - a^2}} \, a \left\{ \frac{1}{r^2} \right.$$

$$= \frac{2 Ne^2}{mV} \int \left(\frac{\cos^2\theta}{a} - \frac{a^2}{b^3 \cos\theta} \right)$$